Primate Brain Evolution

Methods and Concepts

Primate Brain Evolution

Methods and Concepts

Edited by

Este Armstrong

Louisiana State University Medical Center
New Orleans, Louisiana

and

Dean Falk

University of Puerto Rico
Medical Science Campus
San Juan, Puerto Rico

PLENUM PRESS • NEW YORK AND LONDON

Library of Congress Cataloging in Publication Data

Main entry under title:

Primate brain evolution.

"Proceedings of a satellite symposium of the International Primatological Society meeting, held July 4-5, 1980, in Turin, Italy, and of a symposium at the annual meeting of the American Association of Physical Anthropologists, held April 19, 1980, in Niagara Falls, New York"– Verso t.p.

Bibliography: p.

Includes index.

1. Brain – Evolution – Congresses. 2. Primates – Evolution – Congresses. I. Armstrong, Este. II. Falk, Dean, III. International Primatological Society. IV. American Association of Physical Anthropologists. [DNLM: 1. Anatomy, Comparative – Congresses. 2. Brain – Congresses. 3. Evolution – Congresses. 4. Primates – Congresses. WL 300 P952 1980]

QM455.P74 599.8'0188 81-21150

ISBN 0-306-40914-3 AACR2

Based on the proceedings of a satellite symposium of the International Primatological Society Meeting, held July 4-5, 1980, in Turin, Italy, and of a symposium at the Annual Meeting of the American Association of Physical Anthropologists, held April 19, 1980, in Niagara Falls, New York

A Division of Plenum Publishing Corporation
233 Spring Street, New York, N.Y. 10013

Printed in the United States of America

To *Elizabeth C. Crosby*

PREFACE

Given the past decade's explosion of neurobiological and paleontological data and their increasingly sophisticated analyses, interdisciplinary syntheses between these two broad disciplines are of value and interest to many different scientists. The collected papers of this volume will appeal to students of primate and hominid evolution, neuroscientists, sociobiologists, and other behaviorists who seek a better understanding of the substrates of primate, including human, behavior.

Each species of living primates represents an endpoint in evolution, but comparative neurologists can produce approximate evolutionary sequences by careful analyses of representative series. Because nervous tissue does not fossilize, only a comparison of structures and functions among extant primates can be used to investigate the fine details of primate brain evolution. Paleoneurologists, who directly examine the fossil record via endocasts or cranial capacities of fossil skulls, can best provide information about gross details, such as changes in brain size or sulcal patterns, and determine when they occurred. Physical anthropologists and paleontologists have traditionally relied more on paleoneurology, whereas neuroscientists and psychologists have relied more on comparative neurology. This division has been a detriment to the advancement of these fields and to the conceptual bases of primate brain evolution. Both methods are important and a synthesis is desirable. To this end, two symposia were held in 1980--one at the meeting of the American Association of Physical Anthropologists in Niagara Falls, U.S.A., and one at the precongressional meeting of the International Primatological Society in Torino, Italy. Both comparative neurologists and paleoneurologists participated. This volume, which summarizes and assesses major themes in ongoing research on primate brain evolution, is one result of those fruitful dialogues.

For anthropologists, the central thematic statements of brain evolution have sometimes been obscured by specialized terminology. Physical anthropologists interested in primate brain evolution must acquire specialized knowledge in neurobiology. On the other hand, to interpret the findings of comparative neurobiology, the anthropologically minded neuroscientist must understand and grapple with general issues such as locomotion, functional morphology, primate vocalization, and other complex forms of primate behavior, including material forms of culture in the hominid fossil record. Thus the anthropological neurologist must be a generalist in terms of understanding the basic bones and stones of physical anthropology.

Surprisingly, although the central nervous system (CNS) is of primary importance in deciphering primate and human evolution, it has been relegated a rather minor role in sociobiology. Recent sociobiological explorations are highly specific in hypothesizing genetic bases for behavior, yet the CNS interface between genes and culture is greatly oversimplified. Themes of similarities and differences in neural organization and size among taxonomic groups and in concepts of plasticity and structure-function relations need to be thoroughly discussed in an attempt to elucidate biological bases of behavior. Perhaps the role of the nervous system in evolution has been de-emphasized because few scientists can be thoroughly expert in multiple fields. We hope that this volume will serve as a vehicle toward more robust syntheses.

Although the subject matter centers on evolutionary questions, the findings will interest neurobiologists, psychologists, and researchers in the neurological fields of medicine. Because of their phylogenetic closeness to humans, nonhuman primates have traditionally been the subjects of biomedical research. The application of such research to the human condition increases in sophistication and robusticity as our knowledge of neurological similarities and differences among nonhuman primates and humans is refined.

Finally, while all the above reasons show the significance of studies of primate brain evolution for many different academic disciplines, the contributors to this volume share an interest in understanding human nature: the acquisition of culture, language, the ability to produce tools, and that elusive faculty called intelligence. A primary reason that we study primate brain evolution is that it is just plain interesting!

Several contributors review a number of important concepts in primate brain evolution. Campbell discusses the fundamental problems of establishing homologies in nervous tissue, Noback summarizes current information about the neurological substrates for language, and MacLean thoroughly reviews the concept of the triune brain. Questions underlying allometric scaling of brain and body weights are discussed by Holloway and Post, Jerison, Martin, and Radinsky. Falk reviews data on cortical maps.

In addition to review articles, several contributors present new methods, data, and concepts that directly challenge some of the older theories of primate brain evolution.

Some of the chapters present methodological advances. Martin suggests that allometric analyses of brain and body weight relations will improve if the major axis rather than the linear regression technique is used. In reanalyzing brain and body weight data by this method, he concludes that a possible link between the two features is provided by the basal metabolic rate. Bauchot uses volumes of neural structures to construct dendrograms from which phylogenetic conclusions may be drawn. Based on this method, he concludes that *Tarsius* is better classified with prosimians and *Tupaia* is closer to prosimians than to insectivores. A rapid and automatic method for analyzing Nissl-stained material, presented by Zilles, Stephan, and Schleicher, promises to revitalize cytoarchitectonics. These investigators use this method to quantitate differences in the visual cortex and relate them to functional adaptations and phylogenetic relationships. Working in a previously neglected area, Conroy demonstrates the value of applying findings on cerebral vessels and hemodynamics to evolutionary problems.

New data and the new conceptual bases they provide are also presented. Allman documents a *qualitative* difference in the organization of the optic tectum that separates primates from other mammals. Ogren presents data suggestive of a small qualitative difference between macaques and humans in the pulvinar. The gradual reduction of olfactory bulbs during prosimian evolution is challenged by Gurche's data indicating that the reduction had occurred by the Eocene. He suggests, however, that during prosimian evolution the amount of primary sensory-motor cortex devoted to the extremities was increased. Galaburda and Pandya's work indicates that shifts in brain organization, such as the location of the macaque cytoarchitectural field 44 in the inferior limb of the arcuate sulcus rather than in the orbitofrontal cortex as found in the human brain, can occur without being reflected in the gross surface topography. The theory of an evenly distributed increase in numbers of neurons during hominid evolution is challenged by Armstrong's thalamic data, which documents greater relative changes in limbic than in association, motor, or sensory regions. LeMay, Billig, and Geschwind extend our knowledge of asymmetries by noting asymmetric protuberances (petalias) in the occipital region of New World monkeys. An analysis of neonatal brain-to-body weight relationships led Sacher to conclude that the primate growth pattern differs from other mammals and evolved among the Paleocene primates as a mechanism to reduce maternal costs. Leutenegger finds that increased human encephalization is almost exclusively a postnatal growth phenomenon. Radinsky reviews many allometric considerations and then challenges the facile assumption of a relationship between brain size and intelligence.

The approaches summarized in this volume provide the groundwork for anticipated advances during the next decades. Paleoneurology will expand as the number of fossil specimens and their biology is better interpreted. Theoretically, comparative data can be generated from all approaches used to study the structure and function of the CNS. In practice, however, comparative units have remained overwhelmingly morphological. Important contributions from a neurophysiological approach are in this volume, but few comparative neurochemical primate data are now available. In the next decades, comparative neurochemistry, immunology, and neurophysiology will be increasingly used to expand our knowledge of brain evolution in the primate order and in other vertebrates.

This volume was made possible by many people. We thank the personnel at the Editorial Office at Louisiana State University Medical Center, and in particular Virginia Howard for her expert editing and extensive help and Andrea Troxclair for typing the manuscripts and preparing them for publication. Louisiana State University and the University of Puerto Rico are acknowledged for travel support, as are the American Primatological Society and the National Science Foundation, who sponsored a number of American participants to the 1980 International Primatological Society meetings. Professors A. Fasolo and B. Chiarelli are gratefully acknowledged for their generosity in helping to organize the precongressional symposium on primate brain evolution that took place in Italy. We thank all the contributors to this volume. We are saddened to note the death of George Sacher, whose excellent contribution is published posthumously. Finally, we are delighted to acknowledge Elizabeth Crosby, who delivered a superb talk on the primate cortex at the 1980 physical anthropology meetings in Niagara Falls and to whom this volume is dedicated.

ESTE ARMSTRONG
DEAN FALK
October 1981

CONTENTS

ONTOGENETIC PERSPECTIVES

APPROACHES FROM CYTOARCHITECTONICS

A PALEONEUROLOGICAL PERSPECTIVE

THEORETICAL OVERVIEWS

SOME QUESTIONS AND PROBLEMS RELATED TO HOMOLOGY

C.B.G. Campbell*

Comparisons of human brains with those of nonhuman primates and other animals have played a large role in the making of inferences about human origins and evolution. Originally, brain size and gross morphology were the primary objects of comparison. Sir Wilfrid Le Gros Clark, along with some others, was instrumental in placing importance on similarities and differences in internal structure as revealed by descriptive and experimental histology. The concept of homology is central to the making of such comparisons between animals.

This concept, which attempts to define what is the "same" structure in animals of divergent groups, has been the subject of a large and often complex literature. Although several definitions have been proposed, most fall into one of two broad groups or schools of thought. The first takes a "structuralist" approach based on similarity. The second is the "phylogenetic" school, which bases the definition of the relationship on common ancestry. According to the latter school, similarity remains important in inferring that a common ancestry exists. Two structures with a common ancestry may, however, be highly dissimilar in the animals being compared.

The definition of "homology" that I prefer has been presented elsewhere (Campbell and Hodos, 1970). *Structures and other entities are homologous when they could, in principle, be traced back through a genealogical series to a stipulated common ancestral precursor irrespective*

*From the Department of Medical Neurosciences, Division of Neuropsychiatry, Walter Reed Army Institute of Research, Washington, D.C. 20012. This material has been reviewed by the Walter Reed Army Institute of Research, and there is no objection to its presentation and/or publication. The opinions or assertions contained herein are the private views of the author and are not to be construed as official or as reflecting the views of the Department of the Army or the Department of Defense.

of morphological similarity (Ghiselin, 1966). Two other subtypes of homology that apply to special situations are "serial homology" and "field homology" (Smith, 1967). The former refers to the derivation of structures, however similar or dissimilar, from a common anlage, or in other words from the same ontogenetic division of two or more segments, *in any one individual.* The dorsal root ganglia of the spinal cord are an example of serial homology.

Perhaps more important is field homology, the derivation of structures, however similar or dissimilar, from a common anlage, or in other words from the same or different segments, *of any two or more compared individuals or groups of individuals.* This concept has been little used in neurobiology but is potentially important. It may be the only kind of homology that can be established when the organisms compared differ markedly because they are members of two different higher taxa, e.g., different classes.

Associated with the term *homology* are a number of other terms that have been similarly subject to divergences of opinion in regard to their meaning. For further discussion see Campbell and Hodos (1970). The term *homogeny* refers to homologous structures that are also morphologically similar (Lankester, 1870). Particularly important is the term *homoplasy*: structures or entities that are morphologically similar but that cannot, in principle, be traced back to a stipulated common ancestral precursor (Simpson, 1961). There are three sources of homoplastic similarity. *Convergence* is the appearance of similar characters separately in two or more lineages without a common ancestry pertinent to the similarity, but involving adaptation to similar ecological conditions (Simpson, 1961). *Parallelism* is the appearance of similar characters in two or more lineages whose common ancestor lacked the character, but is the product of the interaction of similar genotypes with a common environmental challenge. Parallelism is particularly common in the order Primates. *Chance similarity* refers to resemblances in characteristics that appear in separate groups by independent causes but not involving adaptation to similar ecological conditions (Simpson, 1961). The term *analogy* is applied to correspondence between structures or entities due to similarity in function whether or not they can be traced to a stipulated common precursor. Several workers (e.g., Ghiselin, 1966) emphasized the importance of using a conditional phrase that indicates that the two or more structures are homologous *as something.* The wings of bats, pterodactyls, and birds are homologous *as limbs* but are not homologous *as wings* because their common ancestor lacked wings. This practice has not yet taken hold in comparative neurology, although more attention is now being directed toward homology and homoplasy than formerly. Conditional phrases used in neurobiology might tend to be long and clumsy. An example of the use of conditional phrases would be that the laminae of the lateral geniculate nuclei of marsupial native cats (*Dasyurus*) and placental domestic cats (*Felis*) are homologous *as lateral geniculate neurons*, but are not homologous *as laminae* because their common ancestor almost certainly lacked laminae. In this example the conditional phrases are not complex, but that situation would not always be true. Because less effort is required to assume that others will know what one means, thus avoiding the necessity of constructing conditional phrases, this approach has been the most common course. This course, however, is a potential source of misunderstanding, and an attempt to use conditional phrases is worthwhile.

Most contemporary neurobiologists are concerned with attempting to understand how the human brain works or with discovering broad principles of organization and function applicable to all nervous systems. Most have no serious interest in evolutionary questions, and because evolutionary inferences cannot be readily tested by experiment there has been a lack of respect for evolutionary comparative neurology as a scientific discipline. Consequently, elaborate discussions of definitions of homology and other terms, as well as of criteria for recognition of these relationships, have not had much impact on most neurobiologists. Some workers appear to be uncertain whether the concept even applies to the nervous system. For example, Zeki (1980) stated: "What precisely is meant by 'homology' in the nervous system has never been clear...", as if the nervous system is somehow exempted from the rules that apply to other organ systems.

Uncertainty exists if the user of the term does not specify whether he is using the term in a structuralist or a phylogenetic sense. Once the specification is made, there is now, I believe, reasonable consensus on what is meant in each case. One cannot avoid the effort required to understand the concept (henceforth I refer to the phylogenetic sense of the term exclusively) and criteria for its recognition if one's comparisons are intended for inferring phylogenetic relations or for making generalizations about observed phenomena. Questions and problems related to the concept persist, however, and the organizers of this conference suggested that discussion of some of them might prove useful.

METHODS FOR HOMOLOGY RECOGNITION

So-called criteria for the recognition of homologous relations between different nervous system structures were discussed in a previous paper (Campbell and Hodos, 1970). It was argued there that experimentally determined connectivity patterns were an important source of evidence, along with the more traditional topographic and topologic position of nuclear groups. Because the central nervous system is composed of groups of neurons that have related functions and that communicate with other such groups by means of cell processes, this important feature certainly cannot be ignored. After all, the connectivity relationship is functionally among the most significant. Indeed, some might say it is the *most* significant. Nevertheless, connections were not recommended to be regarded as *the* criterion. Any single criterion may prove misleading in a specific instance. It was suggested that multiple lines of evidence be evaluated before homology is inferred.

Attempts to use a single criterion persist. Winer and Morest (1979) described homologous relations between subgroups of the medial geniculate body in the domestic cat, a placental, and the American opossum, a marsupial. Both species have this diencephalic nucleus divisible into ventral, dorsal, and medial subnuclei. When the Golgi staining method is used, the cell types in each division are similar. Differences exist in the proportions of the different cell types in the various divisions. Two subtypes of Golgi type II (short axon) cells exist in the dorsal division of the cat, whereas there is probably a single type in the opossum. These authors concluded, apparently on the basis of this single mode of comparison, that neuronal structure is the most useful and reliable criterion of homology. Although, technically, neuronal structure includes the axon and its trajectory, these authors are not including connectivity relations in their usage of the term.

The Golgi method in skilled hands demonstrates well the cell body and dendritic tree of individual neurons. Short axons may be seen in their entirety; however, long, highly-branched, complex axons can be followed with difficulty, if at all. The method is capable of providing much information about the population of neurons being studied; however, it does not provide all the necessary information in all cases. For example, when one compares the lateral geniculate body of the domestic cat and opossum, no such striking similarities are found. The nucleus is undifferentiated in the opossum. In the cat it is a highly laminated structure composed of a number of cell types. That the two nuclei are homologous is inferred by other criteria, such as topography and topology, and connectivity.

There is less difficulty in inferring homology when comparing species of the same lower taxon, i.e., of the same family or genus. Species of different orders and, especially, of different classes can be extremely difficult to analyze in this regard. Although the phylogenetic *definition* of homology does not involve similarity, the *recognition* of the homologous relationship requires similarity or a sequence of intermediates between the two structural states. The latter is not usually available to the comparative neurologist. The thalamus of a salamander and that of a rhesus monkey, for example, are in most respects superficially so dissimilar that it is difficult to draw homologies between neuronal populations with confidence. Anamniote brains are usually small and the neuronal cell bodies mostly have not migrated away from the central, fluid-filled ventricle. Amniote brains are usually much larger, composed of many more tightly-packed neurons, and the neuronal cell somata have in many cases migrated away from the ventricles to form highly differentiated nuclear groups and layered cortical formations. A clear understanding of the concept is appropriate in any case, but is most essential in making judgements in situations in which the organisms compared are of different orders or classes.

One cannot always exclude homoplasy (parallelism, convergence, or chance similarity); however, if there are sufficient similarities between two structures and they are in reasonably detailed agreement, then homology can usually be inferred. One should not expect identity in every detail. Some well-known homologues in other systems are quite dissimilar, but the intermediate stages are known. A failure to appreciate some of these features of the concept of homology led Zeki (1980) to argue that the neocortical visual area he described in the rhesus macaque as the "motion area" of the superior temporal sulcus is not homologous with MT, the middle temporal area, found by Allman and Kaas in the owl monkey (Allman and Kaas, 1971).

A comparison of these cortical areas in the two species reveals a great many detailed similarities. The long axis of the temporal lobe in these primates is oriented in an oblique fashion with its longitudinally placed superior temporal sulcus having its posterior extremity superior to its anterior extremity (Fig. 1). In these species MT and the motion area of STS (superior temporal sulcus) are both located in the posterior and superior portion of the temporal lobe. Both are associated with the superior temporal sulcus. The motion area of STS is located in the posterior bank and floor of the sulcus. The middle temporal area, because the less gyrencephalic *Aotus* has a superior temporal sulcus that does not extend so far posteriorly and superiorly, is located on the posterior bank of the sulcus

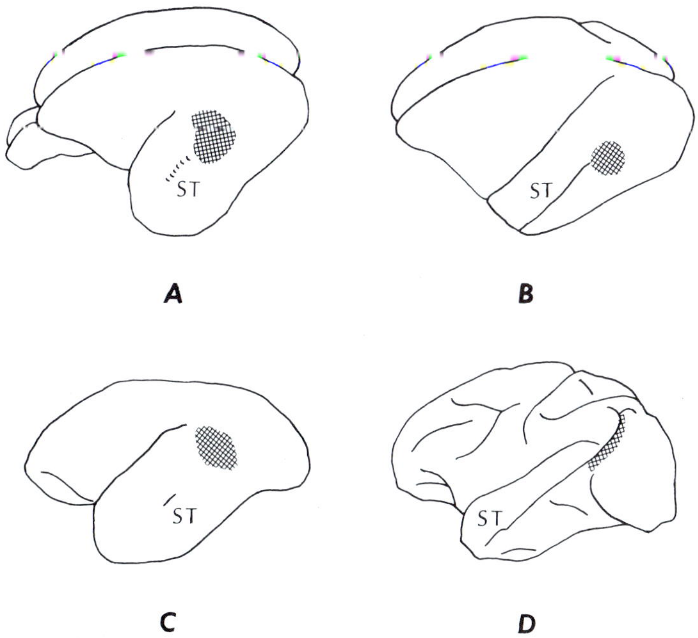

Fig. 1 The middle temporal visual areas (MT) of (A) *Galago*, (B) *Aotus*, (C) *Callithrix*, and the "motion area" of *Macaca* (D). The cortical area is indicated by cross-hatching. ST indicates the superior temporal sulcus or the shallow groove occupying the same position.

and on the lateral cortical surface of the middle temporal gyrus. The sulcus is terminating just as it reaches MT in *Aotus* (Fig. 1). The two areas, however, are topographically extremely similar.

Both MT and the motion area are direct projection zones of Brodmann's area 17 (striate cortex) and 18. Also identical types of neurons, the solitary cells of Meynert, project axons from area 17 to MT and the motion area in both species. The two areas have reciprocal connections with area 17, and the laminar distributions of the connections are the same for both. The contralateral visual field is completely and topographically represented in both areas. A striking band of myelin in the cortex of MT distinguishes it from the surrounding cortex and it is also seen in the motion area of STS. MT and the motion area are both nonhomogeneous architecturally, but they change their appearance in the region representing the far peripheral visual field. Here MT is seen to have more sparsely distributed cells and thinner cortex with less myelin. The shift in cytoarchitecture occurs in the region representing 20-30 degrees from the vertical meridian. These changes are comparable to those found in the motion area. Last, the neurons in both MT and the motion area of STS are highly sensitive to the direction of moving stimuli (Ungerleider and Mishkin, 1979). Zeki (1980) finds these arguments "not compelling."

It is argued (Zeki, 1980) that the term MT stands for middle temporal, whereas the middle temporal gyrus of *Macaca mulatta* is situated far ventrally in the temporal lobe and away from the motion area. Actually, only part of the middle temporal gyrus of *Macaca* is located far ventrally. Its superior extremity reaches to the level of the termination of the lateral sulcus. The motion area of *Macaca* is then partially in the middle temporal gyrus as well as in its superior continuation, the angular gyrus. Both MT and the motion area are located in cortex of the floor and posterior bank of the superior temporal sulcus at its most posterior extremity. More of MT in *Aotus* is exposed on the lateral surface of the hemisphere because this brain is so much less gyrencephalic than that of *Macaca*. Architecturally the two areas are markedly similar.

Zeki (1980) stated that reciprocal connections between striate cortex and the temporal lobe visual areas are also found between striate cortex and other areas of visual cortex. Further, he argued that the fact that both MT and the motion area receive this input from the lower layers of area 17 is somehow negated by the fact that the Clare-Bishop area of the cat receives its input from the upper layers of area 17. The Clare-Bishop area of the cat receives its input from the upper layers of area 17. The Clare-Bishop area of the cat is also a temporal-lobe visual area. That the two primates should be so similar and the cat different is not surprising. The two primates share a more recent common ancestry than either does with the cat. Temporal lobe visual areas are found in other anthropoid primates, e.g., *Saimiri* and *Callithrix* (Fig. 1); and Allman et al. (1975) described a very similar area in a prosimian, *Galago*. Allman (this volume) indicates that this is suggestive that MT was present in the early primates. Temporal lobe visual areas found in nonprimate mammals, e.g., the cat, *Felis domestica*, and the squirrel, *Sciurus carolinensis*, have not been so extensively studied. Whether or not they are homologues of the primate MT is uncertain.

Zeki (1980) found similarities in physiological properties between MT neurons and motion area neurons, as well as what he calls "striking differences." He studied 178 neurons of area MT in *Aotus* and contrasted his results with studies of 155 neurons from the motion area of *Macaca* from a previous study. He found that in his sample of MT neurons 76% were directionally selective, i.e., respond to movement of stimuli in one direction within their receptive field and do not respond to movement in the opposite direction or, more rarely, are inhibited by movement in the opposite direction. Of these neurons, 71% were orientation-selective as well. This means that the orientation of the stimulus (a rectangular slit of light) was critical in determining whether or not the neuron fires.

In the motion area of *Macaca* most cells (about 90%) were directionally selective, but only a very small percentage were also orientation-selective. In other papers (Zeki, 1978a,b), however, a considerably larger percentage of neurons were found to be orientation-selective in the motion area. This raises doubts as to the validity of this difference. Another functional difference used by Zeki to deny homology is the fact that more binocular interactions were seen in MT neurons. Although a majority of cells in both species respond equally well to stimuli delivered to either eye, 42% of MT neurons showed some degree of binocular interaction. He gave the impression that binocular interactions are not found in the motion area of *Macaca*; however, in a previous paper (Zeki, 1978a), he stated that such are present but appear to be rare. He gave no percentages.

In both areas then, a large majority of the neurons are directionally selective and respond equally well to stimuli in either eye. Differences appear to exist in the percentages of neurons that are orientation-selective and display some sort of binocular interaction. These differences are principally quantitative and their significance for visual function are not entirely understood. Again, it must be remembered that homologous structures need not be identical. These species are members of divergent taxa with an appreciable period of independent evolution, and are divergent in activity rhythm, habitat, and lifestyle. One would expect some differences.

Other primates, even a prosimian, have similar areas (Allman et al., 1973). I believe that the evidence is strong that the areas designated as MT (the middle temporal visual area) in *Callithrix*, *Saimiri*, *Aotus*, and *Galago*, as well as Zeki's motion area in *Macaca*, are homologous. Indeed, the evidence reviewed above for *Aotus* and *Macaca*, and originally marshalled by Ungerleider and Mishkin (1979), is one of the better examples of well-founded homology inference in neurobiology. If the motion area is not the homologue of MT, where is MT in *Macaca*? Has this area, apparently present in prosimians and New World monkeys, been lost in Old World monkeys? Is the motion area then a new development in Old World monkeys? What would be the selection pressure for evolution of a new area with such similar functions if MT existed in ancestral prosimians? If the motion area is modified MT, then it *is* the homologue of MT by definition. Homoplasy cannot be absolutely excluded, but little or no evidence for it exists in this instance.

This discussion has not been intended primarily as a criticism of Zeki. It is meant to demonstrate how homologies can be inferred in neurobiology, the nature of neurobiological evidence, and the need to understand the concept thoroughly if it is to be properly applied. Again, it is too much to expect that homology can be differentiated from homoplasy in every instance.

EVOLUTIONARY INFERENCE AS AN AFTERTHOUGHT

Despite the fact that evolutionary questions have not been a primary concern of neurobiologists, there has been an unfortunate tendency for many of them to attempt to place their findings in an evolutionary context. This is often done more or less as an afterthought by workers with only a rudimentary understanding of evolutionary biology, either its concepts or its methods. Numerous examples could be given; however, it is especially common in studies that argue that the human or primate brain has some unique features for which a homologue does not exist in nonhuman primates or nonprimates.

First, the intent is most often not to conduct a true comparative study. Human data, or perhaps experimental data from *Macaca*, are gathered first. Comparative data are then sought from the literature. If such data are lacking, it is stated that there is no evidence of such-and-such in nonhuman primates or nonprimates. Most often if comparative data exist, they have been garnered using older, nonexperimental, or otherwise different techniques. Differences found then are frequently not real. For example, Geschwind's (1965) statements concerning the lack of certain fiber connections to the cortical region of nonhuman primates equivalent to the inferior parietal lobule of man were found to be untrue when modern neural pathway tracing techniques were applied to the cortex of *Macaca* (Pandya and Kuypers, 1969; Galaburda and Pandya, this volume).

Similarly, Geschwind and Levitsky (1968) described left-right asymmetries in the temporal speech region of man and cited work of Pfeifer (1936) and von Economo and Horn (1930) to indicate that no such asymmetries occur in anthropoid apes. Reexamination has shown that similar asymmetries do exist in anthropoids (LeMay and Geschwind, 1975; Yeni-Komshian and Benson, 1976; Falk, this volume; LeMay et al., this volume). The older literature is particularly unreliable in regard to connectivity relations because really reliable techniques for their demonstration have been available only since the mid-1950's. It is a fact of life that as new and better neuroanatomical techniques have been developed, the older literature has been shown to be erroneous or incomplete. This necessitates frequent repetition of previous work.

The principal problem is, however, that few workers have a serious interest in inferring phylogeny in neurobiology. Consequently, one seldom finds persons willing to make the extensive, planned comparisons using identical methods that are necessary to infer homologies and, using that information, to make phylogenetic inferences. Little real progress will be made in our understanding of the evolutionary history of the nervous system so long as this is true.

NEURAL PLASTICITY AND HOMOLOGY

The term *plasticity* has been applied to a heterogeneous group of neurobehavioral reorganizations. All of these reorganizations represent ways in which the functional output or the structure of a neuronal aggregate alter through time when tested by a fixed stimulus or are in some other way reexamined in a standard manner. Anatomical reorganizations are usually produced by surgical intervention early in the life of the organism.

The question arises as to whether or not this aspect of the central nervous system, i.e., its capacity for plasticity, has an impact on our understanding of homology as applied to nervous tissue. It raises the practical problem that neurons might alter their connections with one another more readily than we imagined before. This might then lessen the efficacy of connectivity relations as evidence of homology. This is, perhaps, what Nieuwenhuys and Bodenheimer (1966) feared when they advocated topography and topology of neuronal cell body populations as the most reliable criteria of homology.

Relatively minor, often quantitative variations in pathway disposition have been seen between strains of rodents, but not within strains. This constitutes natural variation. More dramatic pathway variation has been produced by surgically removing the brain region in which a tract terminates, or by removing a matching parallel pathway in the opposite half of the brain. Seemingly, the course an axon takes after it leaves the cell body is not totally determined, because after the surgery axons take an aberrant course. These axons, however, terminate in the same regions and usually in the same layers that they would end in normally, although on the wrong side. Sensory system axons usually even map sensory space in a mirror topographic manner in their new region.

At certain stages of development neurons may project both ipsilaterally and contralaterally. The ipsilateral pathway may normally be lost as development progresses. A genetic change could preserve the ipsilateral pathway in some groups of animals. The "pruning" of inappropriate axons at later stages of development has been shown in a number of instances. Even though axons may be induced to take an aberrant course, there is a strong affinity between axons and their appropriate target cells. This affinity is not absolute. Axons may sometimes terminate in other targets of the same system, but only rarely in uncharacteristic regions, when their usual target is removed (Lund, 1978).

The question is, do phenomena of this sort occur naturally and play a role in evolution? Could such naturally occurring phenomena complicate our efforts to infer homology using axonal connectivity patterns as evidence of the relationship? Selection pressures that produce alterations in peripheral structures could secondarily alter the nuclear groups, both motor and sensory, which innervate these structures. Plasticity effects might then produce changes in central connectivity patterns. I believe that the evidence thus far indicates that connectivity pattern changes are more likely to be quantitative than qualitative. There does not appear to be much evidence for modifications that would make neuronal populations unrecognizable. That such profound changes could occur by this mechanism, however, is not impossible. A great deal of research needs to be done with evolutionary questions allowed to direct the work. The lack of popularity and support of this research area has significantly hindered progress.

Many problems are to be found in applying the concept of homology to studies of the central nervous system, or any morphological system. Homoplasy, including especially parallelism and convergence, will always be a potential source of confusion to us. Nevertheless, we must confront these problems as best we can if we hope to trace the history of the nervous system and its role in the adaptation of organisms.

Acknowledgments: I thank Drs. R. Glenn Northcutt and Ann B. Butler for advice given with enthusiasm and generosity.

REFERENCES

Allman, J.M. and Kaas, J.H., 1971. A representation of the visual field in the caudal third of the middle temporal gyrus of the owl monkey (*Aotus trivirgatus*). Brain Res., 31:85-105.

Allman, J.M., Kaas, J.H., and Lane, R.H., 1973. The middle temporal visual area (MT) in the bushbaby, *Galago senegalensis*. Brain Res., 57:197-202.

Campbell, C.B.G. and Hodos, W., 1970. The concept of homology and the evolution of the nervous system. Brain Behav. Evol., 3:353-367.

Economo, C. von and Horn, L., 1930. Ueber windungsrelief, Masse Rindenarchitektonik der Supratemporalflaeche, ihre individuellen und ihre Seitenunterscheide. Z. Ges. Neurol. Psychiat., 130:678-757.

Geschwind, N., 1965. Disconnexion syndromes in animals and man. Brain, 88:237-294; 585-644.

Geschwind, N. and Levitsky, W., 1968. Human brain: Left-right asymmetries in temporal speech region. Science, 161:186-187.

Ghiselin, M.T., 1966. An application of the theory of definitions to systematic principles. Syst. Zool., 15:127-130.

Lankester, E.R., 1870. On the use of the term homology in modern zoology, and the distinction between homogenetic and homoplastic agreements. Ann. Mag. Natl. Hist., 6:34-43.

LeMay, M. and Geschwind, N., 1975. Hemispheric differences in the brains of great apes. Brain Behav. Evol., 11:48-52.

Lund, R.D., 1978. Development and Plasticity of the Brain, Oxford University Press, New York.

Nieuwenhuys, R. and Bodenheimer, T.S., 1966. The diencephalon of the primitive bony fish *Polypterus* in the light of the problem of homology. J. Morph., 118:415-450.

Pandya, D.N. and Kuypers, H.G.J.M., 1969. Cortico-cortical connections in the rhesus monkey. Brain Res., 13:13-36.

Pfeifer, R.A., 1936. Pathologie der Hörstrahlung und der corticalen Hörsphäre. In, Handbuch der Neurologie, vol. 6., O. Bumke and O. Foerster, eds., J. Springer, Berlin, pp. 533-626.

Simpson, G.G., 1961. Principles of Animal Taxonomy, Columbia, New York.

Smith, H.M., 1967. Biological similarities and homologies. Syst. Zool., 16:101-102.

Ungerleider, L.G. and Mishkin, M., 1979. The striate projection zone in the superior temporal sulcus of *Macaca mulatta*: Location and topographic organization. J. Comp. Neurol., 188:347-366.

Winer, J.A. and Morest, D.K., 1979. What is a homology in the central nervous system? Golgi study of the medial geniculate body of the opossum and cat. Abstr. Soc. Neurosci., 5:147.

Yeni-Komshian, G.H. and Benson, D.A., 1976. Anatomical study of cerebral asymmetry in the temporal lobe of humans, chimpanzees, and rhesus monkeys. Science, 192:387-389.

Zeki, S.M., 1978a. Uniformity and diversity of structure and function in rhesus monkey prestriate visual cortex. J. Physiol., Lond., 277:273-290.

Zeki, S.M., 1978b. Functional specialization in the visual cortex of the rhesus monkey. Nature, Lond., 274:423-428.

Zeki, S.M., 1980. The response properties of cells in the middle temporal area (area MT) of owl monkey visual cortex. Proc. R. Soc., London, Ser. B., 207:239-248.

RECONSTRUCTING THE EVOLUTION OF THE BRAIN IN PRIMATES THROUGH THE USE OF COMPARATIVE NEUROPHYSIOLOGICAL AND NEUROANATOMICAL DATA

John Allman*

In his influential book, *The Antecedents of Man*, Le Gros Clark (1962) maintained that the order Primates cannot be *defined* by specific attributes that uniquely distinguish its members from the members of other orders but rather can only be ***characterized by evolutionary trends*** such as the progressive expansion of the brain. Recent neurophysiological and neuroanatomical data suggest that at least one unique defining feature for the order Primates does exist. In all mammals the retina projects to the optic tectum, but the manner in which the visual field is represented in the optic tectum of Primates differs from that found in all other mammals. In Primates the optic tectum on each side of the brain contains a systematic representation of the *contralateral half of the visual field*, whereas in other mammals the optic tectum of each side contains a systematic representation of the visual field that is viewed by the *contralateral retina*, which is the primitive vertebrate condition found in all non-mammalian vertebrates (see Fig. 1). The drawback in using such neurophysiological criteria in taxonomic definitions is that it is not possible to survey a very large number of species for a particular attribute, and the method cannot be applied to fossils. Nevertheless, neurophysiological mapping of the optic tectum has been conducted in four haplorhine primate species (Cynader and Berman, 1972; Kadoya et al., 1971; Lane et al., 1973; Updyke, 1974), in one strepsirhine primate species (Lane et al., 1973), in a number of other mammalian orders including Lagomorpha (Sitthi-Amorn, personal communication), Rodentia (Siminoff et al., 1966, Lane et al., 1971; Drager and Hubel, 1975; Tiao and Blakemore, 1976), Carnivora (Feldon, et al., 1970; Lane et al., 1974), Insectivora (Lane et al., 1971), marsupials (Lane et al., 1974; Volchan, et al. 1978), and in other classes of vertebrates including birds (Allman, Kaas, Schramm, unpublished map of the optic tectum in the

*From the Division of Biology, California Institute of Technology, Pasadena, California 91125

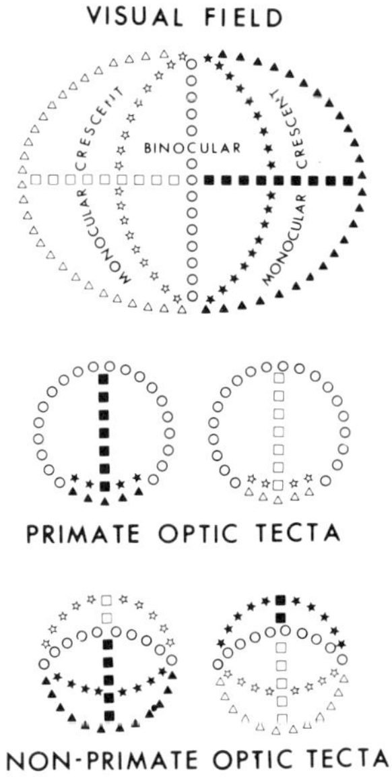

Fig. 1. Schematic plan of the representation of the visual field in the optic tectum of primates and non-primates. Circles indicate the vertical meridian (midline) dividing the two halves of the visual field; squares, the horizontal meridian; triangles, the extreme periphery of the visual field; stars, the division between binocular and monocular portions of the visual field. (From Allman, J.M., 1977. Progr. Psychobiol. Physiol. Psychol., vol. 7. Courtesy of Academic Press.)

seagull), reptiles (Heric and Kruger, 1965), amphibians (Gaze, 1958; Jacobson, 1962), and fish (Jacobson and Gaze, 1965; Schwassmann and Kruger, 1965).

The unique visuotopic organization of the optic tectum in Primates is related to two other distinctive features of Primates: *frontally directed eyes* with a large binocular field and a large *ipsilateral retinotectal projection.* In both strepsirhine and haplorhine primates the ipsilateral retinotectal projection has been variously estimated to be from one third to one half of the total retinotectal projection (Campos-Ortega and Glees, 1967; Giolli and Tigges, 1970; Tigges and Tigges, 1970; Laemle and Noback, 1970; Tigges and O'Steen, 1974; Hubel et al., 1975; Tigges et al., 1977). Cats, whose frontally directed eyes and large binocular overlap approach the primate condition more than any other mammal, conform to the non-primate plan of tectal visual field representation (Lane et al., 1974) and only 5% to 10% of the total retinotectal projection is from the ipsilateral retina (Kanaseki and Sprague, 1974). In all other mammals that have been investigated, including representatives of the artiodactyls, insectivores, lagomorphs, perissodactyls, rodents, marsupials, and monotremes, the ipsilateral retinotectal projection is reported to be small, or absent (Bodian, 1937; Campbell et al., 1967; Cummings and de Lahunta, 1969; Giolli and Guthrie, 1969; Campos-Ortega, 1970; Tigges, 1970; Campbell and Hayhow, 1971; Rockel et al., 1972; Drager, 1974; Pearson et al., 1976).

In tree shrews the visuotopic organization of the optic tectum conforms to the general vertebrate pattern rather than the special primate type (Lane et al., 1971) and the ipsilateral retinotectal pathway has been reported as small or absent (Tigges, 1966; Campbell et al., 1967; Laemle, 1968), which is consistent with the taxonomic studies that exclude tree shrews from the order Primates (Hill, 1953, Campbell, 1966, 1974; Martin, 1968; Cartmill, 1975; Hershkovitz, 1977).

Frontally directed eyes, the unique visuotopic organization of the optic tectum, and the large ipsilateral retinotectal projection in Primates are part of a larger set of changes in the organization of the visual system that occurred when Primates emerged from the ancestral stock of placental mammals. The skull and cranial endocast of an early Eocene primate 55 million years old, *Tetonius homunculus*, are illustrated in Figure 2. *Tetonius* possessed large bony orbits that completely encircled it eyes and a cranium containing a large brain compared with those of its contemporaries of similar size (Jerison, 1973; Radinsky, 1977; Gurche, this volume). The relative size and position of the orbits in *Tetonius* closely resemble living nocturnal prosimians such as *Galago senegalensis* (see Fig. 3). As can be seen in Figure 4, *Galago* possesses large, frontally directed eyes with virtually as much binocular overlap, on the order of 120° to 140°, as is

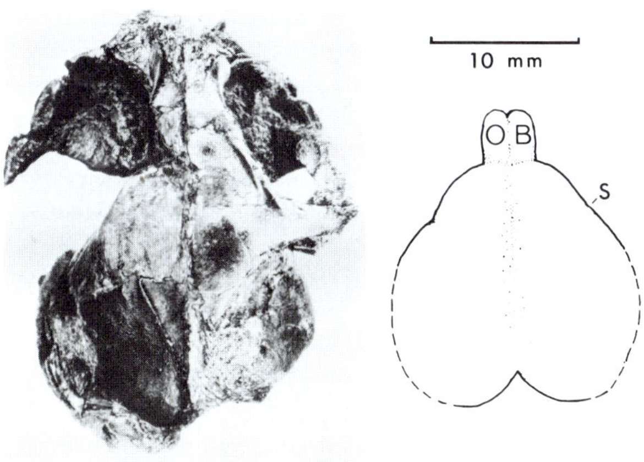

Fig. 2. Left: dorsal view of the skull of *Tetonius homunculus*. A. M. N. H. No. 4194. Right: dorsal view of Radinsky's cranial endocast of *Tetonius*. OB: olfactory bulbs. S: sylvian sulcus. (Both figures reproduced from Radinsky, L.B., 1967. Am. J. Phys. Anthropol., vol. 27. Courtesy of Wistar Press.)

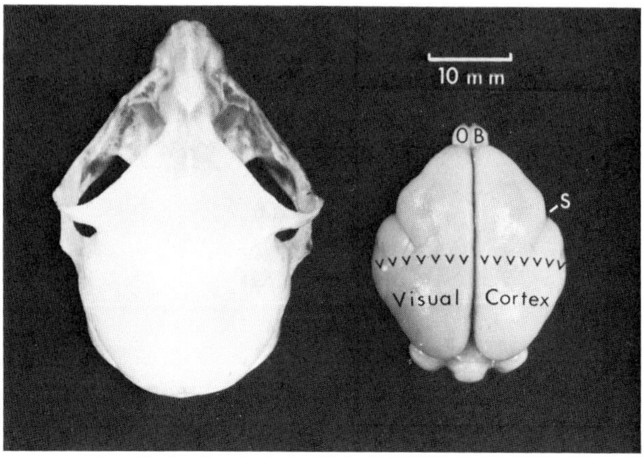

Fig. 3. Left: dorsal view of the skull of *Galago senegalensis*. Right: dorsal view of the brain of *Galago senegalensis*. The visual cortex corresponds to approximately the posterior half of the neocortex. The V's demarcate the anterior border of visual cortex. OB: olfactory bulbs. S: sylvian sulcus. (From Allman, J.M., 1977. Evolution of the visual system in the early Primates. Progr. Psychobiol. Physiol. Psychol., vol. 7. Courtesy of Academic Press.)

present in monkeys, apes, and man (Polyak, 1957; Kaas et al., 1972; Allman and Kaas, unpublished map of the lateral geniculate nucleus in *Galago*). The great similarity in size and position of the orbits in *Galago* and *Tetonius* indicates that *Tetonius* probably also possessed large, frontally directed eyes and was crepuscular or nocturnal. The cranial endocasts of *Tetonius* and other Eocene primates (Radinsky, 1970; Gurche, this volume) show that their brains possessed a conspicuous enlargement of the occipital and temporal lobes, which in modern primates are known to be devoted mainly to the cortical processing of visual information (see Fig. 3 and p. 19). Thus, by the early Eocene, there appeared in Primates the concomitant development of large, frontally directed eyes and an expanded visual cortex. For discussions of the changes in ecological specialization that accompanied these changes in the visual system, see Cartmill (1972, 1974), Allman (1977), and Martin (1979).

In addition to the unique organization of the optic tectum and the developments that can be directly inferred from the skulls and cranial endocasts of Eocene primates, several additional characteristic features of the visual system are found in all primates and probably reflect the ancestral condition. These features include: (1) a high concentration of retinal ganglion cells in the central retina, and the greatly expanded representation of the central retina in the neural maps of the visual field in the brain (Van Buren, 1963; Myerson et al., 1977; DeBruyn et al., 1980); (2) a

Fig. 4. Close-up of the face of *Galago senegalensis*. Note the midline cleft in the upper lip, which is a feature present in all strepsirhine primates but absent in haplorhine primates. (From Allman, J.M., 1977. Progr. Psychobiol. Physiol. Psychol., vol. 7. Courtesy of Academic Press.)

distinctly laminated dorsal lateral geniculate nucleus in which inputs from the two eyes are brought into precise visuotopic register before being relayed to primary visual cortex (Kaas et al., 1972); and (3) a greatly expanded visual cortex containing a number of maps of the visual field (Allman, 1977; Van Essen, 1979; and see Figs. 5 and 6).

The primary visual cortex (V-I) and the adjacent second visual area (V-II) are present in all mammalian species that have been investigated (Allman, 1977). Thus, V-I and V-II probably existed in the early eutherian mammals that were the common ancestors of the various living mammalian orders. The full complement of cortical visual areas varies in different mammals from a minimum of two in a basal insectivore, the hedgehog (Kaas et al., 1970), to a maximum of 12 found recently in cats (Palmer et al., 1978; Tusa and Palmer, 1980). At least 9 cortical visual areas are present in the owl monkey, the primate that has been mapped most completely (see Figs. 5 and 6). Beyond V-I and V-II, homologies are extremely difficult to establish among the visual areas present in mammals belonging to different orders. The last common ancestor of the different mammalian orders lived no more recently than the late Cretaceous period more than 65 million years ago (Romer, 1966), and this ancestral mammal had only rudimentary development of its neocortex (Jerison, 1973). In addition, the adaptive radiation of mammals into different ecological niches with widely divergent behavioral specializations impedes discovery of diagnostic similarities among potentially homologous cortical areas in different mammalian orders (Baker et al., 1980; see Campbell, this volume).

Within the order Primates it is somewhat easier to determine homologous areas in different species (see also Zilles et al., this volume). The middle temporal visual area (MT) is present in strepsirhines (Allman et al., 1973; Tigges et al., 1973) and New World monkeys (Gross et al., 1981; Van Essen et al., 1980), and thus was probably present in the early primates. MT is similar in the different primate species in topographic position (Allman

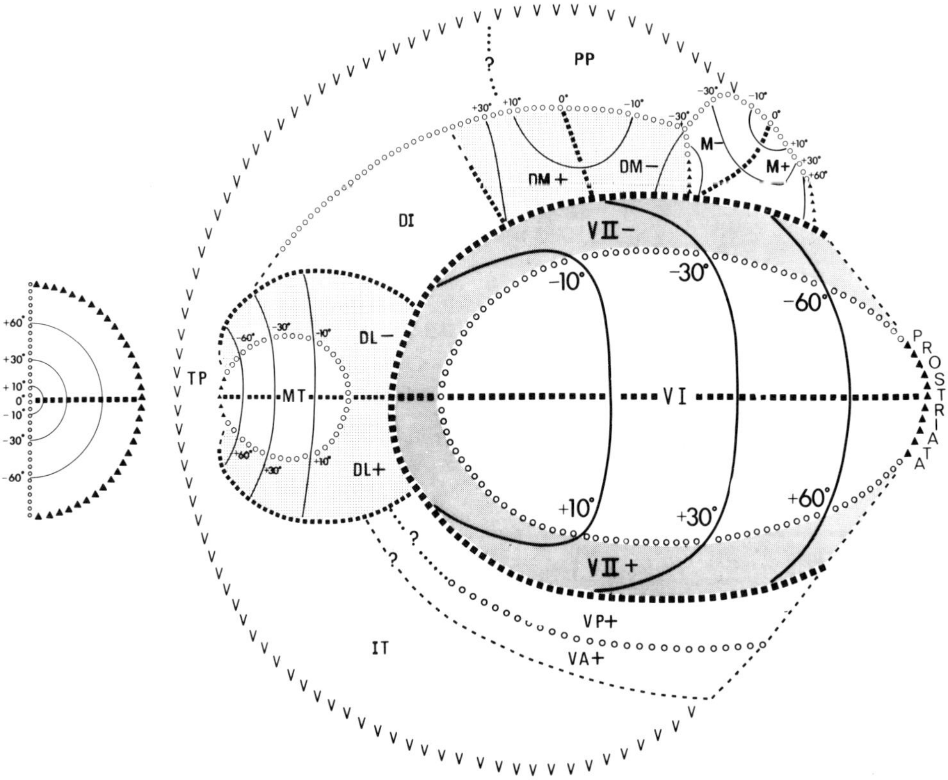

Fig. 5. A schematic unfolding of the visual cortex of the left hemisphere in the owl monkey. The visual cortex corresponds to approximately the posterior third of the entire neocortex. The unfolded visual cortex is approximately a hemispherical surface, which is viewed from above in this diagram. The perimeter chart on the left shows the contralateral (right) half of the visual field. The "prostriata" is histologically and physiologically distinct from V-II (Sanides, 1970; Allman and Kaas, 1971b). Abbreviations and conventions as in Fig. 6. (Figure modified from Allman, J.M., and Kaas, J.H., 1975. Brain Res., vol. 100. Courtesy of Elsevier.)

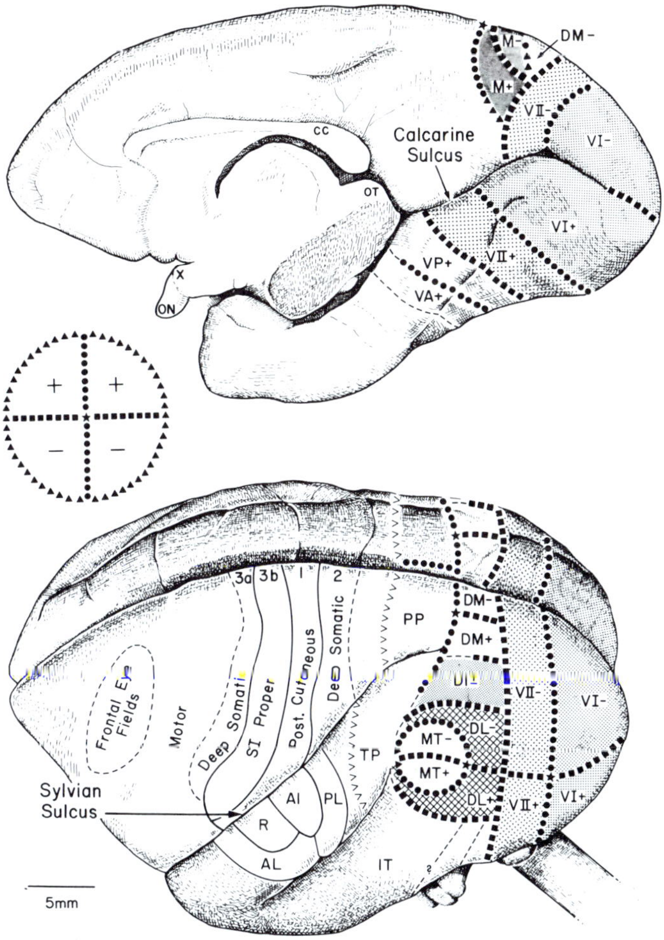

Fig. 6. The representations of the sensory domains in the cerebral cortex of the owl monkey. Above is a ventromedial view; below is a dorsolateral view. On the left is a perimeter chart of the visual field. The symbols in this chart are superimposed on the surface of the visual cortex. Pluses indicate upper quadrant representation; minuses, lower quadrants. The row of V's indicates the approximate border of visually responsive cortex. AI, first auditory area; AL, anterolateral auditory area; CC, corpus callosum; DI, dorsointermediate visual area; DL, dorsolateral crescent visual area; DM, dorsomedial visual area; IT, inferotemporal cortex; M, medial visual area; MT, middle temporal visual area; ON, optic nerve; OT, optic tectum; PL, posterolateral auditory area; PP, posterior parietal cortex; R, rostral auditory area; V-I, first visual area; V-II, second visual area; VA, ventral anterior visual area; VP, ventral posterior visual area; X, optic chiasm. The cortical visual areas were mapped by Allman and Kaas (1971a, b, 1974a,b, 1975, 1976) and Newsome and Allman (1980); the somatosensory areas, by Merzenich et al. (1978); and the auditory areas, by Imig et al. (1977).

and Kaas, 1971a; Spatz and Tigges, 1972; Allman et al., 1973; Gross et al., 1981; Van Essen et al., 1980), characteristic dense myelination of the deeper cortical layers (Allman and Kaas, 1971a; Spatz and Tigges, 1972; Allman et al., 1973; Ungerleider and Mishkin, 1979; Gross et al., 1981; Van Essen et al., 1980), receipt of input from cells in layer IVb and the giant cells of Meynert in V-I (Spatz, 1975; Lund et al., 1976; Lin, Wall and Kaas, unpublished owl monkey data), and characteristic directional selectivity of its neurons (Zeki, 1974; Baker et al., 1981; Zeki, 1980). The directional selectivity of most MT neurons and their strong responsiveness to moving texture fields indicates that the area participates in the *analysis of visual movement* (Zeki, 1974; Baker et al., 1981). Neurons in MT layer V project to the pontine visual nuclei, in which the neurons also are directionally selective and highly responsive to moving texture and which project to the cerebellum, a major center for the control of body and eye movement (Baker et al., 1976; Glickstein et al., 1980).

The dorsolateral visual area (DL) lies adjacent to MT and bears the same topographical relationship to MT as V-II does to V-I (Fig. 5). MT and DL together form a miniature mirror image of the visuotopic organization in V-I and V-II. In DL about 70% of the neurons are selective for the spatial dimensions (length and width) of visual stimuli within excitatory receptive fields that generally are much larger than the preferred stimulus dimensions (Petersen et al., 1980). The dimensional selectivity of DL cells is independent of the sign of contrast in the receptive field, being equally selective to both light-on-dark and dark-on-light stimuli, the amount of contrast, and the position of the stimulus within the excitatory receptive field. DL neurons have a wide range of preferred dimensions from 1° to 30° in length and from 0.25° to 7° in width, and these preferences were independent of each other when both dimensions were tested on the same cell. The dimensional selectivity of DL neurons suggests that DL contributes to *form perception*. This hypothesis is consistent with the observation that DL, of all the visual areas in the owl monkey, has the most expanded representation of the central visual field, where the most acute recognition of form takes place (Allman and Kaas, 1975), and with the recent discovery that DL is the main source of input to inferotemporal cortex (Weller and Kaas, 1980). Inferotemporal cortex has been strongly implicated in the analysis of complex visual stimuli and the learning of visual form discrimination (Gross, 1973; Gross et al., 1981). In an Old World monkey, the macaque, a region adjacent to MT is the main input to inferotemporal cortex (Desimone et al., 1980), and this region, like DL in the owl monkey, emphasizes the representation of the central visual field (Desimone and Gross, 1979). In a strepsirhine, *Galago*, the region corresponding to DL between MT and V-II has been less thoroughly explored than in the owl monkey, but it does appear to contain a similar visuotopic organization and neurons with similar dimensionally selective properties (Allman et al., 1973, and Allman, unpublished data). Thus, DL may be part of the basic complement of areas present in all living primates and existing in the early primates.

The small amount of neocortex possessed by the early placental mammals and the great variation in the number of cortical visual areas reported for different mammalian species suggests that some of the areas beyond V-I and V-II developed at different stages in evolution and independently in different lines of descent. One clear example of variation within the order Primates is among the constituents of the *third tier of visual areas*. In the owl monkey five separate representations of the visual field (medial, M; dorsomedial, DM; dorsointermediate, DI; dorsolateral, DL; and ventral posterior, VP) adjoin the anterior border of V-II and collectively comprise the *third tier* (see Figs. 5 and 6). In a strepsirhine, *Galago senegalensis*, only one visual area, the dorsal (D), is located in the position occupied by three areas, M, DM, and DI, in the owl monkey (Allman et al., 1979). The organization of the comparable region in Old World monkeys has not been completely worked out; however, two areas, V-3 and V-3a, occupy the same locality (Van Essen and Zeki, 1978). On the ventral surface, VP in the owl monkey appears to be homologous with a similarly organized area recently mapped in an Old World monkey (Newsome and Allman, 1980; Newsome et al., 1980); it is not known whether a comparable area is present in strepsirhines. Comparative visuotopic, connectional, and functional data from representative strepsirhine and haplorhine species should eventually establish which areas are homologous in different primates. The existing data suggest that there exists a core of areas including V-I, V-II, MT, DL, and possibly one or two others that are common to all primates, but that there also exist areas present in some species that do not exist in others (see Zilles et al., this volume). Each area probably performs a distinct set of functions in visual perception and visuo-motor coordination, and an area possessed by one species (or larger taxon) and not by another will probably endow its possessor with behavioral capabilities not present in the other. A major task for the future will be to determine what are the distinctive functions of these cortical areas and how they relate to the behavioral and ecological specializations of their possessors.

The early primates probably were small nocturnal predators living in the fine branches of trees (Cartmill, 1972, 1974; Allman, 1977; Martin, 1979); some primates have retained this mode of life, but most have become larger, diurnal folivores or frugivores. Frugivorous diet is correlated positively with brain size and the amount of neocortex relative to body size in Primates (Stephan and Andy, 1970; Clutton-Brock and Harvey, 1980). This association between frugivorous diet and enlarged brain and neocortex may be related to the special demands imposed because a fruit eater's food supply is not constant; different plants bear fruit at different times and at different locations in the complex matrix of the tropical forest (Allman, 1977). Clearly, an animal guided by memory of the locations of fruit-bearing trees can more efficiently exploit the available fruit sources than would otherwise be possible. Thus natural selection would have favored the development in frugivorous primates of capacities for visuospatial memory, which may be localized in a particular area or set of areas.

Another, even more significant behavioral specialization is the development of complex systems of social organization in many primate species. The neural substrate for the mediation of social communication is bound to be an important focus of evolutionary change in the brains of Primates. The order Primates is divided into the strepsirhines (lorises, lemurs, galagos), which tend to have relatively simple forms of social organization, and the haplorhines (tarsiers, monkeys, apes, and humans), in which social organization tends to be much more complex (Hill, 1953; Wilson, 1975). In strepsirhines, as in most mammals, the rhinarium, the space between the upper lip and the nostrils, is a furless, moist mucosal tissue that is tightly bound to the underlying maxillary bone and is divided along the midline by a deep cleft (see Fig. 4). Because strepsirhines share this type of rhinarium with most other mammals, it is highly likely to have been the primitive condition in Primates. By contrast, haplorhines possess a mobile upper lip that is capable of participating in facial expression. Strepsirhines, like most primitive mammals, have scent glands and scent marking behaviors that play a key role in their social communication, and although haplorhines also use olfactory cues to some extent, they rely much more heavily on the use of visually perceived facial expressions and gestures, which allow much more rapid and subtle communication (Schilling, 1979; Sebeok, 1977). Strepsirhines also tend to have much larger olfactory bulbs than do haplorhines (Stephan and Andy, 1970). Thus it appears that as complex systems of social organization evolved in haplorhine primates, social communication was increasingly mediated by the visual channel at the expense of the olfactory. One expression of this evolutionary development is the sensory input to the amygdala, which participates in the control of the neuroendocrine functions of the hypothalamus and thus emotion. In primitive mammals, the main sensory input is from the temporal lobe and particularly from inferotemporal cortex, which is a high-level processor of visual information (Turner et al., 1980; Aggleton et al., 1980). An intriguing possibility is that the temporal cortical neurons projecting to the amygdala correspond to neurons containing the octapeptide cholecystokinin (CCK) (Innis et al., 1979). Neurons responsive to the specific configurations of faces have been recorded in the amygdala and temporal cortex (Sanghera et al., 1979; Perrett et al., 1979; Desimone and Gross, 1980). The clinical condition, prosopagnosia, the inability to recognize familiar faces with relatively little impairment of other visual functions, which is usually associated with lesions located near the occipital temporal junctions, suggests the development of a specialized system for processing the information in faces (Meadows, 1974). Finally, in man another system of social communication, language, has developed along with specialized cortical regions in the temporal and frontal lobes (Penfield and Roberts, 1959; see Noback, this volume; Galaburda and Pandya, this volume).

Acknowledgments: I thank Ms. Leslie Wolcott for drawing Figure 6. This research was supported by National Institutes of Health grants NS-00178 and NS-12131, National Science Foundation grant BNS-77-15605, and the Pew Memorial Trust.

REFERENCES

Aggleton, J.P., Burton, M.J., and Passingham, R.E., 1980. Cortical and subcortical afferents to the amygdala of the rhesus monkey (*Macaca mulatta*). Brain Res., 190:347-368.

Allman, J.M., 1977. Evolution of the visual system in the early primates. In, Progress in Psychobiology, Physiology and Psychology, vol. 7, J.M. Sprague and A.N. Epstein, eds., Academic Press, New York.

Allman, J.M., and Kaas, J.H., 1971a. A representation of the visual field in the caudal third of the middle temporal gyrus of the owl monkey (*Aotus trivirgatus*). Brain Res., 31:84-105.

Allman, J.M., and Kaas, J.H., 1971b. Representation of the visual field in striate and adjoining cortex of the owl monkey (*Aotus trivirgatus*). Brain Res., 35:89-106.

Allman, J.M., and Kaas, J.H., 1974a. The organization of the second visual area (V-II) in the owl monkey: A second order transformation of the visual hemifield. Brain Res., 76:247-265.

Allman, J.M., and Kaas, J.H., 1974b. A crescent-shaped cortical visual area surrounding the middle temporal area (MT) in the owl monkey (*Aotus trivirgatus*). Brain Res., 81:199-213.

Allman, J.M., and Kaas, J.H., 1975. The dorsomedial cortical visual area: A third tier area in the occipital lobe of the owl monkey (*Aotus trivirgatus*). Brain Res., 100:473-487.

Allman, J.M., and Kaas, J.H., 1976. Representation of the visual field on the medial wall of occipital-parietal cortex in the owl monkey, Science, 191:572-575.

Allman, J.M., Campbell, C.B.G., and McGuinness, E., 1979. The dorsal third tier area in *Galago senegalensis*. Brain Res., 179:355-361.

Allman, J.M., Kaas, J.H., and Lane, R.H., 1973. The middle temporal visual area (MT) in the bushbaby, *Galago senegalensis*. Brain Res., 57:197-202.

Baker, J.F., Gibson, A., Glickstein, M., and Stein, J., 1976. Visual cells in the pontine nuclei of the cat. J. Physiol. (Lond.), 255:415-433.

Baker, J.F., Petersen, S.E., Newsome, W.T., and Allman, J.M., 1981. Visual response properties of neurons in four extrastriate visual areas of the owl monkey (*Aotus trivirgatus*): A quantitative comparison of the medial (M), dorsomedial (DM), dorsolateral (DL), and middle temporal (MT) areas. J. Neurophysiol., 45:397-416.

Bodian, D., 1937. An experimental study of the optic tracts and retinal projection of the Virginia opossum. J. Comp. Neurol., 66:133-144.

Campbell, C.B.G., 1966. The relationships of tree shrews: The evidence of the nervous system. Evolution, 20:276-281.

Campbell, C.B.G., 1974. On the phyletic relationships of tree shrews. Mammal Rev., 4:125-143.

Campbell, C.B.G., and Hayhow, W.R., 1971. Primary optic pathways in the echidna, *Tachyglossus aculeatus*: An experimental degeneration study. J. Comp. Neurol., 143:119-136.

Campbell, C.B.G., Jane, J.A., and Yashon, D., 1967. The retinal projections of the tree shrew and hedgehog. Brain Res., 5:406-418.

Campos-Ortega, J.A., 1970. The distribution of retinal fibres in the brain of the pig. Brain Res., 19:306-312.

Campos-Ortega, J.A., and Glees, P., 1967. The termination of ipsilateral and contralateral optic fibers in the lateral geniculate body of *Galago crassicaudatus*. J. Comp. Neurol., 129:279-284.

Cartmill, M., 1972. Arboreal adaptations and the origin of the order Primates, In, The Functional and Evolutionary Biology of Primates, R. H. Tuttle, ed., Aldine, Atherton, Chicago, pp. 97-122.

Cartmill, M., 1974. Rethinking primate origins. Science, 184:436-443.

Cartmill, M., 1975. Primate Origins, Burgess, Minneapolis.

Clutton-Brock, T.H., and Harvey, P.H., 1980. Primates, brains and ecology. J. Zool. (Lond.), 190:309-323.

Cummings, J.F., and de Lahunta, A., 1969. An experimental study of the retinal projections in the horse and sheep. Ann. N.Y. Acad. Sci., 157:293-318.

Cynader, M., and Berman, N., 1972. Receptive field organization of monkey superior colliculus. J. Neurophysiol., 35:187-201.

DeBruyn, E.J., Wise, V.L., and Casagrande, V.A., 1980. The size and topographic arrangement of retinal ganglion cells in the *Galago*. Vision Res., 20:315-328.

Desimone, R., and Gross. C.G., 1979. Visual area: the temporal artery of the macque, Brain Res., 178:363-380.

Desimone, R., and Gross, C.G., 1980. Visual properties of inferior temporal cortex in the macaque. Exp. Brain Res., 41:818-819..

Desimone, R., Fleming, J., and Gross, C.G., 1980. Prestriate afferents to inferior temporal cortex: an HRP study. Brain Res., 184:41-55.

Drager, U., 1974. Autoradiography of tritiated proline and fucose transported transneuronally from the eye to the visual cortex in pigmented and albino mice. Brain Res., 82:284-292.

Drager, U., and Hubel, D.H., 1975. Responses to visual stimulation and relationship between visual, auditory and somatosensory inputs in mouse superior colliculus. J. Neurophysiol., 38:690-713.

Feldon, S., Feldon, P., and Kruger, L., 1970. Topography of the retinal projection upon the superior colliculus of the cat. Vision Res., 10:135-143.

Gaze, R.M., 1958. The representation of the retina on the optic lobe of the frog. Quart. J. Exp. Physiol., 43:209-214.

Giolli, R.A., and Guthrie, M.D., 1969. The primary optic projections in the rabbit. An experimental degeneration study. J. Comp. Neurol., 136:99-126.

Giolli, R.A., and Tigges, J., 1970. The primary optic pathways and nuclei of primates. In, The Primate Brain, C.R. Noback and W. Montagna, eds., Appleton-Century-Crofts, New York, pp. 29-54.

Glickstein, M., Cohen, J., Dixon, B., Gibson, A., Hollins, M., La Bossiere, E., and Robinson, F., 1980. Corticopontine visual projection in the macaque monkey. J. Comp. Neurol., 190:209-230.

Gross, C.G., 1973. Inferotemporal cortex and vision. Prog. Physiol. Psychol., 5:77-115.

Gross, C.G., Bruce, C.J., Desimone, R., Fleming, J., and Gattas, R., 1981. Three visual areas of the temporal lobe. In, Multiple Cortical Areas, C.N. Woolsey, ed., Humana Press, Englewood Cliffs, N.J., in press.

Heric, T.M., and Kruger, L., 1965. Organization of the visual projection upon the optic tectum of a reptile (*Alligator mississippiensis*). J. Comp. Neurol., 124:101-111.

Hershkovitz, P., 1977. Living New World Monkeys (Platyrrhini), vol. 1, Univ. Chicago Press, Chicago.

Hill, W.C.O., 1953. Primates. Comparative Anatomy and Taxonomy. I. Strepsirhini, University of Edinburgh Press, Edinburgh.

Hubel, D.H., LeVay, S., and Wiesel, T.N., 1975. Mode of termination of retinotectal fibers in macaque monkey: An autoradiographic study. Brain Res., 96:25-40.

Imig, T.J., Ruggero, M.A., Kitzes, L.M., Javel, E., and Brugge, J.F., 1977. Organization of auditory cortex in the owl monkey (*Aotus trivirgatus*). J. Comp. Neurol., 171:111-128.

Innis, R.B., Correa, F.M.A., Uhl, G.R., Schneider, B., and Snyder, S.H., 1979. Cholecystokinin octapeptide-like immunoreactivity: Histochemical localization in rat brain. Proc. Natl. Acad. Sci. USA, 76:521-525.

Jacobson, M., 1962. The representation of the retina on the optic tectum of the frog. Correlation between retinotectal magnification factor and retinal ganglion cell count. Quart. J. Exp. Physiol., 47:170-178.

Jacobson, M., and Gaze, R.M., 1965. Types of visual response from single units in the optic tectum and optic nerve of goldfish. Quart. J. Exp. Physiol., 49:199-209.

Jerison, H.J., 1973. Evolution of the Brain and Intelligence, Academic Press, New York.

Kaas, J.H., Guillery, R.W., and Allman, J.M., 1972. Some principles of organization in the dorsal lateral geniculate nucleus. Brain Behav. Evol., 6;253-299.

Kaas, J.H., Hall, W.C., and Diamond, I.T., 1970. Cortical visual areas I and II in the hedgehog: Relation between evoked potential maps and architectonic subdivisions. J. Neurophysiol., 33:595-615.

Kadoya, S., Wolin, L.R., and Massopust, L.C., 1971. Photically evoked unit activity in the tectum opticum of the squirrel monkey. J. Comp. Neurol., 142:495-508.

Kanaseki, T., and Sprague, J.M., 1974. Anatomical organization of pretectal nuclei and tectal laminae in the cat. J. Comp. Neurol., 158:319-338.

Laemle, L.K., 1968. Retinal projections of *Tupaia glis*. Brain Behav. Evol., 1:473-499.

Laemle, L.K., and Noback, C.R., 1970. The visual pathways of the lorisid lemurs (*Nycticebus coucang* and *Galago crassicaudatus*). J. Comp. Neurol., 138:49-62.

Lane, R.H., Allman, J.M., and Kaas, J.H., 1971. Representation of the visual field in the superior colliculus of the grey squirrel (*Sciurus carolinensis*) and the tree shrew (*Tupaia glis*). Brain Res., 26:277-292.

Lane, R.H., Allman, J.M., Kaas, J.H., and Miezin, F.M., 1973. The visuotopic organization of the superior colliculus of the owl monkey

(*Aotus trivirgatus*) and the bushbaby (*Galago senegalensis*). Brain Res., 60:335-349.

Lane, R.H., Kaas, J.H., and Allman, J.M., 1974. Visuotopic organization of the superior colliculus in normal and Siamese cats. Brain Res., 70:413-430.

Le Gros Clark, W.E., 1962. The Antecedents of Man, University of Edinburgh Press, Edinburgh.

Lund, J.S., Lund, R.D., Hendrickson, A.E., Bunt, A.H., and Fuchs, A.F., 1976. The origin of efferent pathways from the primary visual cortex, area 17, of the macaque monkey as shown by retrograde transport of horseradish peroxidase. J. Comp. Neurol., 164:287-304.

Martin, R.D., 1968. Towards a new definition of primates. Man, 3:377-401.

Martin, R.D., 1973. Comparative anatomy and primate systematics. Symp. Zool. Soc. London, 33:301-337.

Martin, R.D., 1979. Phylogenetic aspects of prosimian behavior. In, The Study of Prosimian Behavior, G.A. Doyle and R.D. Martin, eds., Academic Press, New York.

Meadows, J.C., 1974. The anatomical basis of prosopagnosia. J. Neurol. Neurosurg. Psychiatr., 37:489-501.

Merzenich, M.M., Kaas, J.H., Sur, M., and Lin, C.S., 1978. Double representation of the body surface within cytoarchitectonic areas 3b and 1 in SI in the owl monkey (*Aotus trivirgatus*). J. Comp. Neurol., 181:41-74.

Myerson, J., Manis, P.B., Miezin, F.M., and Allman, J.M., 1977. Magnification in striate cortex and retinal ganglion cell layer of owl monkey: a quantitative comparison. Science, 198:855.

Newsome, W.T., and Allman, J.M., 1980. The interhemispheric connections of visual cortex in the owl monkey, *Aotus trivirgatus*, and the bushbaby, *Galago senegalensis*. J. Comp. Neurol., 194:209-234.

Newsome, W.T., Maunsell, J.H.R., and Van Essen, D.C., 1980. Areal boundaries and topographic organization of the ventral posterior area (VP) of the macaque monkey. Soc. Neurosci. Abstr., 6:579.

Palmer, L.A., Rosenquist, A.C., and Tusa, R.J., 1978. The retinotopic organization of lateral suprasylvian visual areas in the cat. J. Comp. Neurol., 177:237-256.

Pearson, L.J., Sanderson, K.J., and Wells, R.T., 1976. Retinal projections in the ringtailed possum *Pseudocheirus peregrinus*. J. Comp. Neurol., 170:227-240.

Penfield, W., and Roberts, L.A., 1959. Speech and Brain Mechanisms, Princeton Univ. Press, Princeton.

Perrett, D.I., Rolls, E.T., and Caan, W., 1979. Temporal lobe cells of the monkey with visual responses selective for faces. Neurosci. Lett., 53:358.

Petersen, S.E., Baker, J.F., and Allman, J.M., 1980. Dimensional selectivity of neurons in the dorsolateral visual area of the owl monkey. Brain Res., 197:507-511.

Polyak, S.D., 1957. The Vertebrate Visual System, Univ. of Chicago Press, Chicago.

Radinsky, L.B., 1967. The oldest primate endocast. Am. J. Phys. Anthropol., 27:385-388.

Radinsky, L.B., 1970. The fossil evidence of prosimian brain evolution. In, The Primate Brain, C.R. Noback and W. Montagna, eds., Appleton-Century-Crofts, New York, pp. 209-224.

Radinsky, L.B., 1977. Early primate brains: Facts and fiction. J. Hum. Evol., 6:79-86.

Rockel, A.J., Heath, C.J., and Jones, E.G., 1972. Afferent connections to the diencephalon in the marsupial phalanger and the question of sensory convergence in the "posterior group" of the thalamus. J. Comp. Neurol., 145:105-130.

Romer, A.S., 1966. Vertebrate Paleontology, 3rd ed., Univ. Chicago Press, Chicago.

Sanghera, M.K., Rolls, E.T., and Roper-Hall, A., 1979. Visual responses of neurons in the dorsolateral amygdala of the alert monkey. Exp. Neurol., 63:610-626.

Sanides, F., 1970. Functional architecture of motor and sensory cortices in primates in light of a new concept of neocortical evolution. The Primate Brain, C.R. Noback and W. Montagna, eds., Appleton-Century-Crofts, New York, pp. 137-208.

Schilling, A., 1979. Olfactory communication in prosimians. In, The Study of Prosimian Behavior, G.A. Doyle and R.D. Martin, eds., Academic Press, New York.

Schwassmann, H.O., and Kruger, L., 1965. Organization of the visual projection upon the optic tectum of some fresh water fish. J. Comp. Neurol., 124:113-126.

Sebeok, T.A., 1977. How Animals Communicate, Indiana University Press, Bloomington.

Siminoff, R., Schwassman, H.O., and Kruger, L., 1966. An electrophysiological study of the visual projection to the superior colliculus of the rat. J. Comp. Neurol., 127:435-444.

Spatz, W.B., 1975. An efferent connection of the solitary cells of Meynert. A study with horseradish peroxidase in the marmoset *Callithrix*. Brain Res., 92:450-455.

Spatz, W.B., and Tigges, J., 1972. Experimental-anatomical studies on the "Middle Temporal Visual Area (MT)" in primates. I. Efferent cortical connections in the marmoset (*Callithrix jacchus*). J. Comp. Neurol., 146:451-463.

Stephan, H., and Andy, O.J., 1970. The allocortex in primates. In, The Primate Brain, C.R. Noback and W. Montagna, eds., Appleton-Century-Crofts, New York, pp. 109-135.

Tiao, Y.-C., and Blakemore, C., 1976. Functional organization in the superior colliculus of the golden hamster. J. Comp. Neurol., 168:483-504.

Tigges, J., 1966. Ein experimenteller Beitrag zum subkortikalen optischen System von *Tupaia glis*. Folia Primatologia, 4:103-123.

Tigges, J., 1970. Retinal projection to subcortical optic nuclei in diurnal and nocturnal squirrels. Brain Behav. Evol., 3:121-134.

Tigges, J., and O'Steen, W., 1974. Termination of retinofugal fibers in squirrel monkey: A re-investigation using autoradiographic methods. Brain Res., 79:489-495.

Tigges, J., Bos, J., and Tigges, M., 1977. An autoradiographic investigation of the subcortical visual system in chimpanzee. J. Comp. Neurol., 172:367-380.

Tigges, J., Tigges, M., and Kalaha, C.S., 1973. Efferent connections of area 17 in *Galago*. Am. J. Phys. Anthropol., 38:393-398.

Tigges, M., and Tigges, J., 1970. The retinofugal fibers and their terminal nuclei in *Galago crassicaudatus* (primates). J. Comp. Neurol., 138:87-102.

Turner, B.H., Mishkin, M., and Knapp, M., 1980. Organization of the amygdalopetal projections from modality-specific cortical association areas in the monkey. J. Comp. Neurol., 191:515-543.

Tusa, R.J., and Palmer, L.A., 1980. The retinotopic organization of area 20 and 21 in the cat. J. Comp. Neurol., in press.

Ungerleider, L.G., and Mishkin, M., 1979. The striate projection zone in the superior temporal sulcus of *Macaca mulatta*: Location and topographic organization. J. Comp. Neurol., 188:347-366.

Updyke, B.V., 1974. Characteristics of unit responses in superior colliculus of the *Cebus* monkey. J. Neurophysiol., 37:896-909.

Van Buren, J.M., 1963. The Retinal Ganglion Cell Layer, Charles C Thomas, Springfield.

Van Essen, D.C., 1979. Visual cortical areas. In, Ann. Rev. Neurosci., W. M. Cowan, ed., Ann. Reviews, Inc., Palo Alto.

Van Essen, D.C., and Zeki, S.M., 1978. The topographic organization of rhesus monkey prestriate cortex. J. Physiol. Lond., 277:193.

Van Essen, D.C., Maunsell, J.H.R., and Bixby, J.L., 1980. The organization of extrastriate visual areas in the macaque monkey. In, Multiple Cortical Areas, C.N. Woolsey, ed., Humana Press, Englewood Cliffs, N.J., in press.

Volchan, E., Rocha-Miranda, C.E., Lent, R., and Gawryszewski, L.G., 1978. The retinotopic organization of the superior colliculus in the opossum (*Didelphis marsupialis aurita*). In, Opossum Neurobiology, C.E. Rocha-Miranda and R. Lent, eds., Academia Brasileira de Ciencias, Rio de Janeiro.

Weller, R.E., and Kaas, J.H., 1980. Connections of the dorsolateral visual area (DL) of extrastriate visual cortex of the owl monkey (*Aotus trivirgatus*). Soc. Neurosci. Abstr., in press.

Wilson, E.O., 1975. Sociobiology, Harvard University Press, Cambridge, Mass.

Zeki, S.M., 1974. Functional organization of a visual area in the posterior bank of the superior temporal sulcus of the rhesus monkey. J. Physiol. Lond., 236:549.

Zeki, S.M., 1980. The response properties of cells in the middle temporal area (area MT) of owl monkey visual cortex. Proc. Roy. Soc. Lond., B. 207:239.

SOME CAUTIONARY NOTES ON MAKING INFERENCES ABOUT RELATIVE BRAIN SIZE

Leonard Radinsky*

The purpose of this paper is to call attention to some of the assumptions that are involved in analyzing and interpreting evolution of relative brain size. These assumptions are well known to the few people who generate and analyze primary data in this area, but may not be apparent to the larger audience that reads and repeats their conclusions. The assumptions I wish to consider are those involved in (1) estimating body weights of fossil species, (2) comparing relative brain sizes of different taxa or of different groups of taxa, and (3) attributing biological significance to differences in brain size.

FOSSIL BODY WEIGHTS: CHOICE OF MODELS

Relative brain size is usually considered as brain size relative to body weight, although other standards of comparison have been suggested (e.g., cross-sectional area of the corpus callosum or the medulla: Anthony, 1938; weight of spinal cord: Krompecher and Lipak, 1966; foramen magnum area: Radinsky, 1967; and medulla weight: Passingham, 1975). Brain sizes of extinct species can be estimated by water displacement of endocasts or by double graphic integration of lateral and dorsal view drawings of endocasts (for the latter technique, see Jerison, 1973, p. 50). Body weight, however, is more difficult to estimate for extinct species. A common approach is to use the body weight of a modern species that matches the fossil in preserved parts. I used this method to estimate body weights of several early prosimians (Radinsky, 1977), most of which are known only from skulls. The problem with that approach is its lack of demonstration that the modern species chosen for comparison is the best model, or even a reasonably accurate model to use for estimating body weight of the fossil. The latter

*From the Department of Anatomy, University of Chicago, Chicago, Illinois 60637

may have different proportions (e.g., relative tooth or skull size, or body length to body weight) from the modern model, or there may be a wide range of variation in body weight among modern species of a given skeletal dimension (e.g., tooth size, skull length, or total body length).

A more rigorous approach for estimating fossil body weights is to extrapolate from skeletal measurement-body weight relationships in a *group* of modern species chosen for their similarity in size and proportions to the fossil. For example, Gingerich (1977) estimated body weight of *Aegyptopithecus* and Proconsul (= *Dryopithecus*) by extrapolating from the regression equations describing tooth length-body weight relationships in modern hominoids and in modern primates, and Jerison (1979) estimated body weights of early prosimians from skull length-body weight regressions in modern prosimians and modern ceboid monkeys. Both authors give 95% confidence limits for their weight estimates, but because those confidence limits are derived from a group of species means rather than from a random sample of individuals drawn from a natural population, there is no reason to expect normal distribution and the predicted confidence limits. That is, there is no reason to expect some species to have body weight-skull length or tooth length relationships that lie over two standard deviations from the means of those relationships in the group of species sampled. Perhaps a more meaningful expected range for body weight estimates could be derived from the observed range of body weights for a given skeletal measurement, or the equation-derived range of body weights from the samples of modern species (Gurche, this volume).

A more important consideration concerns the choice of modern species from which the regression equations are derived. Fossil body weight estimates may seem like hard data, particularly when confidence limits are provided, but one should consider whether evidence is available that the fossil species, usually incompletely known even with respect to skeleton, was proportioned similarly to the modern species in the sample used. Gingerich (1977) recognized this problem, and noted that body weight estimates derived from tooth size would be inappropriate for some obviously relatively large-toothed fossil hominids. Jerison (1979) also recognized this problem and provided body weight estimates from skull length-body weight equations derived from both modern prosimian and modern ceboid samples. However, when it came to calculating relative brain size, Jerison (1979) chose the ceboid model body weights, arguing that fossil omomyid prosimians were phylogenetically closer to modern ceboids than to modern strepsirhine prosimians. This choice made a significant difference for the larger fossil genera, such as *Rooneyia*, which has an estimated body weight of 400 gm from the prosimian regression and 330 gm from the ceboid regression. Thus, use of the ceboid regression resulted in a lower body weight estimate and therefore a higher relative brain size index for *Rooneyia* than would have been true had the prosimian regression been used. Choice of an appropriate model may be a subjective decision, but it seems clear that similarity in body proportions is more important than phylogenetic proximity. No complete skeletons are known for omomyid prosimians, so it is not obvious whether the ceboid skull length to body weight regression equation gives more accurate omomyid body weight estimates than does the prosimian regression. One could look at other aspects of the skeleton, e.g., tooth size, foramen magnum area, aspects of postcranial anatomy, to see what body weight estimates would be derived from them. In the absence of such independent extrapolations, it is not possible to judge which modern primate sample is the better model for omomyid body weights, and both estimates should be used for calculating relative brain size. In summary, a critical assumption to examine in evaluating fossil body weight estimates is that the modern model used was indeed appropriate.

RELATIVE BRAIN SIZE COMPARISONS: DIFFERENCES IN SLOPES

Because brain size scales with negative allometry with respect to body weight, some sort of regression analysis (as opposed to a straight brain weight-body weight ratio) must be used to compare relative brain size in species of different body size. Bauchot and Stephan (1966, 1969) used a regression equation based on brain weight-body weight relationships in primitive modern insectivorans (called basal insectivores), with a slope of 0.63, to derive their Progression Indices. Jerison (1973) used a regression equation based on a large sample of modern mammals from several orders, with a slope of 0.67, to calculate his Encephalization Quotients. In both cases, the indices are derived by calculating the expected brain size for a species of a given body weight, and then dividing its actual brain size by the "predicted" brain size. It is arbitrary which of those two base lines is used, but problems may arise in cases where brain weight-body weight relationships scale significantly differently from 0.63 to 0.67. For example, among individuals of a given species, regression slopes usually fall between 0.2 and 0.4, and one sometimes gets similarly low slopes for groups of individuals from closely related taxa. If one calculates relative brain size indices in such groups from regression equation slopes between 0.6 and 0.7, the smaller members will have higher indices than the larger members, simply from the lower intragroup slope. (See Holloway and Post, this volume; Martin, this volume; and Gould, 1975, for a fuller discussion of this problem.)

This situation is relevant in analysis of relative brain size in great apes and australopithecines. Among the great apes, brain size vs. body weight slopes have been estimated at 0.18 (Clutton-Brock and Harvey, 1980) and 0.34 (Pilbeam and Gould, 1974). (The difference may arise because the latter analysis includes the pygmy chimp whereas the former does not.) Relative brain size indices derived from equations with 2/3 slopes place the great apes within the range of relative brain size of monkeys, with the gorilla relatively low, even within the monkey range (see Fig. 1). In response to this unexpected result, i.e., finding apes of the same relative brain size as monkeys, various workers have devised other indices for comparing relative brain size in these primates. Hemmer (1971) used a slope of 0.23 to calculate a Cephalisationskonstante for modern primates, and got indices showing the great apes above monkeys in relative brain size.

Leutenegger (1973) used Hemmer's regression equation to derive relative brain size indices for australopithecines, and found *A. africanus* and *A. robustus* similar in relative brain size when the lower (0.23) slope was used. (Comparisons with a 2/3 slope result in higher relative brain size indices for the smaller *A. africanus.*) Pilbeam and Gould (1974) calculated a great ape brain weight-body weight regression slope of 0.34, and an *Australopithecus africanus-A. robustus-A. boisei* slope of 0.33, and concluded, as did Leutenegger (1973), that the robust australopithecines were similar to the gracile species in encephalization.

The problem with those studies is that the lower slopes are substituted in an *ad hoc* way that may reflect the investigators' preconceived beliefs rather than objective analysis. Most workers have a preconceived belief that apes are higher on the "scale of nature," or at least smarter, than monkeys, and because of that expect to find relatively larger brains in apes than monkeys. When an initial analysis does not produce that result, i.e., higher relative brain size indices for the great apes, there is a temptation to search for another analysis that will produce results more in conformity with previous expectations. Although *intraspecific* brain weight-body weight relationships do indeed scale to 0.2 to 0.4 slopes, it is not a general rule that such slopes hold for members of different species or different genera. Among carnivores, species of the genera *Felis, Panthera, Canis,* and *Vulpes* have brain weight-body weight relationships scaling with slopes of about 2/3 within their respective genera (Radinsky, unpublished data). Thus, the problem is how do we decide in a rigorous way which groups of species or genera should have their brain weight-body weight relationships characterized by a 1/3 slope, and which by a 2/3 slope?

Clutton-Brock and Harvey (1980) developed an innovative approach to this question by calculating a brain weight-body weight regression equation for each major family of primates, and then using those equations to calculate relative brain size indices (their Comparative Brain Size index) for genera within their respective families. The only difficulty with this is that, for families with few species, one has less confidence in the calculated regression slope. Thus Clutton-Brock and Harvey found brain weight-body weight slopes of 0.17 and 0.18 for hylobatids (two species included) and pongids (three species included) respectively, compared with slopes ranging from 0.49 to 0.79 for five other primate families with larger numbers of species. With such small numbers of species for the apes, how can one decide between the following two possibilities: the low slopes reflect the true brain weight-body weight relationships among the apes, and the large species (e.g., gorilla, siamang) are as encephalized as their smaller confamilial relatives; or, the larger species are indeed less encephalized than their smaller relatives, and we have lower slopes only because so few species are available for comparison. Obviously there is no 'correct' solution to this problem. The choice of regression line slope in such cases as these is arbitrary, and one can only be aware of that and be sensitive to biases from preconceived beliefs, when the final results (i.e., indices) are considered.

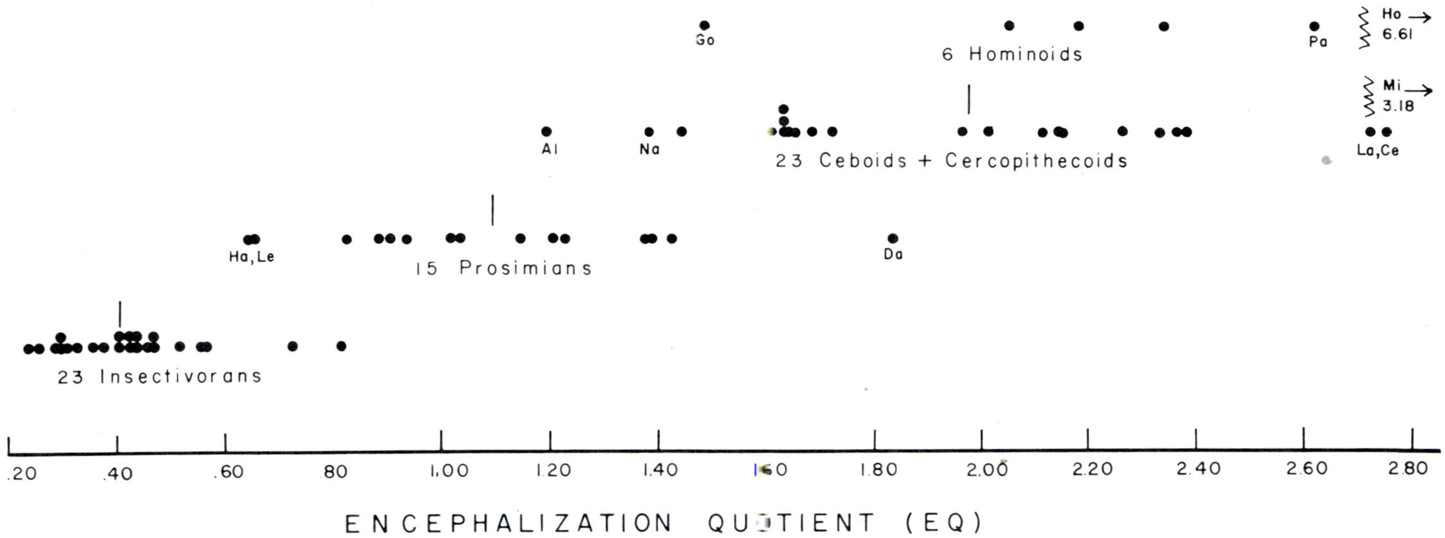

Fig. 1. Relative brain size in modern insectivorans and primates, calculated from data in Bauchot and Stephan, 1966 and 1969. The Encephalization Quotient is the brain size (grams or cubic centimeters) of a given species divided by the brain size expected for the average living mammal of that species' body weight (grams):

$$EQ_i = \frac{\text{brain size}_i}{0.12 \text{ body weight}_i^{0.67}}$$

Letters identify prosimian and anthropoid extremes of EQ. Al, *Alouatta*; Ce, *Cebus*; Da, *Daubentonia*; Go, *Gorilla*; Ha, *Hapalemur*; Ho, *Homo*; La, *Lagothrix*; Le, *Lepilemur*; Mi, *Miopithecus talapoin*; Na, *Nasalis*; Pa, *Pan*.

Relative Brain Size Comparisons: Group Diversity and Sampling

There is a considerable diversity in relative brain size among higher taxonomic groups of modern mammals (see Fig. 1), and there is no reason to believe that considerable diversity in that feature did not exist among extinct groups (e.g., Jerison, 1973; Radinsky, 1978). Therefore, one should be cautious about making generalizations from samples of only one or a few species. For example, Wilson and Szalay (1972) compared relative brain size in the Eocene prosimian *Notharctus* with that of the modern prosimian *Propithecus*, and concluded that *Notharctus* had a relatively smaller brain than do modern prosimians, and Martin (1973) compared relative brain size of the Eocene prosimian *Adapis* with that of two species of *Lemur*, and reached a similar conclusion for *Adapis*. Those conclusions are probably correct, although the differences in relative brain size between the Eocene prosimians and modern prosimians as a group are less than was suggested by those studies, because there are considerably less encephalized modern prosimians than *Lemur* and *Propithecus* (Bauchot and Stephan, 1966).

The brains of modern anthropoids are relatively large (although not uniquely so) compared with those of other modern mammals, and it is commonly believed that the brains of primates have always been relatively large compared with those of their contemporaries (e.g., Le Gros Clark, 1971; Jerison, 1973,1979). However, from the first evolutionary radiation of primates, the Paleocene Plesiadapiformes (about 65 to 55 m.y.), we have only *one* very incomplete endocast (Radinsky, 1977) from only one of the approximately 20 to 25 known genera. The endocast is from *Plesiadapis*, one of the most derived of the Paleocene primates, and even if we could make accurate estimates of its relative brain size, it would not be sufficient information to allow generalizing about relative brain size in the earliest primates.

Our record for the second radiation of primates, from about 55 to 35 m.y., is better and includes endocasts of three genera of adapids and three genera of omomyids (see Table 1). Adapid relative brain size was below that of modern prosimians, but at least the latest of the omomyids fell within the range of modern prosimians for relative brain size. Because of the differences between adapids and omomyids, and between omomyids of different ages, one cannot generalize about relative brain size in the second radiation of primates. Further, comparison of relative brain size between these early prosimians and their carnivore and ungulate contemporaries (Table 1) indicates that relatively large brains were not a distinctive feature of the Eocene prosimians.

BIOLOGICAL SIGNIFICANCE OF RELATIVE BRAIN SIZE

A common belief is that relative brain size correlates with intelligence. There is a large body of literature describing differences in performance of a variety of species on a variety of "intelligence" tests (e.g.,

TABLE 1. Relative Brain Size (EQ) in Early Primates, Carnivores and Ungulates

10^6 years ago	Primates[a]		Carnivores and Ungulates[b]	
24				
M. - L. Oligocene	*Aegyptopithecus*	.66 -.97	(18)	.46 - 1.08
28				
E. Olig	*Rooneyia*	.81 - 1.23	(1)	.98
37				
L. Eocene	*Necrolemur*	.56 - .94	(4)	.44 - .77
	Adapis	.39 - .53		
46				
M. Eocene	*Smilodectes*	.47 - .53	(5)	.37 - .77
49				
E. Eocene	*Tetonius*	.42 - .71	(2)	.46 - .59
54				

	Modern EQs
23 insectivorans	.41 (.24 - .82) (excluding tupaiids and macroscelidids)
15 prosimians	1.10 (.65 - 1.84)
28 anthropoids	1.98 (1.20 - 3.18) (excluding *Homo* = 6.61)

[a]Range of estimates from Radinsky, 1977; Jerison, 1979; and Gurche, this volume
[b]Data from Radinsky, 1978.

Glickman and Sroges, 1966; Ehrlich et al., 1976; Riddell and Corl, 1977), most of which shows species with larger brains doing better on the tests. However, as has been often pointed out in recent years, the results of most cross-species "intelligence" tests are equivocal because of species differences in sensory capacities and the lack of evidence that experimental conditions are equivalent for different species. (A striking example of the significance of differences in sensory specializations is the demonstration by Slotnick and Katz (1974) that rats tested with olfactory cues perform as well on learning sets as primates tested with the usual visual stimuli.) For further discussion of this problem, and an attempt to analyze qualitative rather than quantitative differences, see Warren, 1973, 1974.

Even if we could devise "intelligence" tests that were equivalent for different species, the significance of the results would depend on the resolution of another crucial problem, one that has so far been little explored. That is the question of the relationship between the behavioral capacities being tested in the lab and behavior that is important in the real world (i.e., under natural conditions) for reproductive success. To demonstrate the hypothesis that the adaptive significance of enlarged brains is increased intelligence, one must provide evidence that increased intelligence leads to increased fitness (i.e., increases the chances of having offspring in the next generation). Jerison (1979) has asserted that increase in brain size enhances "information-processing capacity," and that the latter contributes to fitness. Although that may seem intuitively obvious, I know of no evidence showing that either larger brains or enhanced information processing capacity do indeed contribute to fitness. The fact that relative brain size has increased independently many times during evolution of birds and mammals suggests that there was a common selective pressure involved, but just what it was remains an enigma.

In recent years, other interesting factors have been discovered that correlate with relative brain size. These include: maximum potential longevity in mammals (Sacher, 1975); diets and foraging strategies in bats (Pirlot and Stephan, 1970; Eisenberg and Wilson, 1978); and diet and home range size in primates (Clutton-Brock and Harvey, 1980). Although a statistically significant correlation does not necessarily imply anything about causality, studies such as those just cited, which consider natural condition behavioral/ecological factors, are a rich source of suggestions for possibly testable hypotheses on the functional significance of differences in brain size.

REFERENCES

Anthony, R., 1938. Essai de recherche d'une expression anatomique approximative du degré d'organisation cérébrale autre que le poids de l'encéphale comparée au poids du corps. Bull. Soc. Anthropol. (Paris), 9 (ser. 8):17-67.

Bauchot, R., and Stephan, H., 1966. Données nouvelles sur l'encéphalisation des insectivores et des prosimiens. Mammalia, 30:160-196.

Bauchot, R., and Stephan, H., 1969. Encéphalisation et niveau évolutif chez les simiens. Mammalia, 33:225-275.

Clutton-Brock, T.H., and Harvey, P., 1980. Primates, brains and ecology. J. Zool. London, 190:309-323.

Ehrlich, A., Fobes, J.L., and King, J.E., 1976. Prosimian learning capacities. J. Human Evol., 5:599-617.

Eisenberg, J.F., and Wilson, D.E, 1978. Relative brain size and feeding strategies in the Chiroptera. Evol., 32:740-751.

Gingerich, P.D., 1977. Correlation of tooth size and body size in living hominoid primates, with a note on relative brain size in *Aegyptopithecus* and *Proconsul*. Am. J. Phys. Anthrop., 47:395-398.

Glickman, S.E., and Sroges, R.W., 1966. Curiosity in zoo animals. Behaviour, 26:151-188.

Gould, S.J., 1975. Allometry in primates, with emphasis on scaling and the evolution of the brain. In, Approaches to Primate Paleobiology Contrib. Primat., Vol. 5, F. Szalay, ed., Karger, Basel, pp. 244-292.

Hemmer, H., 1971. Beitrag zur Erfassung der progressiven Cephalisation bei Primaten. Proc. 3rd Int. Congr. Primat., Zurich 1970, 1:99-107.

Jerison, H.J., 1973. Evolution of the Brain and Intelligence. Academic Press, N.Y, 482 pp.

Jerison, H.J., 1979. Brain, body and encephalization in early primates. J. Human Evol., 8:615-635.

Krompecher, S., and Lipak, J., 1966. A simple method for determining cerebralization. J. Comp. Neurol., 127:113-120.

Le Gros Clark, W.E., 1971. The Antecedents of Man, 3rd. ed., Quadrangle Books, Chicago.

Leutenegger, W., 1973. Encephalization in australopithecines: a new estimate. Folia Primat., 19:9-17.

Martin, R.D., 1973. Comparative anatomy and primate systematics. Symp. Zool. Soc. Lond, 33:301-337.

Passingham, R.E., 1975. The brain and intelligence. Brain, Behav. Evol., 11:1-15.

Pilbeam, D., and Gould, S.J., 1974. Size and scaling in human evolution. Science, 186:892-901.

Pirlot, P., and Stephan, H., 1970. Encephalization in Chiroptera. Canadian J. Zool., 48:433-444.

Radinsky, L., 1967. Relative brain size: a new measure. Science, 155:836-838.

Radinsky, L., 1977. Early primate brains: facts and fiction. J. Human Evol., 6:79-86.

Radinsky, L., 1978. Evolution of brain size in carnivores and ungulates. Am. Natur., 112:815-831.

Riddell, W.I., and Corl, K.G., 1977. Comparative investigation of the relationship between cerebral indices and learning abilities. Brain, Behav. Evol., 14:385-398.

Sacher, G., 1975. Maturation and longevity in relation to cranial capacity in hominid evolution. In, Primate Functional Morphology and Evolution, R. Tuttle, ed., Mouton, The Hague, pp. 417-441.

Slotnick, B.M. and Katz, H.M., 1974. Olfactory learning-set formation in rats. Science, 185:796-798.

Warren, J.M., 1973. Learning in vertebrates. pp. 471-509, In, Comparative Psychology, D.A. Dewsbury, and D.A. Rethlingshafer, eds., McGraw-Hill, New York.

Warren, J.M., 1974. Possibly unique characteristics of learning by primates. J. Human Evol., 3:445-454.

Wilson, R.W., and Szalay, F.S., 1972. New paromomyid primate from middle Paleocene beds, Kutz Canyon area, San Juan Basin, New Mexico. Am. Mus. Novitates, 2499:1-18.

ALLOMETRIC APPROACHES TO THE EVOLUTION OF THE PRIMATE NERVOUS SYSTEM

R. D. Martin*

Both reconstruction of phylogenetic relationships between organisms and the related exercise of classification depend on assessment of similarities and differences. These undertakings are complicated by the fact that the organisms compared may differ (perhaps considerably) in body size. Maintenance of any functional constancy in species spanning a range of body sizes often obliges other features to change in a systematic nonlinear fashion with body size (allometry) (e.g., Gould, 1966, 1975a,b; Huxley, 1932; Jerison, 1973, 1977; Martin, 1975, 1979, 1980; McMahon, 1973; Radinsky, 1967; Schmidt-Nielsen, 1970, 1972; Stahl, 1962, 1963, 1967; Stephan, 1972). Most dimensional characters do not vary in direct proportion to body size (isometry), and brain size in vertebrates generally does not increase in simple proportion with increasing body size. Instead, the ratio of brain size to body size decreases systematically with increasing body size within any given taxonomic group (negative allometry). Such "scaling" of brain size to fit any given body size appears to follow a biological law to maintain some functional constancy and is, therefore, broadly predictable. Thus, whenever we conduct comparisons between organisms of different body sizes, we must take into account any recognizable biological laws governing the scaling of characters to body size.

Where a comparison involves any two organisms that differ in body size, differences that reflect functional scaling to body size must be distinguished from those that reflect fundamental differences in biological adaptation (Martin, 1980). Of course the simplest procedure for excluding the scaling effects of body size in a comparison between organisms is to take species of the *same* body size, in which case any difference can be

*Department of Anthropology, University College, Gower Street, London WCI, England

taken as directly indicating differential adaptation. (This is, incidentally, also a useful test to apply in any cases where theoretical inferences have previously been made about scaling effects.)

In most cases, however, comparison among organisms of the same body size is ruled out and, therefore, some indirect method for recognizing and accounting for the scaling effects of body size is used. In virtually all cases, the method used is *empirical*, in that the overall relationship between a given dimensional character and body size is surveyed in a group of species spanning a range of body sizes, and the resulting pattern is interpreted in terms of allometric variation. Such interpretation depends on the recognition of approximate concordance between the data (or subsets of the data) and some mathematical function, and in practice considerable success has been obtained with the use of the standard *allometric formula*: $Y = k.X^{\alpha}$ (where Y is the dimensional character considered, such as brain size, α is the "allometric exponent," and k is the "allometric coefficient"). This empirical formula has the advantage of being relatively simple and of becoming linear in form when subjected to logarithmic conversion: $\log Y = \alpha \log X + \log k$. Accordingly, the dimensional data involved in an allometric comparison are customarily presented in the form of a logarithmic plot, and the resulting pattern is interpreted in terms of apparent linear relationships. In some cases, all of the data may appear to fit a single linear relationship, whereas in others separate linear groups of points may be identified. In the latter case some subjectivity may be involved in interpreting the data, but if the separate linear relationships have the same slope (α) and therefore differ only in the value of log k, then, reasonably, each group is following the same allometric law. A good example is provided by logarithmic plots of neonatal body weight against maternal body weight in primates (Leutenegger, 1973; Martin, 1975; Sacher, this volume). Strepsirhine primates (lemurs and lorises) differ considerably from haplorhine primates (tarsiers, monkeys, apes, and man) in that at any given body size the latter have considerably bigger neonates (about three times bigger) than do strepsirhines. Allometric lines of very similar slope can be fitted to the separate data sets for strepsirhines and haplorhines, and there is no overlap between the two distributions. Very similar slopes are also found in logarithmic plots of brain size against body size in insectivores and primates (Stephan et al., 1970; Jerison, 1973; Martin, 1979), but in this case there is considerable overlap between the categories (e.g., "basal" insectivores vs. "advanced" insectivores vs. strepsirhines vs. haplorhines).

As has been pointed out (Martin, 1980), the vertical separation of linear allometric data sets can be regarded as representing a 'grade' differentiation (Fig. 1). In other words, the differences between two sets of data in the values of a dimensional character at any given body size can be attributed to a fundamental difference in adaptation, reflected in some aspect of bodily organization. In the case of adaptive radiation, characters will generally be scaled to body size to maintain approximate functional equivalence in vital characteristics, whereas in long-term, progressive evolution a shift through successive grades of organization will often be

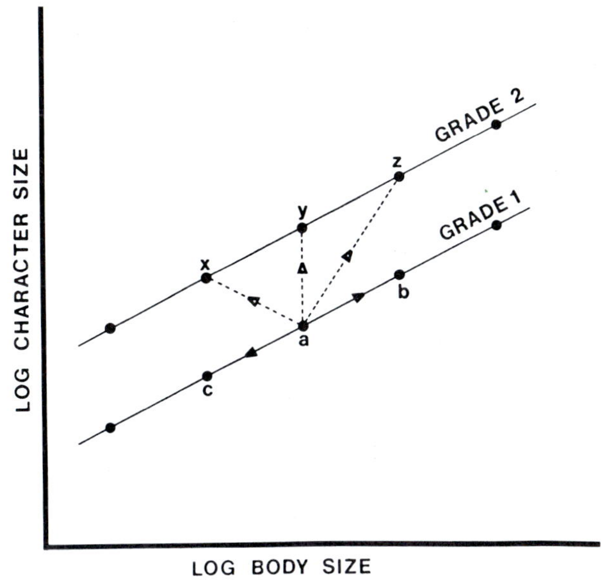

Fig. 1. Illustration of the concepts of scaling to body size and "grades." In each of the allometric series c-a-b and x-y-z, the character concerned is scaled to body size according to the same natural law (identical slope = same allometric exponent). Some difference in functional organization within each "grade" (1 = c-a-b; 2 = x-y-z) leads to differential expression of the character at any given body size, reflected in the different intercepts of the lines on the ordinate. The dotted lines indicate how in "phylogenetic allometry" involving a grade-shift over time, the slope of the line (a-x; a-y; a-z) depends on a combination of change in grade and change in body size. (After Martin, R.D., 1980. Z. Morph. Anthrop., vol. 71. Courtesy of E. Schweizerbart'sche Verlag.)

involved. For this reason (Fig. 1; see also Martin, 1980), allometric lines should not be fitted to data derived from fossil lineages (or approximations thereof) because this can confuse body size changes and grade shifts, thus making the key parameters (slope, α; intercept, log k) functionally meaningless. Hence, the concept of "phylogenetic allometry" (see Gould, 1966) is best avoided (See also Holloway and Post, this volume).

For attempts at establishing standard allometric relationships, organisms represented in any data set should belong to the same general grade of organization (e.g., mammals vs. reptiles; primates vs. insectivores; haplorhines vs. strepsirhines). Following the standardization, differential adaptation can be recognized in vertical departures from the overall linear relationship. As Gould (1975) succinctly stated: "We must establish the allometric regularities of functional scaling in interspecific, "mouse-to-elephant" plots in order to provide criteria for the recognition of special adaptations unrelated to the requirement of size." Special attention should be directed toward the value of the allometric exponent (α), because it is directly related to the functional constancy maintained by scaling. For instance, it is at present widely accepted that the allometric exponent for brain-to-body size relationships approximates the value of 2/3 (0.67) in various groups of vertebrates (Stephan, 1972; Jerison, 1973). This is consistent with the idea that, as body size increases, brain size keeps pace with body surface area rather than with body volume, and this might appear logical because both sense organs and effectors tend to be organized over surfaces rather than through the bulk of body tissue. As will emerge later, however, this apparent "biological law" may not be as reliable as has generally been assumed, and modification of the accepted value for the allometric exponent can have far-reaching implications.

STATISTICAL CONSIDERATIONS

In the empirical recognition of allometric relationships, the techniques used to fit lines to logarithmically transformed data play a central part. A fundamental problem with all such techniques is that requirements relating to normality of distribution are unlikely to be met by biological data where the points represent average conditions for individual species. Nevertheless, the use of standard statistical techniques to describe overall trends in such biological data is probably justifiable in the absence of any more reliable method. A much more serious problem arises, however, in the widespread use of *linear regressions* to determine the basic parameters (α ; log k) of allometric relationships. Fitting a linear regression to a given set of data assumes that only one of the two variables (usually the Y - variable) is subject to "error." In other words, a linear regression takes into account natural variation in only one variable. Yet there is no reason whatsoever to assume that body size should be any less subject to "error" than, for example, brain size. Kermack and Haldane (1950) long ago pointed out that, for this reason, the linear regression is generally unsuitable for biological data, and they proposed instead the *reduced major axis*, which takes into account errors in both variables (see also Gould, 1975; Jolicoeur, 1973). An alternative technique is to use the *major axis* (= principal axis), which similarly takes into account errors in both variables, but involves a different mathematical approach. In practice, it does not matter which of the three line-fitting techniques (linear regression; reduced major axis; major axis) is used when there is little scatter of the points around the lines (i.e., where the correlation coefficient, r, has a value of 0.98 or more). As the scatter of the points increases, however, so does the discrepancy between the parameters determined by the linear regression and those determined by either the reduced major axis or the major axis. The discrepancy is systematic in that the slope (i.e., the value determined for the allometric exponent, α) is always lower for a regression of Y on X than for the reduced major axis or the major axis. Even with correlation coefficient values as high as r = 0.95, the slope for the linear regression will be much lower than for the reduced major axis or principal axis. Thus, when one is interpreting the functional significance of the empirically determined allometric exponent (α), care must be taken if the correlation coefficient (r) is less than 0.98. As a general rule, the reduced major axis or the major axis is safer to use in determining empirical values for allometric relationships.

The discrepancy between results obtained with a linear regression and, say, those obtained with the major axis is of considerable importance whenever the value of the slope (i.e., the allometric exponent, α) is taken as a basis for detailed interpretation. Two cases in point are (1) the analysis of functional implications of the apparent allometric relationship and (2) the use of the empirical values determined for the allometric formula to predict the "expected" value for a dimensional character at any given body size. In the latter case, the best-fit line determined for a given set of data plotted on logarithmic coordinates can be used as a "standard" against which the deviations of individual plots can be assessed. Thus, Stephan (1972) used the allometric relationship between brain size and body size in "basal" insectivores as a standard for calculating "encephalization indices" of individual primate species, whereas Jerison (1973) used the brain-to-body size relationship in mammals generally as a basis for establishing "encephalization quotients" for individual mammal species. Both authors took a value for α of about 0.67 for computing their indices. Obviously, if the value determined for α is too low (e.g., as a result of using a linear regression instead of the major axis for data with a correlation coefficient of less than 0.98), the result will be that either index will overestimate the degree of encephalization in large-bodied species. Hence, the relevance of any conclusions about the relative degree of brain development in species of different body size is directly dependent on the reliability of the line-fitting technique used (also see Holloway and Post, this volume). As a general rule, then, linear regressions should not be used for computing indices or for any similar operations.

One of the best established, and most far-reaching, allometric relationships in mammals is that existing between basal metabolic rate and body size (Brody, 1945; Kleiber, 1947, 1961, 1969). This is a negative allometric relationship and it was originally believed (e.g., Rubner, 1883) that the basal metabolic rate was directly linked to body surface area so that the allometric exponent (α) had a value of 0.67. Kleiber (1932, 1947) was able to show that the allometric exponent is actually significantly greater than this, approximating the value of 0.75 (see also Brody, 1945). Subsequently, Hemmingsen (1950, 1960) provided evidence that a similar value (about 0.73) applies not only to mammals but to living organisms in general, though there are grade relationships involved. For instance, homeothermic vertebrates (birds and mammals) exhibit far greater basal metabolic rates than poikilotherms (fish, amphibians and reptiles) at any given body size, but each group (homeotherms; poikilotherms) follows an allometric relationship with the slope of about 0.73. The standardized data considered by Kleiber (1961) for basal metabolic rate and body size in 26 selected mammalian taxa provide a good illustration of a convincing allometric relationship. Recalculation of the data (Martin, 1981a) shows that the correlation coefficient is extremely high ($r = 0.998$) and there is accordingly relatively little difference between the parameters determined with the different statistical procedures:

(1) Linear regression: $\log M = 0.756 \log W - 0.442$
(2) Reduced major axis: $\log M = 0.757 \log W - 0.446$
(3) Major axis: $\log M = 0.757 \log W - 0.446$

(where M = basal metabolic rate in Kcal/day; W = body weight in kg; logarithms to base 10 are used throughout unless otherwise stated).

From these empirically determined formulae and the 95% confidence limits of the major axis (0.740 - 0.774), the allometric exponent for these mammals is clearly significantly greater than 0.67. Hence, in assessing the functional significance, one can reliably conclude that there is *not* a straightforward relationship between metabolism and body surface area. In this case, rejection of the simple surface area hypothesis is based on the availability of strictly comparable data covering an adequate range of body sizes (Kleiber, 1961), rather than on the particular statistical technique used to describe the allometric relationship. The example clearly illustrates the importance of accurately determining the value of the allometric exponent (α).

All of the above considerations prompted me to re-examine the relationship between brain size and body size in vertebrates, especially because of the widespread acceptance of the allometric exponent value of 0.67, indicating a body-surface area relationship. Surprisingly, the allometric relationship between brain size and body size in mammals and other vertebrates does not seem to have been rigorously subjected to appropriate analysis, despite general reliance on the value of 0.67. Once again, the value accepted for the allometric exponent is of importance both in functional terms and with respect to computation of encephalization indices or quotients. As Jerison (1973) stated in his comprehensive treatment of brain size allometry in vertebrates: "We should note that an exponent of 2/3 implies a surface-to-volume relationship and may, therefore, be the basis for theorizing on the significance of the brain size." Jerison in his analysis of the vertebrate brain evolution, used a mixture of conventional regression analyses and fitting of principal axes to "minimum convex polygons" in allometric plots to justify a general allometric exponent of 0.67, but no supporting formulae are provided. In fact, the formula given by Jerison (1973) for the calculation of "encephalization quotients" in mammals ($B=0.12. W^{2/3}$, where B = brain weight in grams and W = body weight in grams) is based on the *a priori* assumption that the slope *is* 0.67, rather than on some empirically determined formula.

When major axes are fitted to large samples of data for mammals, birds and reptiles, no statistical justification for taking the value of the allometric exponent as 0.67 for *any* of these groups emerges. In the case of the placental mammals, analysis of the original set of data provided by Crile and Quiring (1940) and used by Jerison (1973) does indicate a slope of 0.67 for the logarithmic plot. The major axis determined for these data (N = 97 species) is: $\log B = 0.687 \log W + 2.025$, with 95% confidence limits on the slope of 0.647 - 0.728. The correlation coefficient is high (r = 0.960), but the linear regression still gives a somewhat lower slope: log B = 0.669 log W + 2.095. When the sample size is considerably increased (from N = 97 to N = 309 species), however, by the inclusion of additional data (Bauchot and Stephan, 1966, 1969; Count, 1947; Stephan et al., 1970; von Bonin, 1937) giving a more representative range, the situation is radically changed. Although the correlation coefficient remains about the same (r = 0.963), the slope of the major axis (Fig. 2) is greatly increased:

$$\log B = 0.755 \log W + 1.774.$$

Once again, the linear regression (as expected) gives a somewhat lower slope:

$$\log B = 0.734 \log W + 1.848.$$

The new 95% confidence limits on the slope of the major axis (0.732 -0.779) now rule out a value of 0.67 as appropriate for the allometric relationship

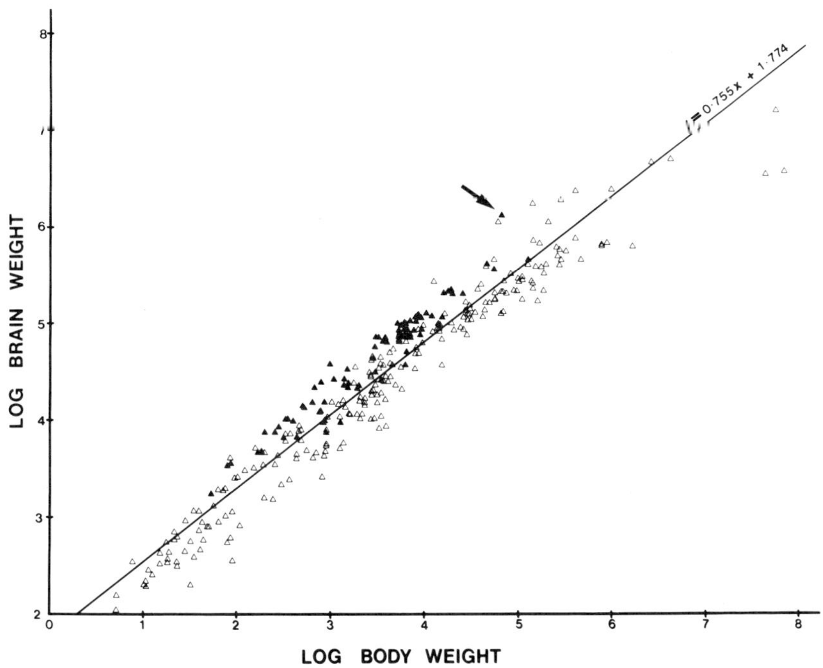

Fig. 2. Logarithmic plot of brain weight against body weight for a sample of 309 mammals. The major axis (continuous line) for the overall sample has a slope of 0.755, indicating an allometric exponent close to 3/4. Primates (black triangles) tend to have large brains for their body size, but there is considerable overlap with other mammals (white triangles). The closest species to man (arrowed) are, in fact, cetaceans. (From Martin, R.D., 1981b. Nature, vol. 293. Courtesy of MacMillan Journals.)

and suggest some connection between mammalian brain size and basal metabolic rate rather than a link between brain size and body surface. In both cases (brain size vs. body size; basal metabolic rate vs. body size) the allometric exponent is significantly greater than 0.67 and approximates 0.75. Considerable evidence has now been collected to suggest that brain size is intimately linked to the overall energetic coordination of the body (Sacher and Staffeldt, 1974; Rudder, 1979; Sacher, this volume), and it might, therefore, be more profitable to seek explanations for relationships between brain size and body size in mammals in terms of metabolic constraints rather than in terms of surface-to-volume interactions.

With birds and reptiles, the picture is considerably different. For these two vertebrate groups, there has never been any real evidence for the often-assumed relationship between brain size and body surface area (i.e., $\alpha = 0.67$), and analysis of available data now points to the surprising conclusion that the allometric exponent is actually significantly *less* than 0.67 for birds and reptiles. For birds, the original sample (N = 52 species) of brain and body weights provided by Crile and Quiring (1940) actually gives a major axis formula of:

$$\log B = 0.531. \log W + 2.256$$

with 95% confidence limits on the slope of 0.477 - 0.586 (r = 0.936). Expansion of these data (to N = 180 species) by inclusion of values calculated from the tables provided by Portmann (1946, 1947) gives a similar revised major axis formula of:

$$\log B = 0.576. \log W + 2.112$$

with 95% confidence limits on the slope of 0.544 - 0.608 (r = 0.932). In both cases, the 95% confidence limits exclude a figure of 0.67 for the allometric exponent as being too high. Similarly, for reptiles the original sample (N = 19 species) cited by Crile and Quiring (1940) gives a major axis formula of:

$$\log B = 0.610. \log W + 0.943$$

with 95% confidence limits on the slope of 0.544 - 0.680 (r = 0.971). Increasing the sample size to (N = 59 species) with data from Platel (1979) gives a somewhat lower value for the allometric exponent with a major axis formula of:

$$\log B = 0.540. \log W + 1.222$$

with 95% confidence limits on the slope of 0.500 - 0.584 (r = 0.955). Here, the revised value for the allometric exponent obtained with the larger sample is, once again, significantly below the "expected" value of 0.67. Therefore, for both birds and reptiles available data indicate that brain size is not scaled to the 2/3 power of body size, but to some lesser power close to 0.56.

At first sight, this unexpected finding with brain-to-body size relationships in birds and reptiles may seem to rule out the possibility of any link with basal metabolic rate. Birds and reptiles typically reproduce, however, by means of the cleidoic egg, whereas mammals are viviparous. In mammals, nutrients for the developing fetal brain can be continually supplied by the maternal uterus throughout gestation and the brain weight of the neonate may, therefore, directly reflect the mother's metabolic turnover. Provided that adult brain weight is related to neonatal brain weight in a proportional manner (e.g., see Rudder, 1979, for data on primates), adult brain weight might be expected to reflect the maternal basal metabolic rate (i.e., with the 3/4 power of adult body weight). In birds and reptiles, nutrients for the developing brain of the offspring must be provided in the egg at the outset of development. If the size of the egg is assumed to be related to the basal metabolism of the adult, such that:

$$\log E = 0.75 \log W + \log k'$$

(where E = egg weight, and W = adult body weight),

and the brain size of the hatchling is assumed to be related to the metabolic rate of the egg such that:

$$\log C = 0.75 \log E + \log k''$$

(where C = brain weight of the hatchling),

then brain size of the hatchling should be related to the body weight of the adult by the formula:

$$\log C = (0.75 \times 0.75) \log W + \log k'''$$

$$= 0.563 \log W + \log k'''$$

If a proportional relationship exists between the brain weight of the hatchling (C) and the brain weight of the adult (B), then the following relationship can be predicted:

$$\log B = 0.563 \log W + \log k''''$$

This theoretically expected value of 0.563 for the allometric exponent agrees well with the empirically determined values of 0.576 for reptiles and 0.540 for birds, and lies well within the 95% confidence limits in both instances. Concerning egg weight (in grams) in birds, the theoretical expected relationship with adult body weight (in grams) is fairly confirmed by the major axis for the data sample (N = 127 species) provided by Heinroth (1930):

$$\log E = 0.793. \log W + 2.338$$

with 95% confidence limits on the slope of 0.751 - 0.838 ($r = 0.954$). Obviously it will be necessary to confirm that a similar relationship between egg weight and adult body size exists in reptiles and that metabolic turnover in eggs follows the usual relationship between basal metabolic rate and body size determined for whole animals. The suggested proportionality link between hatchling brain size and adult brain size also remains to be confirmed, but at least there may be sound theoretical reasons why mammalian brain size scales to the 0.75 power of body size, whereas in birds and reptiles it scales to the 0.56 power of body size. Even if this theoretical explanation does not survive further testing, the hypothesis that brain size scales according to body surface area in mammals, birds, or reptiles clearly cannot be supported. Analysis of increased sample sizes of species using the major axis shows that the value of 0.67 for the allometric exponent is excluded in all three groups of vertebrates.

As a final note on this topic (see also Radinsky, this volume), in studies of brain-to-body size relationships in vertebrates, the value of the allometric exponent progressively decreases as the taxonomic level for the species included in the comparison decreases, reaching its lowest magnitude with the limiting case of intraspecific variation (see Gould, 1975a, for a general discussion of this phenomenon). The reasons for this remain obscure, but obviously any inferences regarding the value of the allometric exponent must be restricted to the taxonomic level at which analysis is conducted. I must emphasize, therefore, that the above discussion of a possible link between basal metabolic rate and brain size applies only at the level of comparison between the classes Mammalia, Aves, and Reptilia. Within a smaller taxonomic unit, such as the order Primates, more detailed analysis is required.

BODY WEIGHT ESTIMATION AND THE PROBLEM OF INTERVENING VARIABLES

A further problem that can arise in the analysis of relative brain size in primates and other mammals stems from the use of *intervening variables* as indicators of body size. This procedure is often unavoidable where fossil species are included in comparisons, because there is rarely sufficient fossil material to permit reconstruction of the entire skeleton (see also Radinsky, this volume). Special problems of interpretation are encountered where indirect measures of body size are used and these can be illustrated with reference to two such measures: (1) foramen magnum area, and (2) cheek-tooth dimensions.

Radinsky (1967, 1970) suggested the use of foramen magnum area as an indirect measure of body size in fossil mammals. He found a good correlation between foramen magnum area and body size within a number of different orders of placental mammals (Radinsky, 1967) and went on to relate cranial endocast size to foramen magnum area in living and fossil prosimian primates (Radinsky, 1970). From the latter comparison, the relative brain size in Eocene prosimians such as *Adapis parisiensis* seemed similar to the average condition for modern prosimians. The overall dimensions of the skull of *Adapis parisiensis*, however, are those generally comparable to the modern mongoose lemur, *Lemur mongoz* (Martin, 1973), and the latter has a considerably greater cranial capacity (17.6 ml vs 8.8 ml). Thus, either *Adapis parisiensis* had a larger head relative to body size than *Lemur mongoz*, or some error is involved in inferring body size from foramen magnum area in this fossil primate species. Radinsky (1977, this volume) went on to point out, rightly, that individual species comparisons of this kind are somewhat unreliable and that comprehensive surveys of living primate species should be used as a basis for the assessment of fossils. In line with this, a survey was conducted on various cranial and dental measures in a representative sample of 36 nonhuman primate species (Martin, 1980) to assess their relative worth as indicators of body size. The following parameters were found to be highly correlated (r = 0.98 in all cases) with body weight on logarithmic coordinates:

1. Maximal skull length (SL)
2. Maximum width across the zygomatic arches (BZW)
3. A measure of occipital condyle area (COA)
4. A measure of foramen magnum area (FMA)

One might therefore assume that any one of these four variables might be equally reliable for inferring body weight in a fossil primate species; however, when the major axis formulae calculated for the living nonhuman primate sample are applied to these skull parameters for *Adapis parisiensis* (N = 2; Martin, 1980; Gingerich and Martin, in preparation), considerable discrepancy is found between the body weight indicated by the first three variables and that suggested by foramen magnum area. SL, BZW, and COA together indicate an average body weight of 2.3 kg (range 2.2 -2.5 kg), whereas FMA indicates a body weight of 0.4 kg--a sixfold difference! When the relative brain size of *Adapis parisiensis* is determined using a body weight of 2.3 kg, the degree of brain development in this Eocene primate is found to be considerably less than in any modern prosimian species. The same result is found with *Adapis magnus* (Gingerich and Martin, in preparation).

As has been pointed out from three independent sources (Jerison, 1973; Martin, 1973, 1980; Gould 1975a), the basic problem involved in the use of foramen magnum area as an indicator of body size is that it is also correlated with brain size in living primates. For the sample of 36 nonhuman primate species (Martin, 1980), foramen magnum area was found to be as highly correlated with cranial capacity as it is with body size (r = 0.98 in both cases). When the evolution of the brain through time is considered, an additional factor comes into play (Fig. 3). If the foramen magnum area were to be an accurate indicator of body size throughout the evolutionary history of the primates (hypothesis 1), then the girth of the spinal cord would remain unchanged as the brain expanded, relative to body size. On the other hand (hypothesis 2), as brain size expanded relative to body size as in primate evolution, the spinal cord might also have undergone further development such that the foramen magnum area increased. *Adapis parisiensis* and *Adapis magnus* apparently possessed not only small brains but also relatively small-calibre spinal cords compared with those of modern

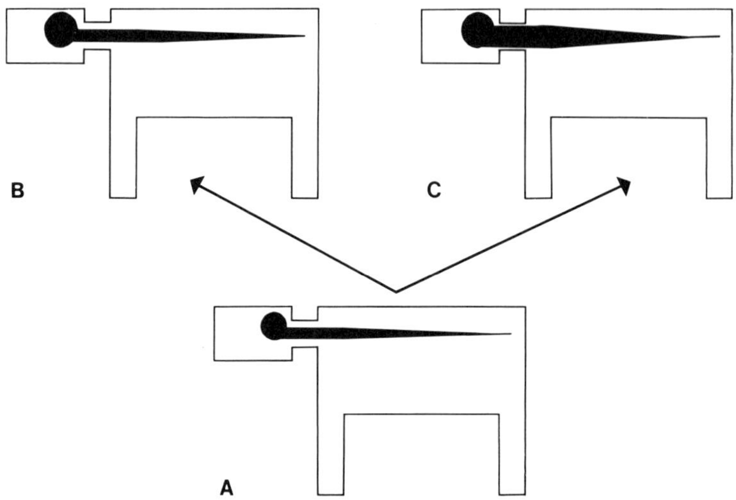

Fig. 3. Diagram illustrating two alternative hypotheses for the evolution of the brain and spinal cord.
Hypothesis 1 (A --B) : The brain expands relative to body size in evolution, but the spinal cord does not. The girth of the spinal cord may therefore be a reliable indicator of body size.
Hypothesis 2 (A --C) : Both the brain and the spinal cord expand relative to body size in evolution. The girth of the spinal cord, in an evolutionary sequence, is therefore more reflective of brain size than of body size.

prosimian primates, which agrees with the second hypothesis (Martin, 1980; Gingerich and Martin, in preparation) In this case, use of foramen magnum area to measure body size in fossil primates is inherently unreliable because of the possibility that the foramen magnum has itself been modified. Because of the complex relationship between brain size, foramen magnum area, and body size, great care must be exercised in the interpretation of brain-size to foramen-magnum-area relationships even in living mammals. For example, Findley's study (1969) of brain-volume to foramen-magnum-area relationships in bats cannot be simply equated with an analysis of brain-to-body size relationships without more detailed investigation.

The other main example in literature of the use of an intervening variable to infer body size in fossil primates involves individual or collective measures of the cheek teeth. A number of studies have been conducted on the allometric relationships of single molar teeth or the total cheek-tooth area both in primates specifically (e.g., Kay, 1975; Gingerich, 1977; Pirie, 1978; Gingerich and Schoeninger, 1979; Gingerich, et al., 1980; Martin, 1979, 1980; Kay and Simons, 1980), and in mammals generally (Gould, 1975b; Creighton, 1980). Two distinct problems can be identified in the use of such dental measurements as a basis for investigating allometric relationships. First, cheek-tooth measurements are not sufficiently highly correlated with body size across the order Primates to permit really reliable inference of body weights. In other words, there are different grades of cheek-tooth to body-size relationships within the primates, which confuse the issue when any other variable (e.g., brain size) is considered. Second, when the relationship between cheek-tooth dimensions and body size are examined to relate dental adaptations to metabolic requirements, errors may be inadvertently introduced if an intervening variable (e.g., maximum skull length) is used as an indicator of body size.

When a survey is conducted of empirical formulae for the relationship between cheek-tooth dimensions and overall body size (Table 1), the average value for the correlation applies whether the dental measure used is a linear dimension of a single tooth or an index of cheek-tooth area calculated from the maximum length and the maximum width of the upper cheek-tooth row ("upper-cheek-tooth-area index"). Now a correlation coefficient of $r = 0.95$ may seem to be high enough to provide a good predictive basis for determining body weight from an empirical formula. Nevertheless, of all the skull measures I have examined (Martin, 1980), the upper-cheek-tooth-area index gave the lowest correlation with body weight; other measures gave values of $r = 0.98$. Even with these other relatively high correlation coefficients, there is considerable scatter of points around the major axis and a fair range of error is involved in body weight predictions based on these nondental measures. It may, therefore, be preferable (as I advocated elsewhere, 1980) to use a panel of measures for predicting body weight, taking the average of all predictions in each case. When the data provided by Swindler (1976) for 19 nonhuman primate species are computed to obtain empirical relationships between body weight and two different cheek-tooth measures (M_2 length; upper-cheek-tooth-area index - Table 1), derived "predictions" of body weight for individual species in the sample differ by an average of almost 40% from the original measures. In individual extreme cases, the reliability of dental predictions of body weight is considerably worse than this. For *Papio cynocephalus*, the predicted body weight is more than twice the actual figure (+160%), whereas for *Ateles geoffroyi* the predicted value is less than half of the real value (-58%). With a correlation coefficient value as low as $r = 0.95$, use of a regression formula (as is the usual procedure in the literature) will lead to systematic underestimation of body weight for large-bodied species and over-estimation for small-bodied species.

In a recent paper by Creighton (1980), the correlation coefficient for a sample of 288 mammals giving the relationship of lower-first-molar (M_1) area to head and body length is 0.952, and the accompanying figure showed considerable scatter around the calculated regression line. Estimations indicate that the head + body length predicted for a given dental measurement may be out by a factor of 1.6 at the boundaries as indicated by the 95% confidence limits, which would correspond to a fourfold overestimation or underestimation of body weight from the overall allometric formula. Because grade relationships in the allometric variation of a given dimensional character with body size (see Fig. 1) often involve a twofold or threefold

TABLE 1. Formulae Relating Cheek Tooth Measures to Body Weight in Various Primate Samples

Authors		Modern Primate Sample		r		Empirical Formula
KAY (1975)		10 nonhuman primate species, ranging from *Microcebus murinus* to *Gorilla gorilla*.		?		†$\log_e A = 0.59. \log_e W - 4.65$ (where A = the summed products of the lengths and breadths of cheek teeth)
GINGERICH (1977)	(i)	38 primate species, excluding Cercopithecoidea	(i)	0.963	(i)	†$\log W = 3.289. \log L + 1.085$ (where L = M_2 length)
	(ii)	7 hominoid primate species, including man	(ii)	0.942	(ii)	†$\log W = 2.989. \log L + 1.462$ (where L = M_2 length)
MARTIN (1980 and in prep.)		36 nonhuman primate species (data for ♂♂ & ♀♀ averaged)		0.947		*$\log W = 2.061. \log L + 0.998$ (where A = upper cheek-teeth area index)
from SWINDLER (1976) calculations conducted on tabulated data		19 nonhuman primate species (data for ♂♂ & ♀♀ averaged)	(i)	0.948		*$\log L = 0.361. \log W - 0.563$ (where L = M_2 length)
			(ii)	0.943		*$\log A = 0.612. \log W - 0.042$ (where A = upper cheek-teeth area index)
GINGERICH, SMITH and ROSENBERG (1980)		45 primate species (♂♂ & ♀♀ treated separately)	(i)	0.93	(i)	†$\log W = 1.84. \log A + 0.81$ (where A = M^1 area)
			(ii)	0.95	(ii)	†$\log W = 1.67. \log A + 1.30$ (where A = M_1 area)
KAY and SIMONS (1980)		106 primate species, excluding Callitrichidae		0.949		†$\log 10 = 2.86. \log L + 1.37$ (where L = M_2 length)

* = principal axis
† = regression

difference between the grades in the value for the character predicted at a given body size, the relatively poor correlation between dental dimensions and body size in primates (Table 1) is obviously a cause for concern. If, for example, brain size is plotted against a dental dimension to investigate "relative brain size" in fossil forms, the grade relationships may be partially or completely obscured by grade relationships between the intervening variable itself (i.e., the dental measure) and body size.

A related problem arises when dental dimensions are related to skull dimensions in an attempt to test hypotheses about relationships between dietary intake and basal metabolism in primates (Gould, 1975; Pirie, 1978; Martin, 1979). The theoretical basis for such attempts is provided by the following three formulae:

(i) Body weight (W) is related to some linear measure of body size, such as skull length (SL) by the formula:

$$W = k.' (SL)^3$$

(ii) Metabolic requirements (M) are related to body size by the formula:

$$M = k.'' (W)^{3/4}$$

from which equation (i) gives:

$$M = k.''' (SL)^{9/4}$$

(iii) If some dental measurement (D) is directly related to metabolic requirements, then:

$$D = k.''''(SL)^{9/4}$$

If this reasoning is correct, then the allometric coefficient for the relationship between metabolically relevant dental measurement and skull length should be 2.25 (i.e., 9/4). Three separate publications (Gould, 1977; Pirie, 1978; Martin, 1979 - *mea culpa*) showed allometric relationships between postcanine tooth area and skull length with a value of the allometric coefficient close to the expected 2.25, with the conclusion that this provides evidence of a link between the area of the cheek teeth and metabolic requirements. Kay (1975) found, however, with a sample of ten primate species ranging from the mouse lemur to the gorilla, that the allometric exponent for the relationship between postcanine tooth area and actual body weight is 0.59, not 0.75. This apparent discrepancy is explained by the fact that body weight does not scale according to the cubed power of skull length in accordance with the theoretical expectation (above). For a sample of 36 nonhuman primate species, I found (Martin, 1980) the major axis for the relationship between body weight in grams (W) and maximal skull length in mm (SL) to be:

$$\log W = 3.889 \log SL - 4.085 \ (r = 0.983)$$

Therefore the allometric exponent for the relationship between post-canine tooth area and actual body size should be *less* than 0.75, which agrees with the empirical formula determined by Kay (1975). Once again, use of an intervening variable instead of actual body size in the analysis of allometric relationships has led to confusion over functional relationships.

Acknowledgments: The development of the concepts underlying this paper was significantly influenced by discussions with Ms. L. Aiello, Dr. B.C.C. Rudder, and Dr. P.H. Harvey, to all of whom I owe my gratitude. Valuable advice on statistical aspects was provided by Dr. M. Hills of the British Museum (Natural History) and considerable practical help in the preparation of the paper was provided by Ms. A. MacLarnon. Dr. J.M.M. Rollinson assisted with the measurements of the primate skulls that were used to derive some of the empirical formulae in the paper.

REFERENCES

Bauchot, R., and Stephan, H., 1966. Données nouvelles sur l'encéphalisation des insectivores et des prosimiens. Mammalia, 30:160-196.

Bauchot, R., and Stephan, H., 1969. Encéphalisation et niveau évolutif chez les simiens. Mammalia, 33:225-275.

Brody, S., 1945. Bioenergetics and Growth, Rheinhold Pub. Co., New York.

Count, E.W., 1947. Brain and body weight in man: their antecendents in growth, and evolution. Ann. N. Y. Acad. Sci., 46:993-1122.

Creighton, G.K., 1980. Static allometry of mammalian teeth and the correlation of tooth size and body size in contemporary mammals. J. Zool. (Lond.)., 191:435-443.

Crile, G.W., and Quiring, D.P., 1940. A record of the body weight and certain organ and gland weights of 3690 animals. Ohio J. Sci., 40:219-259.

Findley, J.S., 1969. Brain size in bats. J. Mammal., 50:340-344.

Gingerich, P.D., 1977. Correlation of tooth size and body size in living hominoid primates, with a note on relative brain size in *Aegyptopithecus* and *Proconsul*. Am. J. Phys. Anthrop., 47:395-398.

Gingerich, P.D., and Schoeninger, M.J., 1979. Patterns of tooth size variability in the dentition of primates. Am. J. Phys. Anthrop., 51:457-465.

Gingerich, P.D., Smith, B.H. and Rosenberg, K., 1980. Patterns of allometric scaling in the primate dentition and prediction of body size from tooth size. Am. J. Phys. Anthrop., 52:231-232.

Gould, S.J., 1966. Allometry and size in ontogeny and phylogeny. Biol. Rev., 41:587-640.

Gould, S.J., 1975a. Allometry in primates with emphasis on scaling and the evolution of the brain. Contrib. Primatol., 5:244-292.

Gould, S. J., 1975b, On the scaling of tooth size in mammals. Am. Zool., 15: 351-362.
Heinroth, O., 1930. Angaben zur Fortpflanzungsbiologie der Vögel und Säugetiere. In, Tabulae Biologicae, vol. 6, C. Oppenheimer, and L. Pincussen, eds., W. Junk, Berlin, 716-741.
Hemmingsen, A.M., 1950. The relation of standard (basal) energy metabolism to total fresh weight of living organisms. Rep. Steno. Mem. Hosp., 4:7-58.
Hemmingsen, A.M., 1960. Energy metabolism as related to body size and respiratory surfaces, and its evolution. Rep. Steno. Mem. Hosp., 9:1-110.
Huxley, J.S., 1932. Problems of Relative Growth. Methuen, London.
Jerison, H.J., 1973. Evolution of the Brain and Intelligence, Academic Press, New York.
Jerison, H. J., 1977. The theory of encephalization. Ann. N. Y. Acad. Sci., 299:146-160.
Jolicoeur, P., 1973. Imaginary confidence limits of the slope of the major axis of a bivariate normal distribution: a sampling experiment. J. Am. Stat. Assoc., 68:866-871.
Kay, R.F., 1975. The functional adaptations of primate molar teeth. Am. J. Phys. Anthrop., 43:195-215.
Kay, R.F., and Simons, E.L., 1980. The ecology of Oligocene African Anthropoidea. Int. J. Primatol., 1:21-37.
Kermack, K.A., and Haldane, J.B.S., 1950. Organic correlation and allometry. Biometrika, 37:30-41.
Kleiber, M., 1932. Body size and metabolism. Hilgardia, 6:315-353.
Kleiber, M., 1947. Body size and metabolic rate. Physiol. Rev., 27:511-541.
Kleiber, M., 1961. The Fire of Life: An Introduction to Animal Energetics, John Wiley, New York.
Kleiber, M., 1969. Further consideration of the relation between metabolic rate and body size. In, Energy Metabolism of Farm Animals, K.L. Blaxter, J. Kielanowski, and G. Thorbek, eds., Oriel Press, Newcastle-upon-Tyne, 505-511.
Leutenegger, W., 1973. Maternal-fetal weight relationship in primates. Folia Primatol., 20:280-293.
McMahon, T., 1973. Size and shape in biology. Science, 179:1201-1204.
Martin, R.D., 1973. Comparative anatomy and primate systematics. Symp. Zool. Soc. Lond., 33:301-337.
Martin, R.D., 1975. The bearing of reproductive behavior and ontogeny on strepsirhine phylogeny. In, Phylogeny of the Primates, W.P. Luckett, and F.S. Szalay, eds., Plenum Pub. Co. New York, pp. 265-297.
Martin, R.D., 1979. Phylogenetic aspects of prosimian behavior. In, The Study of Prosimian Behavior, G.A. Doyle, and R.D. Martin, eds., Academic Press, New York, pp. 45-77.
Martin, R.D. 1980. Adaptation and body size in primates. Z. Morph. Anthrop., 71:115-124.
Martin, R.D., 1981a. Field studies of primate behaviour. Symp. Zool. Soc. Lond., 46:287-336.

Martin, R.D., 1981b. Relative brain size and basal metabolic rate in terrestrial vertebrates. Nature, 293:57-60.

Pirie, P.L., 1978. Allometric scaling in the post-canine dentition with reference to primate diets. Primates, 19:583-591.

Platel, R., 1979. Brain weight-body weight relationship. In, Biology of the Reptilia, vol. 9, C. Gans, ed., Academic Press, London, pp. 147-171.

Portmann, A., 1946. Etudes sur la cérébralisation chez les oiseaux. Alauda, 14: 2-20.

Portmann, A., 1947. Etudes sur la cérébralisation chez les oiseaux. Alauda, 15:1-15.

Radinsky, L.B., 1967. Relative brain size: a new measure. Science, 155:838.

Radinsky, L.B., 1970. The fossil evidence of prosimian brain evolution. In, The Primate Brain, C.R. Noback, and W. Montagna, eds., Appleton-Century-Crofts, York, pp. 209-224.

Radinsky, L.B., 1977. Early primate brains: facts and fiction. J. Hum. Evol., 6:79-86.

Rubner, M., 1883. Über den Einfluss der Körpergrösse auf Stoff- und Kraftwechsel. Z. Biol., 19:535-562.

Rudder, B.C.C., 1979. The Allometry of Primate Reproductive Parameters, Ph. D. Thesis, University of London.

Sacher, G.A. and Staffeldt, E.F., 1974. Relation of gestation time to brain weight for placental mammals: implications for the theory of vertebrate growth. Am. Nat., 108:593-616.

Schmidt-Nielsen, K, 1970. Energy metabolism, body size and problems of scaling. Fed. Proc., 29:1524-1532.

Schmidt-Nielsen, K., 1972. Locomotion: energy cost of swimming, flying and running. Science, 177:222-228.

Stahl, W.R., 1962. Similarity and dimensional method in biology. Science, 137:205-212.

Stahl, W.R., 1963. The analysis of biological similarity. Adv. Biol. Med. Phys., 9:355-464.

Stahl, W.R., 1967. Scaling of respiratory variables in mammals. J. Appl. Physiol., 22:453-460.

Stephan, H., 1972. Evolution of primate brains: a comparative anatomical investigations. In, Evolutionary Biology of Primates, R. Tuttle, ed., Aldine-Atherton, Chicago, pp. 155-174.

Stephan, H., Bauchot, R. and Andy, O.J., 1970. Data on size of the brain and of various brain parts in insectivores and primates. In, The Primate Brain, C.R. Noback and W. Montagna, eds., Appleton-Century-Crofts, New York, pp. 289-297.

Swindler, D.R., 1976. Dentition of Living Primates, Academic Press, New York.

von Bonin, G., 1937. Brain-weight and body-weight of mammals. J. Gen. Physiol., 16: 379-389.

THE RELATIVITY OF RELATIVE BRAIN MEASURES AND HOMINID MOSAIC EVOLUTION

Ralph L. Holloway*
David G. Post

Several developments in recent years have renewed interest in finding measures or parameters to characterize brain and body size relationships both quantitatively and objectively. Among three developments the most outstanding are 1) a deepening appreciation and understanding of all metric relationships in evolutionary change and of formal comparisons among extant species (Bauchot and Stephan, 1969; Gould, 1975, Holloway, 1976; 1979; Jerison, 1973; Passingham, 1973; Passingham and Ettlinger, 1975; Sacher, 1975; Stephan et al., 1970; Hemmer, 1971; Leutenegger, 1973); 2) the publication of Jerison's (1973) volume on brain size and intelligence in evolutionary perspective; 3) the attempts to quantify body size parameters of certain early hominid specimens by physical anthropologists, McHenry (1974,1975) and Holloway (1975); and 4) the recent designation (Johansen and White, 1979) of the Hadar and Laetoli (Ethiopia and Tanzania, respectively) hominids to a new taxon, *Australopithecus afarensis*, a taxon clearly having a small ape-sized brain, yet with other morphological attributes indicative of true hominid status (e.g., bipedality, dentition).

Frequently a variety of neural values, foremost among them encephalization quotient scores (E.Q.) and "extra neurons," are determined for particular fossil hominids (e.g., Tobias, 1971), and reports of such determinations are accompanied by discussions of the role of neural or brain evolution in the overall dynamics of human evolution (see above references, plus White, 1980). Although some of us who work directly with brain evolution data are aware of the relativistic nature of such measures (e.g., Jerison, 1973; Hemmer, 1971; Leutenegger, 1973; Martin, this volume; Radinsky, this volume), we observe that many scientists do not appear to appreciate that most of these measures are *relative to the data base chosen*,

*From the Department of Anthropology, Columbia University, New York, New York 10027

and that little progress has been made in establishing criteria to determine which data base should be selected for particular purposes.

During the past several years, one of us (RH) has had the opportunity to prepare brain endocasts from original hominid fossils and to reconstruct those brains, both to find the most accurate volume estimates possible and to learn what the actual fossil evidence suggests about the evolution of the brain. In the course of these researches, encephalization quotients were seen to be a useful preliminary heuristic device to compare stages of brain size development in hominids, once some relatively accurate estimation of body size (weight) had been made; however, E.Q. scores were also observed not to be an invariant, inherent property of a given fossil specimen (or extant species) and therefore, functional or taxonomic judgments based on these scores have to be made with caution. Our intent here is twofold: (a) to demonstrate the often unappreciated dependence of E.Q. scores on the specific form of the reference equation and data bases used for their computation and (b) to spark some necessary discussion and debate regarding the best criteria for selecting particular data bases.

ENCEPHALIZATION

Allometric techniques have played a major role in the analysis of brain evolution, starting with the pioneering studies of Snell (1892) and of Eugene Dubois (1897). As "the study of size and its consequence" (Gould, 1966), allometry has unarguable relevance for our attempts to interpret the diversity of brain size both among extant organisms of different body weights and within phyletic lineages. We will not review the insight this approach has provided, nor comment on many of the more controversial aspects of this method (e.g., which regression model is most approprate for curve-fitting purposes, the interpretation of y-intercepts,the treatment of outliers--see Smith, 1980; Martin, this volume; Radinsky, this volume). We will focus instead on the use of empiricially-determined regression lines as "criteria of subtraction" (Gould, 1975) wherein deviations from the line are interpreted as size-independent, functional adaptations.

Encephalization coefficients exemplify the "criteria of subtraction" approach. An encephalization coefficient or quotient is simply a measure of the vertical residual or deviation of the observed brain size of an individual or species from a bivariate regression line constructed using brain and body size values for a group of organisms (e.g., all mammals, insectivores, primates). That is, on the assumption that brain-body relationships can be described by a power equation of the form:

$$V = k\,W^{\alpha}$$

where V = brain size, W = body weight, and α = power to which body weight is raised.

k = empirically determined parameters

a linear equation of the following form is fit, generally by the least-squares or major-axis techniques, to these data points:

$$\log V = \log k + \alpha \log W$$

The E.Q. for any specimen i (EQ_i) is then computed as the ratio of observed-to-expected brain weight, or

$$EQ_i = \frac{V_i}{k\, W_i^{\alpha}}$$

where V_i and W_i are the brain and body weights, respectively, for specimen i.

Obviously, each set of data points chosen will provide a somewhat different reference equation, with different values for the parameter k and α, and the E.Q. score of any particular specimen will change accordingly; the consequences of this fact will be examined below. Regardless of the data base selected, two salient facts are apparent when looking at primate and fossil hominid E.Q.'s: (1) modern *Homo sapiens* has the highest E.Q., and (2) fossil hominids are always intermediate between extant pongids and modern *Homo*, *but the degree of intermediacy depends on the data base selected.*

EMPIRICAL CONSIDERATIONS

Table 1 gives a sample of reference equations for the computations of E.Q.'s. The formulas are derived from least-squares regressions of Bauchot and Stephan (1969) in which the data base (89 primate species including modern *Homo*) is a combination of empirical determinations and results of their review of the literature. The equations for the various taxa or combinations thereof were determined from the SPSS scattergram and regression packages (Nie et al., 1975). In addition, #10, the "Homocentric" equation, was derived simply by using Bauchot and Stephan's values for *Homo sapiens*, and completing the regression line through "origin," i.e., zero-body weight and zero-brain weight. The "Homocentric" equation is derived from the premise that *Homo sapiens* is the highest encephalized animal known and has the advantage that an E.Q. score derived from it is *directly* expressed as a percentage of the modern *Homo* value. Table 2 is a listing of hypothetical brain and body weights for a sample of 20 early fossil hominids, in which the brain weights (volumes) are accurate and the body weights were chosen to cover a reasonable range from light to heavy.

TABLE 1. Ten Empirically Derived Equations Resulting From Plotting Log_{10} Brain Weight Against Log_{10} Body Weight For Different Combinations of Primate Taxa

Equation No.	Equation	Exponent Values	Correlation Coefficient R	Data Base	Data Source
1	Brain y = .12 x Body	.66	N.A.[a]	All mammals, N=198	Jerison 1973
2	= .0429 Body	.63	.98	Insectivores, N=22	Bauchot & Stephan 1969
3	= .0407 "	.64818	.99329	Basal insectivores, N=11	"
4	= .0893 "	.77698	.96937	All primates, including *Homo*, N = 89	"
5	= .0991 "	.76237	.97062	All Primates, No *Homo*, N = 88	"
6	= .1482 "	.65683	.94980	Prosimians, N = 24	"
7	= .0963 "	.79855	.98931	New World monkeys, N = 21	"
8	= .5477 "	.57750	.91439	Old World monkeys, N = 36	"
9	= .6216 "	.58152	.98135	Pongids, N = 7	"
10	= 1.0 "	.64906[b]		*Homo sapiens*, N=1	Holloway

[a] N.A. = Not available (fitted by convex polygon method, Jerison 1973).

[b] This exponent is the result of choosing a brain weight of 1330 grams, and a body weight of 65,000 grams. The line passes through origin.

TABLE 2. Hypothetical Brain and Body Weights for a Sample of 20 Early Fossil Hominids and Known Weights for 4 Extant Hominoids

	Hominid No.	Brain Volume (ml)	Body Weight (grams)[a]
	1	400	20000
	2	400	18144
	3	442	28020
	4	442	23900
	5	442	21700
	6	585	20412
	7	485	22680
	8	485	24947
	9	485	27215
	10	485	29483
	11	485	34019
	12	485	54359
	13	530	36500
	14	530	31500
	15	530	34019
	16	530	29483
	17	530	27215
	18	508	22680
	19	508	34019
	20	508	45359
(Gorilla)	21	465	165000
(Chimpanzee)	22	420	46000
(Orangutan)	23	370	55000
(Homo)	24	1,330	65000

[a] The occasional appearance of *gram* figures in the body weights reflects conversion from ranges initially rounded in *pound* units.

Hominids #1-5, for example, could represent small-brained australopithecines, because the value of 400 to 442 g more or less represents the Taung specimen. The 485 g values of hominids #6-12 represent the volume for the STS5 australopithecine, and the 530 g value of #13-16 represents the OH5 and SK1585 robust australopithecines. The 508 g value is that for KNM-ER-1813, a small East African hominid.

Table 3 provides the E.Q. scores for chimpanzee, gorilla, orangutan, *Homo sapiens*, and the 20 hominids, and their raw percentages of the modern human value based on the equations in Table 2.

Table 4 shows the difference between the *highest and lowest* percent of *Homo sapiens* values for the 10 equations listed in Table 1, which is always between equations #7 and #8, that is, between the equations for New

TABLE 3. Encephalization Quotient (E.Q.) Scores and Their Raw Percentages of the Modern Human Value[a,b]

HOMINOID #	EQUATION NUMBER 1	2	3	4	5	6	7	8	9	10
1	4.55	18.19	16.02	2.04	2.12	4.04	1.53	2.40	2.03	.65
	65.90	63.20	64.60	75.10	74.00	65.20	77.20	59.40	59.40	65.00
2	4.86	19.35	17.06	2.20	2.29	4.30	1.65	2.53	2.15	.69
	70.30	67.20	68.80	81.10	79.70	69.50	83.30	62.80	63.20	69.00
3	4.02	16.26	14.22	1.73	1.81	3.57	1.29	2.18	1.84	.57
	58.20	56.50	57.30	63.90	63.20	57.70	65.10	54.00	54.20	57.40
4	4.47	17.97	15.77	1.96	2.05	3.97	1.46	2.39	2.02	.64
	64.70	62.40	63.50	72.30	71.40	64.10	73.70	59.20	59.50	63.60
5	4.77	19.10	16.79	2.11	2.20	4.29	1.58	2.53	2.14	.68
	69.00	66.30	67.70	77.90	76.80	68.30	79.80	62.60	62.90	67.70
6	5.45	21.78	19.17	2.43	2.53	4.83	1.82	2.87	2.43	.77
	78.90	75.60	77.30	89.70	88.30	78.00	91.90	71.20	71.50	77.30
7	5.08	20.38	17.90	2.24	2.34	4.51	1.67	2.70	2.29	.72
	73.50	70.80	72.10	82.60	81.50	72.80	84.30	67.00	67.30	72.20
8	4.76	19.19	16.83	2.08	2.17	4.23	1.55	2.56	2.16	.68
	69.00	66.70	67.80	76.70	75.80	68.40	78.30	63.40	63.60	68.00
9	4.50	18.17	15.91	1.95	2.03	4.00	1.45	2.43	2.06	.64
	65.10	63.10	64.10	71.70	70.90	64.60	73.20	60.30	60.50	64.20
10	4.26	17.28	15.10	1.83	1.91	3.79	1.36	2.32	1.96	.61
	61.70	60.00	60.90	67.40	66.70	61.30	68.70	57.60	57.80	60.90
11	3.88	15.79	13.76	1.64	1.72	3.45	1.21	2.14	1.81	.55
	56.10	54.80	55.50	60.30	59.80	55.80	61.10	53.00	53.10	55.50
12	2.84	11.75	10.16	1.14	1.20	2.54	.83	1.63	1.37	.41
	41.10	40.80	40.90	41.90	41.70	41.00	41.90	40.40	40.50	40.90
13	4.04	16.50	14.37	1.69	1.78	3.60	1.25	2.24	1.89	.58
	58.50	57.30	57.90	62.40	62.00	58.20	63.10	55.60	55.70	57.90
14	4.46	18.11	15.81	1.90	1.99	3.97	1.41	2.44	2.04	.64
	64.60	62.90	63.70	70.00	69.30	64.10	71.20	60.50	60.70	63.80
15	4.24	17.25	15.04	1.79	1.88	3.77	1.32	2.34	1.97	.61
	61.30	59.90	60.60	65.90	65.40	61.00	66.70	57.90	58.10	60.70
16	4.66	18.88	16.50	2.00	2.09	4.15	1.48	2.54	2.15	.66
	67.50	65.60	66.50	73.70	72.90	67.00	74.70	62.90	63.10	66.60
17	4.91	19.85	17.38	2.13	2.22	4.37	1.58	2.66	2.25	.70
	71.20	69.00	70.00	78.40	77.50	70.60	79.80	66.00	66.10	70.10
18	5.32	21.35	18.75	2.35	2.45	4.72	1.75	2.83	2.39	.76
	77.00	74.10	75.60	86.60	85.40	76.30	88.40	70.20	70.50	75.60
19	3.35	13.80	12.00	1.37	1.44	2.99	1.01	1.90	1.60	.48
	48.50	47.90	48.40	50.50	50.30	48.40	51.00	47.00	47.10	88.20
20	3.35	13.79	11.96	1.37	1.44	2.99	1.01	1.90	1.60	.48
	48.50	47.90	48.20	50.50	50.30	48.40	51.00	47.00	47.10	48.20
21 (Gorilla)	1.56	6.67	5.68	.57	.61	1.41	.41	.97	.81	.23
	22.60	23.20	22.90	21.00	21.30	22.70	20.70	24.00	23.90	22.90
22 (Chimp)	.74	11.30	9.80	1.12	1.18	2.45	.82	1.55	1.31	.39
	39.70	39.30	39.50	41.30	41.20	39.60	41.40	38.60	38.60	39.50
23 (Orang)	2.15	8.90	7.69	.86	.91	1.92	.63	1.24	1.04	.31
	31.10	30.90	31.00	31.70	31.60	31.00	31.80	30.60	30.70	31.00
24 (*Homo*)	6.91	28.80	24.81	2.71	2.87	6.19	1.98	4.03	3.40	1.00
	100.00	100.00	100.00	100.00	100.00	100.00	100.00	100.00	100.00	100.00

[a]Based on the equations in Table 1 and hominoids in Table 2.
[b]The top number is the actual E.Q. value, and the bottom value is the percentage of modern Homo's E.Q. (e.g., Hominid #7, E.Q. of 5.08 represents 73.5% of 6.91, the modern *Homo sapiens'* 6.91 E.Q.)

TABLE 4. Sample Percentage Differences Between Extremes in Sets of Equations[a]

		Differences between equations	
	Hominoid number	#4 and #8 (all primates, N.W. monkeys)	#7 and #8 (N.W. and O.W. monkeys)
	1	15.7	17.8
	2	18.1	20.5
	3	9.9	11.1
	4	13.1	14.5
	5	15.3	17.2
	6	18.5	20.7
	7	15.7	17.3
	8	13.3	14.9
	9	11.4	12.9
	10	9.8	11.1
	11	7.3	8.1
	12	1.5	1.5
	13	6.8	7.5
	14	9.4	10.7
	15	8.0	8.8
	16	10.7	11.8
	17	12.5	13.8
	18	12.4	18.2
	19	7.6	4.0
	20	3.5	4.0
Gorilla	21	-2.9	-3.3
Chimpanzee	22	2.7	2.8
Orangutan	23	1.0	1.2
Homo sapiens	24	0.0	0.0

[a]Based on the percentage of modern *Homo* E.Q. values in Table 3.

World and Old World monkeys, a consequence of the fact that these equations have the highest and lowest exponents, respectively (see below). We have included the differences between equations #4 and #8 to show that the difference between the "usual" E.Q.'s, i.e., those used by Jerison (1973) and Stephan, et al., (1970) or the "Homocentric" one, are minimal.

Table 5 gives rank-order correlations for the E.Q. scores of 89 species when different reference equations are used. Although the correlations are generally extremely high (and all are significantly different from zero), their nonunity implies that the rank order of these species is not necessarily invariant under a change of reference equation (see below).

TABLE 5. **Rank-Order Correlations for the E.Q. Scores of 89 Species When Different Reference Equations from Table 1 are Used**[a,b]

	1	2	3	4	5	6	7	8	9	10
1	---	.98	.99	.76	.82	1.00	.68	.95	.95	.99
2	.90	---	.99	.66	.73	.99	.57	.99	.99	.99
3	.95	.95	---	.71	.77	1.00	.62	.97	.97	1.00
4	.60	.50	.55	---	.99	.74	.99	.57	.57	.71
5	.66	.56	.61	.94	---	.80	.97	.64	.65	.78
6	.97	.93	.97	.57	.63	---	.65	.96	.96	1.00
7	.52	.42	.47	.92	.86	.49	---	.47	.48	.63
8	.82	.92	.87	.42	.48	.85	.34	---	1.00	.97
9	.82	.92	.88	.43	.49	.85	.35	.99	---	.97
10	.95	.95	1.00	.55	.61	.98	.47	.87	.87	---

[a]The figures in the top right-hand portion of the table are Spearman's rho's, those in the bottom left portion are Kendall's taus. For example, equation 5 shows the highest correlation with equation 4 in both Kendall's tau (.94) and Spearman's rho (.99).

[b]Although the E.Q.'s of primates are in close agreement for all 89 species, they are not perfect, indicating changes in rank order depending on the equations used.

THEORETICAL CONSIDERATIONS

The "reversibility" of the E.Q. rank orders of different taxa when different reference equations are used, documented on the above empirical data base, presents a serious problem in the interpretation of E.Q. scores, particularly because there seems to be little informed agreement (or, indeed, little debate) concerning which reference equation is most appropriate for examining brain-body relationships within any particular group of related organisms. We will now briefly examine the general behavior of E.Q. scores as a function of a small number of parameters; in particular, we will show that the *relative* magnitude of the E.Q.'s change and thus can be, in many cases, arbitrarily manipulated to provide almost any rank order the researcher prefers.

Let

V_i = brain volume (or weight) of species i
V_j = brain volume (or weight) of species j
W_i = body weight of species i
W_j = body weight of species j

Using an equation of the form

$$y = kW^{\alpha}$$

we can compute the E.Q.'s of species i and j (=EQ_i and EQ_j respectively) as:

(1a) $EQ_i = V_i / kW_i^{\alpha}$ and

(1b) $EQ_j = V_j / kW_j^{\alpha}$

Thus, $EQ_i = EQ_j$ if and only if

$$V_i / kW_i^{\alpha} = V_j / kW_j^{\alpha} \text{ or}$$

(2) $V_i / V_j = (W_i / W_j)^{\alpha}$

By taking logarithms of equation 2, we get the following expression for the conditions under which $EQ_i = EQ_j$:

$$\log (V_i/V_j) = \alpha \log (W_i/W_j)$$

Changing the latter equation to an inequality, we find that:

(3a) if $\alpha < \log (V_i/V_j)/\log (W_i/W_j)$

$EQ_i > EQ_j$ <u>if</u> $W_i/W_j > 1$ and $EQ_i < EQJ_j$ if $W_i/W_j < 1$

(3b) if $\alpha > \log V_i/V_j \log (W_i W_j)$

$EQ_i > EQ_j$ <u>if</u> $W_i/W_j < 1$ and $EQ_1 < EQ_j$ if $W_i W_j > 1$

(Note that the above expressions are undefined if $W_i = W_j$, in which case $EQ_i/EQ_j = V_i/V_j$ for all values of α).

The parameter k, which determines the y-intercept of the logarithmically transformed version of the brain-body equation, has no effect whatsoever on the *relative* magnitudes of the two species' E.Q.'s (although it will, of course, influence the absolute magnitude of those scores).

Thus, using equations 3a and 3b and brain and body weights, we can define for any pair of taxa a value for the exponent of the reference equation (α^*) above which one species has the higher E.Q. and below which the other is judged more "highly cephalized." The values of α^* for each pair of hypothetical hominids in Table 2 are given in Table 6. Although most values of α^* lie outside the range of exponent values normally encountered in a "reasonable" set of potential reference equations approximately ($.4 < \alpha < .8$), a substantial number do not (these have been *<u>underlined</u>* in Table 6). In these cases the choice of a reference equation from among a "reasonable" set of potential reference equations will be critical in forming our

TABLE 6. Values of α^* For Each Pair of Hypothetical Hominids[a]

	12	20	13	11	19	15	14	10	16	3	9	17	8	4	7	18	5	6	1	2
12	—	-.26	-.22	0	-.10	-.19	-.16	0	-.15	.14	0	-.13	0	.11	0	-.05	.10	0	.19	.18
20		--	-.20	.16	0	-.15	-.12	.11	-.10	.29	.09	-.08	.08	.22	.07	0	.19	.06	.29	.26
13			--	1.26	.60[b]	0	0	.42	0	.69	.30	0	.23	.43	.19	.09	.35	.15	.47	.40
11				--	x[c]	x	-1.15	0	-.62	.48	0	-.40	0	.26	0	-.11	.21	0	.36	.31
19					--	x	-.55	.32	-.30	.72	.21	-.19	.15	.39	.11	0	.31	.09	.45	.38
15						--	0	.62	0	.94	.40	0	.29	.51	.22	.10	.40	.17	.53	.45
14							--	1.34	0	1.55	.61	0	.38	.66	.27	.13	.49	.20	.62	.51
10								--	x	1.82	0	-1.11	0	.44	0	-.18	.30	0	.50	.40
16									--	3.57	1.11	0	.53	.86	.34	.16	.59	.24	.73	.58
3										--	-3.18	-6.23	-.80	0	-.44	-.66	0	-.29	.30	.23
9											--	x	0	.71	0	-.25	.41	0	.63	.48
17												--	1.02	1.40	.49	.23	.80	.31	.91	.69
8													--	2.17	0	-.49	.67	0	.87	.61
4														--	-1.77	-2.66	0	-.59	.56	.36
7															--	x	2.10	0	1.53	.86
18																--	3.15	.44	1.90	1.07
5																	--	-1.52	1.22	.56
6																		--	9.45	1.64
1																			--	0
2																				

[a]Entries are values of α (α^*) below which the row hominid will have a higher encephalization coefficient than the column hominid.

[b]Values of α^* between 0.4 and 0.8 are underlined. See text for details.

[c]Entries of "x" indicate hominid pairs with the same body weight.

judgments concerning the relative encephalization of that particular pair. Without some explicit *a priori* criterion for choosing the "best" equation, those judgments are, to a large extent, a matter of taste. Thus, we take issue with Gould's claim that the use of E.Q.'s "avoids a reliance upon very precise (and probably incorrect) values for allometric exponents" (1975). The above suggests that the "choice" of exponent is ultimately dependent on the data base chosen, and affects the relative E.Q. scores of different taxa and hence our interpretations of evolutionary relationships among them. This will present a serious problem *unless* one of the two following conditions hold:

(a) An allometric exponent could be chosen on the basis of some well-established *a priori* functional hypothesis. That is, if some exponent were known, from structural or "engineering" principles, to represent functional equivalence within the interspecific brain/body weight allometric relationship, that exponent could be used with confidence for the calculation of the E.Q.'s of various taxa, functioning as a "criterion of subtraction." The scaling exponent of 0.66 comes immediately to mind, for it appears frequently in empirical studies and, more important, it implies a functional determination of brain weight by body surfaces rather than volumes. As Gould (1975) argued, however, the underlying causes of this surface-dependence have never been adequately explained, and "we cannot ever be sure that the presumed surface-dependency is more than coincidental." Furthermore, the scaling could vary in different taxa for different reasons (see also Martin, this volume).

(b) In the absence of an understanding of the functional requirements for size increase, and if, regardless of the data base chosen, all exponents varied within narrow limits, perhaps some average could be taken for the reference equation. As can be seen in Table 1, the exponents for the various combinations of primate taxa vary from .577 in Old World monkeys (N=36) to .798 in New World monkeys (N=21). Although an "average" value of around .66 is suggested, the observed variability in the scaling exponents is not simply a function of small-sample fluctuations, as can be seen in Figure 1 where 95% confidence limits for the exponents have been plotted. We are aware that any attempt at a functional or adaptive interpretation of this variability in scaling parameters, which is in any case beyond the more limited scope of this report, must come to grips with several serious methodological problems. For example, within the New World monkeys, the slopes for the Callithricidae and Cebidae are both close to .66, yet given the relative placement of these two groups with respect to brain and body weight, the combination of the two families within one higher-order taxon yields a slope of .798. At which taxonomic level are comparisons of slope meaningful?

In this paper, we have concentrated on equations based on intergeneric or higher-order taxonomic groupings, because these have been used most often. Hemmer (1971) and Leutenegger (1973) used intrageneric data, which yield slopes of roughly .2 to .3 in most cases. For the same reasons

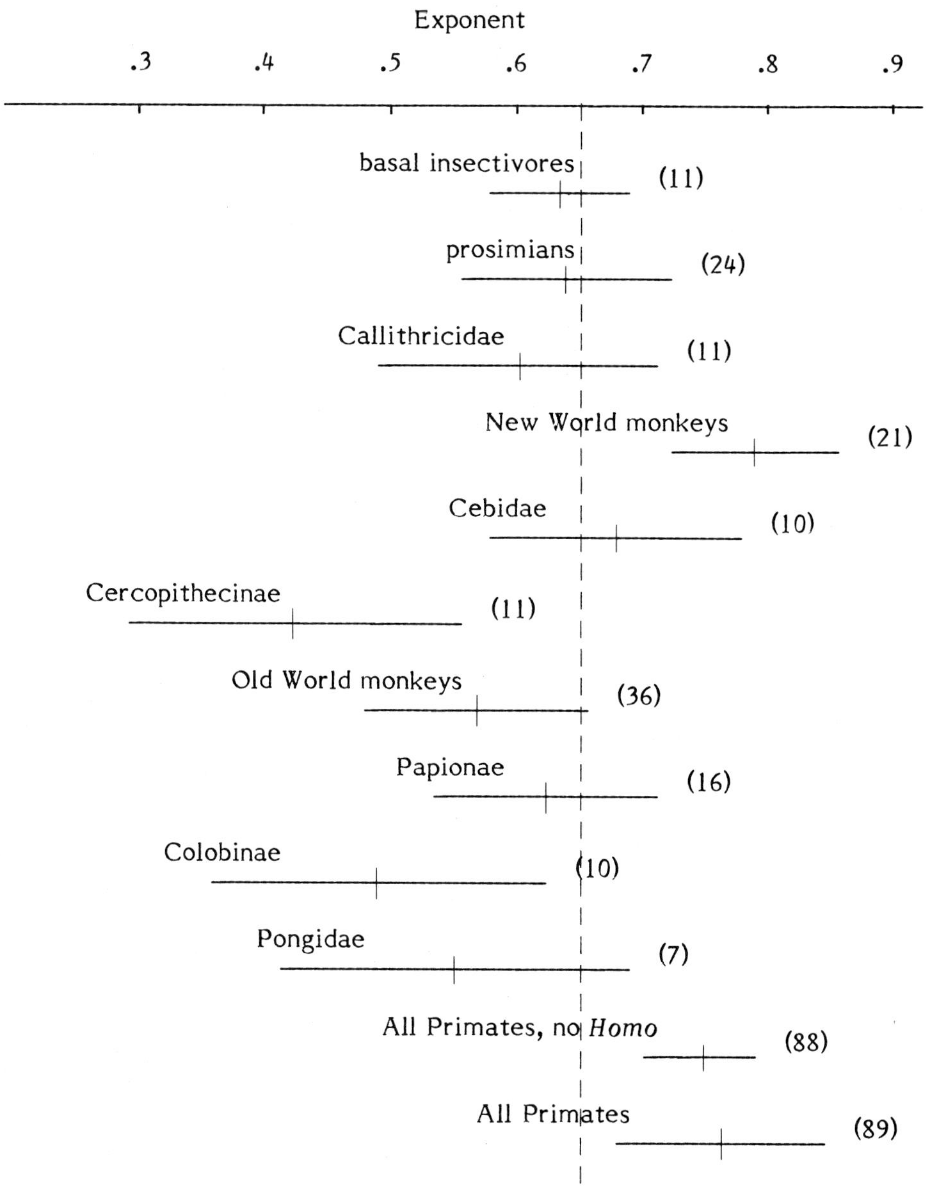

Fig. 1. The mean and 95% confidence intervals for different Primate taxa. The values in parentheses are the number of species in each group. The dashed line gives the exponential value of 0.66.

indicated here, we are not convinced that any one particular intrageneric allometric curve is any less selective or meaningful than for higher-order taxa combinations. Nor are we certain that the data bases for such intrageneric groups are any more accurate or adequate than the others. Slopes of values lower than .6 do have the effect of lowering E.Q. scores for smaller-brained species. Using a slope of .23, for example, gives *Homo sapiens* an E.Q. of 103.96, and *Pan troglodytes* a value of 35.55. *Gorilla* has a value of 29.34.

The australopithecine E.Q.'s are relatively lower than those given in Table 3, varying from about 41 to 45. The Hadar specimen provides an E.Q. of about 40.5. These values are still significantly higher than any pongid value, but respectively lower than those based on an exponent of roughly .6. This illustrates, again, two major points: (1) the australopithecine E.Q.'s remain advanced over pongid values, and (2) the values as a percentage of modern *Homo* are sensitive to the slopes chosen. We frankly do not see any empirically derived reasons or physiological ones to convince us that intrageneric slopes are more appropriate.

Variability of slopes exists and no one reference equation has such a high frequency that it appears regardless of the data points on which the equation is based.

HUMAN MOSAIC EVOLUTION

As the above data should make clear, one's interpretation of where any particular hominid (or choice of brain and body weight combinations) fits in relationship to modern *Homo* will depend not only on the choice of weights, but on the data base used for calculating E.Q.'s. If "brain evolution" is equated with brain size increase and thereby encephalization, one could conclude that brain evolution was the terminal phase of hominid evolution (McHenry, 1974, 1975; White, 1980), following dental, locomotory, and other musculoskeletal changes. Alternatively, one could regard high encephalization scores indicative of early brain evolutionary changes, at least with respect to size. Thus, as some have argued, the australopithecines had apelike brains, or as others of us have argued (Holloway, 1976, 1979; Leutenegger, 1973), they show truly hominid characteristics, *including their E.Q.'s*. Of course, hypotheses about early human evolution are speculative, and vary according to the weight placed on functional systems evolving through time, e.g., bipedalism, the brain, tool-use and tool-making, manual dexterity, hunting and gathering, and sexual division of labor.

The question of weighting and interpretation takes on even greater significance the further back into prehistory we go, and the most recent discoveries of the Hadar and Laetolil specimens require, we believe, careful consideration of which data bases are most appropriate for the eventual judging of whether or not the brain had an important role in the earliest phases of human evolution.

WHICH DATA BASES FOR WHAT?

The question really revolves about the extrapolation from extant living forms to fossil species. Is it logical to compute E.Q.'s for hominids from an equation based on extant living species ranging from mouse to whale in size, and including order upon order of mammals only distantly related to hominids? Or is it more logical to compute E.Q's from some data base more closely related to actual phylogenetic relationships, e.g., hominids from pongids, Old World primates, or just modern *Homo*, as in equation 10? Or, should intrageneric or specific exponents of .2 to .3 be used? Obviously, the only secure data bases available are from living species, as we can never be certain about the body weights for fossil specimens. If we desire a purely comparative, phyletically uniform paradigm, then either the Jerison (1973) or Bauchot and Stephan (1969) and Stephan et al. (1970) equations (#1 and #2) make sense. We prefer the latter, because it represents a group of animals that seems overall most similar to the original ancestors of many of the mammalian groups, and to the primates in particular. Because equation #10, the "Homocentric" equation, allows the E.Q. to be immediately expressed in a percentage of the modern human value, and because we *are* the most highly cephalized creature, this equation serves the same descriptive purpose of equation #1, except that we ourselves become the standard for comparison.

But what of phylogenetic realtionships? Why not compute E.Q.'s for the Prosimii from a "basal insectivore" line? Why not compute E.Q.'s of New World monkeys from a prosimian line, or those of Old World monkeys from prosimians, or indeed, given the small-sized and primitive Oligocene "pongids" (Simons, 1972), compute E.Q.'s for extant pongids from the Prosimian or Old World (or New World) forms? Should the hominids' E.Q.'s be computed from the extant pongid equation, from "basal" insectivores, from a composite of mice, whales, giraffes, lions, walruses, bovids, and dogs, or from a general intrageneric equation that will also vary depending on genera and species chosen?

Two assumptions are inherent in the above discussion: (1) E.Q.'s are useful measures, and we should continue to grind them out; and (2) we really do know something about the ancestral lines from which various taxa (at whatever level) are derived. Both assumptions require challenging and much more thought than they have received previously. E.Q.'s are useful as a composite indicator of brain-body size relationships, and if accurate brain and body weights and absolute dates are available, they could provide a more refined indication of evolutionary dynamics through time. And yes, we do believe we know something about ancestral relationships of extant and even fossil species, however brightly or dimly in particular groups.

Figure 2 shows graphically the results of calculating E.Q.'s for each major primate taxon on the basis of the popular conception of what approximates *the next* lowest taxonomic grade. In other words, prosimian E.Q.'s have been calculated from a "basal insectivore" line, and so forth.

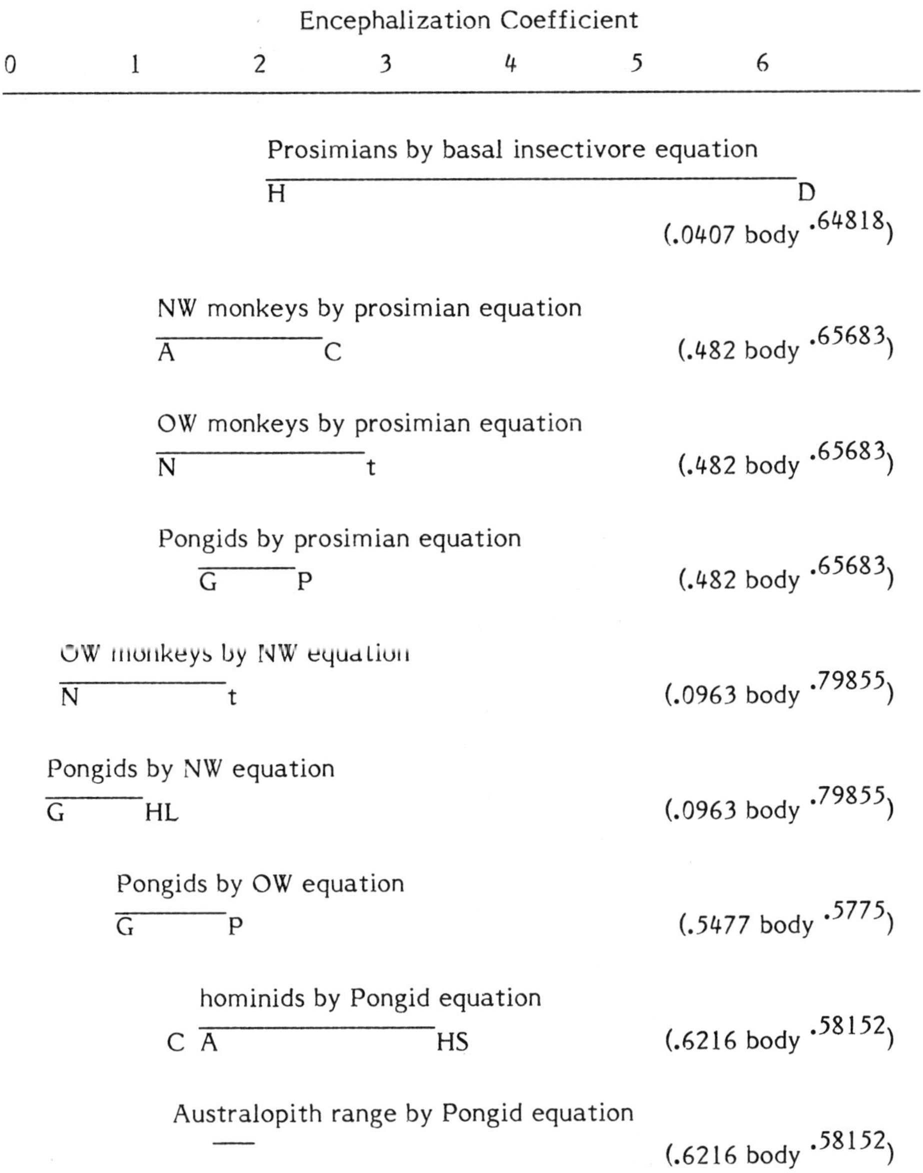

Fig. 2. E.Q.'s computed for different primate taxa based on next-lower grouping. Figures in parentheses are the power equations used to obtain E.Q. values. H = *Hapalemur*; D = *Daubentenoia* (aye-aye); A = *Aotus*; C = *Cebus* (capuchin); N = *Nasalis*; t = talapoin monkey; G = *Gorilla*; P = *Pan troglodytes*; HL = *Hylobates lar*; HS = *Homo sapiens*.

The authors are *not* claiming that these taxa necessarily evolve from their "lower" groups. The figure is purely for descriptive purposes. Indeed, one could justifiably derive the pongids from either a prosimian--or New World monkey--like ancestral base, for example, and both New World and Old World monkeys from a prosimian-like base. The most striking aspect of the figure is that almost all groups are considerably overlapping in their E.Q. ranges when approached in this manner. The exceptions are the Old World monkeys and pongids from their respective base lines and the Prosimian range based on an insectivore line. One *could* conclude that the Prosimians show the highest relative E.Q. scores of all primates, including modern *Homo*.

We are not claiming this approach to be more logical than using equations based on distantly related species, as in equations 1 or 2, or 10. Indeed, the lie is given away when one places some of the newest, and one supposes more accurate, estimations of Gurche's (1978, this volume) brain and body weights for certain fossil prosimians along the prosimian-from-basal insectivore regression equation. Only *Rooneyia* falls within the extant prosimian range, suggesting that ancestral groups differed in brain-body scaling relationships. The derivation of hominid E.Q. scores from the extant pongid line is interesting in that even the heaviest of the hypothetical STS 5 *Australopithecus africanus*, with a body weight of 54 kg (119 pounds), does not overlap the highest encephalized pongid, the chimpanzee.

Perhaps a more specific and recent example would help in this question. The newly discovered Hadar 133-45 adult cranium (Johansen and White, 1979) has a reconstructed endocast whose volume is about 500 ml (RH, unpublished). If the brain/body weights ratio were the same as the chimpanzee average, body weight would be 55,555.5 grams, or 122.5 pounds, a value considerably above published estimates. With a figure of 75 pounds for body weight and 500 ml for brain volume, the E.Q. scores for the Hadar adult range from 54.7 to 63.1% of the E.Q. for modern *Homo sapiens*. The two extreme values are for equations #8 and #7 respectively, or Old World and New World monkeys. In Table 3 the chimpanzee E.Q. percent of the modern *Homo sapiens* values are 38.6 and 41.4% respectively.

Three points should be emphasized here. First, assuming a body weight of 75 pounds for the Hadar adult, the E.Q.'s are well advanced over the pongid range. Second, the differences between values (based on different equations) increase as the brain/body weights ratios increase (Table 3). Third, if the body weight for the Hadar 133-45 adult were lighter, the percentage of modern human E.Q. score would increase. White's (1980) discussion of hominid mosaic evolution overlooks the clear possibility that relative brain size had undergone evolutionary change by 3.5 million years ago. *For all we know at the present time, relative brain size could have preceded full bipedal locomotion in hominid evolution, and such an interpretation could be dependent on the data base chosen to calculate the E.Q. scores.*

This point is not an attack on the *mosaic* nature of human evolution (McHenry, 1975; White, 1980). If evidence suggested that brain/body ratios were higher than those of pongids during the earlier phases of hominid evolution, we would suggest that selection pressure on growth and development had altered in the past, with concomitant effects on such factors as postnatal dependency time and brain maturation, all of which occurred within a social-environmental matrix. One of us (Holloway, 1975, 1976, 1979) has consistently explored these ideas. The point we are making is that a strict adherence to any one "just-so" mosaic theory of human evolution could dim our perceptions regarding the importance of the brain throughout our evolutionary history.

These statistical exercises show that relative brain size parameters, such as E.Q.'s, are relative to the data bases chosen, and that although the rank-order correlations between E.Q.'s for some 89 primate species are high regardless of the chosen base lines, rank-orders do shift, and placements of hominids with regard to modern *Homo*, or chimpanzee, can vary by almost 20%. These kinds of shifts can effect interpretations of time-related shifts in the nature of hominid mosaic evolution, leading to a polarization of conceptual viewpoints regarding the role of the brain in human evolution. The tendency to reify these primitive heuristic devices (E.Q.'s) needs examination and correction, so that E.Q.'s remain a useful "first-step" tool in analyzing evolutionary trends, rather than being enshrouded with an unrelativistic and objective "mystique." E.Q.'s are residuals from a log log regression line between two variables, brain and body weights, and although surely some biologically relevant connection exists between the two, perhaps in terms of "minimal brain mass," or body surface area to volume, relationships to behavioral evolution or other biologically relevant factors are neither known nor explained. Furthermore, one should not assume (except for purposes of testing hypotheses) that the same set of biological relationships is homogeneous throughout all taxa of animals. That is, different taxa could scale differently for different reasons.

Finally, there is a critical need for evaluating the data bases to be chosen when using these E.Q. scores, as well as articulating their purposes, whether purely descriptive or phylogenetic. Until this is done, these heuristic devices really tell us nothing about brain-behavioral evolution.

SPECULATIVE COMMENTS

This paper has been purposefully lacking in speculation about brain evolution in the Hominidae. Our queries regarding appropriate data bases for calculation of E.Q. scores, and the various tables presented in this paper should not be construed as a blanket acceptance of allometry being a sufficient explanatory base for hominid brain evolution. Indeed, as indicated elsewhere (Holloway, 1979, 1976, 1975), brain-size changes through time are viewed as but one phenotypic facet that requires synthesis with other levels of neurobiological structuring, such as reorganization and heirarchy (or timing differentiation, development, and growth of the brain and its parts). Here we show that, insofar as gross brain and body size units are to be manipulated to measure neural evolutionary change, i.e., E.Q.'s or other measurements of cortical cephalization (e.g., Jerison's N_C's or "extra cortical neurons," 1973), the relational position of any given hominid (or primate) will vary according to the data base. Viewpoints on the nature of mosaic human evolution, meaning the interpretation of major evolutionary "events" (e.g., bipedalism, canine reduction, tool-making and manual dexterity, brain evolution) will covary with the weighting of E.Q. values. Whatever the "true" australopithecine or habiline configurations of brain and body weights, those figures and the resulting E.Q.'s do not address such important issues as reorganization and/or hierarchy during the evolution of the brain. E.Q.'s are not currently related to hemispheric asymmetry and specialization in modes of cognitive functioning. Yet comparative studies are showing very distinct asymmetrical (petalial) patterns between pongid and hominid endocasts (Holloway and de Lacoste, in preparation; LeMay et al., this volume). Nor are E.Q.'s correlated with endocast stereoplotting studies (Holloway, 1981), which suggests that once allometric changes are taken into consideration, hominid and pongid taxa significantly differ in endocast surface shape, particularly in the inferior parietal region. Nor have E.Q.'s yet been used to investigate the extremely thorny issues of the position of the lunate sulcus (Falk, this volume) and its significance for differentiating between pongid and hominids, both structurally and functionally.

The relativism in relative brain size should be kept in perspective, as *one* element in brain evolutionary change. Allometric explanations are interesting initial attempts to provide a window on complex changes that occurred through time. Ultimately, the alterations will be linked to changes in regulatory DNA, that part of the genome that initiates, terminates, and sustains complex molecular events at the level of structural DNA, and that controls the unfolding of sequences in interaction within a bodily and social environment.

A preoccupation with what is *presently* quantifiable can blind us in seeking more holistic syntheses between comparative and paleoneurological evidence or mask the probable complexity of evolutionary change in primate brains.

Acknowledgments: Many of the quantitative analyses were performed while the senior author (R.H.) was supported by NSF grants BNS 7805651, 74-20149, and GS-92231X for which support he is grateful.

REFERENCES

Bauchot, R. and Stephan, H., 1969. Encéphalisation et niveau évolutif chez les simiens. Mammalia, 33:235-275.

Dubois, E., 1897. Sur le rapport du poids de l'encéphale avec la grandeur du corps chez le mammifères. Bull. Soc. Anthrop., Paris, 8:337-376.

Gould, S.J., 1966. Allometry and size in ontogeny and phylogeny. Biol. Rev., 41:587-640.

Gould, S.J., 1975. Allometry in primates, with emphasis on scaling and the evolution of the brain. In, Contributions to Primates, F. Szalay, ed., Vol. 5, Karger, Basel, pp. 244-292.

Gurche, J., 1978. Early primate brain evolution. Master's thesis, University of Kansas.

Hemmer, H., 1971. Beitrag zur Erfassung der progressiven Cephalisation bei Primaten. In, Proc. 3rd Congress Primatology, Zurich 1970, H. Biegert and W. Leutenegger, eds., vol. 1, Karger, Basel, pp. 99-107.

Holloway, R.L., 1975. 43rd James Arthur Lecture on the Evolution of the Human Brain: The Role of Human Social Behavior in the Evolution of the Brain. Am. Mus. Nat. Hist., N.Y.

Holloway, R.L., 1976. Some problems of hominid brain endocast reconstruction, allometry, and neural reorganization. In, Colloquium VI of the IX Congress of the U.I.S.P.P., Nice, 1976 Congress, Pretirage, pp. 69-119.

Holloway, R.L., 1979. Brain size, allometry and reorganization: toward a synthesis. In, Development and Evolution of Brain Size, M.E. Hahn, B.C. Dudek, and C. Jensen, eds., Academic Press, New York, pp. 59-88.

Holloway, R.L., 1981. Exploring the dorsal surface of hominoid brain endocasts by stereoplotter and discriminant analysis. Phil. Trans. Soc. Lond., B292:155-166.

Jerison, H., 1973. Evolution of the Brain and Intelligence. Academic Press, New York.

Johansen, D.C., and White, T.D., 1979. A systematic assessment of early African hominids. Science, 203:321-329.

Leutenegger, W., 1973. Encephalization in Australopithecines: A new estimate. Folia Primatologia, 19:9-17.

McHenry, H.M., 1974. Fossil hominid body weight and brain size. Nature, 74:686-688.

McHenry, H.M., 1975. Fossils and the mosaic theory of human evolution. Science, 190:425-431.

Nie, N.H., Hull, C.H., Jenkins, J.G., Steinbrenner, K., and Bent, D.H., 1975. S.P.S.S. (Statistical Package for the Social Sciences), 2nd ed., McGraw-Hill, New York.

Passingham, R.E., 1973. Anatomical differences between the cortes of man and other primates. Brain, Behavior and Evolution, 7:337-359.

Passingham, R.E., and Ettlinger, G., 1975. A comparison of cortical functions in man and other primates. International Review Neurobiology, 16:233-299.

Sacher, G.A., 1975. Maturation and longevity in relation to cranial capacity in hominid evolution, In, Primate Functional Morphology and Evolution, R. Tuttle, ed., Mouton Press, The Hague.

Simons, E.L., 1972. Primate Evolution, MacMillan Company, New York.

Smith, R.J., 1980. Rethinking allometry, J. Theor. Biol., 87:97-111.

Snell, O., 1892. Die Abhängigkeit des Hirngewichts von dem Körpergewicht und den geistigen Fähigkeiten. Arch. Psychol., 23: 436-445.

Stephan, H., Bauchot, R., and Andy, O.J., 1970. Data on brain size of the brain and various brain parts in insectivores and primates, In, The Primate Brain, C.R. Noback and W. Montagna, eds., Appleton-Century-Crofts, New York, pp. 289-297.

Tobias, P.V., 1971. The Brain in Hominid Evolution, Columbia University Press, New York.

White, T.D., 1980. Evolutionary implications of Pliocene hominid footprints. Science, 208:175-176.

ALLOMETRY, BRAIN SIZE,

CORTICAL SURFACE, AND CONVOLUTEDNESS

Harry J. Jerison*

The size of the mammalian brain may be directly and closely related to the amount of information that the brain can handle. My purpose is to show how and why I think this is the case and to apply the results to comparisons among primates. Two measures of size are emphasized here: the gross weight of the undissected brain and the total surface area of the cerebral cortex. A third measure, the convolutedness, or extent of fissurization, is closely related to the first two measures.

To minimize confusion about most of the terms and hypotheses of the first paragraph, I shall define as many terms as possible. The bases of the hypotheses have been discussed in detail before (Jerison, 1973, 1977, 1979). The relationships among size, surface, and convolutedness are the subject of the present report. *Information* is used in a more or less formal sense, as changes of state in the neural machinery of the brain per unit time. The measures of weight, volume, and surface are always in the grams-centimeter system. *Cortical surface* refers to the entire surface, including parts hidden (buried) in sulci and fissures. *Cortical* includes paleocortical as well as neocortical cerebral structures. The analysis in this report is based on data reported by Elias and Schwartz (1971) and by Brodmann (1913); the former authors are explicit in describing their measures of surface as including paleocortical structures, and this usage is implicit in Brodmann's report. The consistency of the data from the two reports suggests that they were measuring the same surface areas.

The amount of fissurization was measured by Elias and Schwartz (1971), who used a simple method. After preparing serial sections of whole brains they counted the number of separate gyri in each cross-section. A

*From the Department of Psychiatry and Biobehavioral Sciences, UCLA Medical School, Los Angeles, California 90024

gyrus was defined as a *convexity* on the surface, whereas a dimpling or sulcus would appear as a concave region separating gyri. The measure was undertaken explicitly for application to the study of fossil endocasts, and the lengths of gyri that were measured were restricted to those on the external cerebral surfaces. Midline gyri such as the cingulate gyrus were excluded from the measurements. The count of gyri on cross-sections of the whole brain was converted into a measure of the total length of gyri by multiplying the gyral number by the number of sections and the thickness of each section. The definition is, thus, purely quantitative and does not concern details of the fissural pattern. The total length of the gyri multiplied by their average width should exactly equal the exposed surface area of a brain as it might be measured on an endocast.

THE SIGNIFICANCE OF SURFACE AREA

Among the many technical and scientific advances in modern neurology has been the beginning of a new understanding of the microstructure and statistics of the mammalian cerebral cortex. The entire cortex seems to be made up of orderly columns, or modules, built on a common plan, and each made up of several hundreds or thousands of neurons (Eccles, 1979; Mountcastle, 1978; Szentagothai, 1978). More generally, the structure of the cerebral *cortex* is becoming recognized as having a basic uniformity that is almost as regular as that of the structure of the cerebellum, although the regularity is not as striking in anatomical preparations (Rockel et al., 1980). Comparable regularity of structure is identifiable in some subcortical regions (Scheibel and Scheibel, 1970). Although the details of the interpretation of this columnar organization are controversial (Towe, 1975), some kind of columnar organization clearly is present. Eccles' (1979) interpretation of that organization points to the measure of the cerebral surface area as being proportional to its information-processing capacity. The modules are conceptually in the role of elements ("chips") in microprocessors.

Szentagothai (1978) indicated that individual cortical modules are about 0.25 mm in diameter. Although the modular depth is related to cortical thickness, the total number of neurons per module is surprisingly constant in the nonvisual mammalian cortex (Rockel et al, 1980). Taking the module as the unit of processing and following Eccles' (1979) interpretation, we can estimate the number of modules from the cortical area. A human brain with 2000 cm^2 of *cortex* would contain 4 million modules. A dolphin with 3000 cm^2 would have 6 million modules. A chimpanzee with 800 cm^2 of cortex would have 1.6 million modules. And a mouse with 4 cm^2 would have 800 modules.

It would be a mistake to treat the numbers as exact measures of processing capacity, but it is not a mistake to recognize that the exact measures are definable. Cerebral cortical area may be used in a working definition of processing capacity, especially if the biological analysis of its modular structure, such as that suggested by Szentagothai (1978) and other authors cited earlier, proves to be correct.

SURFACE/VOLUME RELATIONS IN THE MAMMALIAN BRAIN

The identification of a fundamental role for cortical surface in processing capacity antedated modern neuroanatomy. It is discussed in Baillarger's (1853) short report, in which he also discussed the basic geometric constraint on surface/volume relationships in the brain. His report is curious and worth a digression because it anticipates the kinds of errors still made in handling data on this issue.

Baillarger noted that intelligence in animals had been associated (by others) with the amount of surface area, and he further recognized that if one wished to compare species with brains of different gross sizes, size had to be taken into account or controlled for in some way. The natural adjustment was to take a simple surface/volume ratio, which Baillarger did for man and rabbit. His ratios were $1700 cm^2/900 g$ (cerebral hemisphere only) for humans and $24 cm^2/5 g$ for rabbits. He did the same adjustment for the brains of a cat, sheep, dog, and hog, and found a similar paradoxical result. The animals had relatively more cortical surface than humans. To resolve the problem, Baillarger used Galileo's principle of "similitude," noting that the volume of a structure must be proportional to the cube of its length whereas its surface is proportional to the square (see D'Arcy Thompson, 1942). In modern terms we write the equation:

$$\underline{S} = \underline{k}\,\underline{V}^{2/3} \tag{1}$$

$\underline{S}$ = surface, $\underline{V}$ = volume and $\underline{k}$ is a constant.

Equation 1 defines the shape of any solid object in that the value of $\underline{k}$ is fixed for all objects of a given shape. All spheres have $\underline{k} = 4.84$. All cubes have $\underline{k} = 6.0$. (The units of surface $\underline{S}$ and volume $\underline{V}$ have to be dimensionally consistent, and in this report we are using the centimeter-gram-second system.)

If Baillarger had analyzed the surface-area-to-volume relationships in the brain dimensionally using Eq. 1 for the human and rabbit data, he might have noted that the rabbit $\underline{k}$ was 8.2, whereas in man it was 18.2. Man had much more surface per unit volume. Baillarger did not make that calculation, however; he noted only that man had no less *relative* surface area than a rabbit and that a proper statement would have to take into account surface/volume relationships.

It is possible to make better comparisons now, because we now have a rather good picture of the nature of the surface/volume relationships in the mammalian brain. Good data were first presented, without analysis, by Brodmann (1913). More than a half century later, Elias and Schwartz (1971) presented additional data with a purely empirical "regression" analysis. My purpose here is to emphasize fundamentals by summarizing these two sets of data graphically (Fig. 1).

The relationship displayed in Figure 1 can be discussed on several levels. Empiricists might treat the relationship as displaying the role of brain weight as a statistic, that is, a number to estimate a parameter. The parameter in this case is the cortical surface area, and the empiricist's satisfaction is that a measure as gross as brain weight can be used to estimate the important parameter of cortical surface area. Elias and Schwartz (1971) took the argument to this level and wrote a regression equation based on 15 species, which I have recalculated on the basis of the data from 48 species:

$$\log \underline{S} = 0.91 \log \underline{E} + \log 3.74 \quad (2)$$

$\underline{S}$ is surface in cm^2 and $\underline{E}$ is weight in grams.

This result is the empiricist's basic goal, that is, to describe a relationship. He can go further and analyze it statistically. The correlation coefficient is $r = 0.995$; to two significant figures this is a perfect correlation. In view of this impressive statistical result, it is obviously appropriate to look at the association represented by Eq. 2 more closely, to try to understand its meaning.

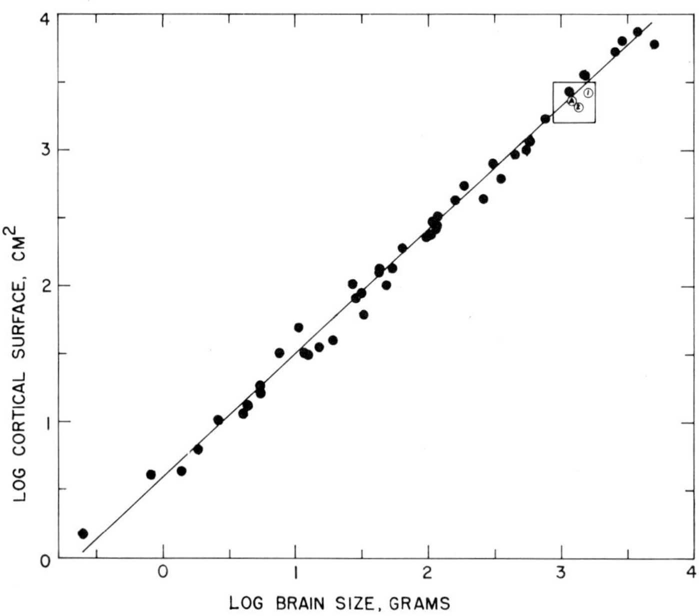

Fig. 1. Cortical surface as function of brain size in 48 species of mammals (data from Brodmann, 1913; Elias and Schwartz, 1971).

Baillarger's (and Galileo's) dimensional approach is helpful. Recast into a dimensionally meaningful format, Eq. 2 is:

$$\underline{S} = 3.74\,\underline{E}^{0.91} \quad (3)$$

The difference in the exponent between Eq. 1 and Eq. 3 is important. The empirical exponent, .91, which is significantly greater than 2/3, suggests that mammalian brains "behave" as if they were solid bodies that change their shape in an extraordinarily orderly way as they increase in size. The change is the increase of surface area beyond that expected for geometrically similar bodies of different volumes. The primary mechanism of the increase is, of course, fissurization. Equation 3 may explain the amount of observed fissurization. It predicts that measured fissurization should be determined almost entirely by brain size, that is, to about the same extent that surface area is so determined. This prediction is confirmed (see below), but more fundamental things may be learned from the dimensional approach.

The dimensional analysis implied by Eq. 1 and Eq. 3 can lead to important insights if the dimensional implications of several additional functions of brain size are considered. I have reviewed these functions (Jerison 1973, 1977, 1979), and they are cortical volume ($\underline{U}$), number of cortical neurons ($\underline{N}$), neuronal density ($\underline{N}/\underline{U}$), cortical thickness ($\underline{T}$) and brain weight ($\underline{E}$). The important feature is always the exponent, and the equations are:

$$\underline{U} = \underline{k_1}\,\underline{E}^{1.08} \quad (4)$$

$$\underline{N} = \underline{K_2}\,\underline{E}^{2/3} \quad (5)$$

$$\underline{N}/\underline{U} = k_3\,E^{-1/3} \quad (6)$$

$$\underline{T} = \underline{k_4}\,\underline{E}^{1/6} \quad (7)$$

With the exception of the exponent in Eq. 4, the exponents are approximations from empirical analyses, to suggest the dimensional constraints of brain size on these several variables. Equation 4 might be written with an exponent of 1.0 from a dimensional perspective, to suggest a dimensionally balanced system with respect to volume. But the empirically determined exponent (Jerison, 1977) was actually 1.08, and the analysis represented by Eq. 3 suggests that cortical volume cannot be balanced dimensionally against total brain volume. The empirical exponent of Eq. 4 appears to be a sensible one, as is evident if we combine Eqs. 3 and 7:

$$\begin{aligned} \underline{U} &= \underline{S}\ \underline{T} \\ &= \underline{k_o}\underline{k_4}\,\underline{E}^{.91+.17} \\ &= \text{constant} \times \underline{E}^{1.08} \end{aligned}$$

The system is reasonably self-consistent.

The insights come from the implications of all the equations. The 2/3 exponent of Eq. 5 shows that larger brains should be able to accommodate their number of cortical neurons and maintain geometric similarity. The disproportionate increase of cortical surface area indicated by Eq. 3 and its .91 exponent, therefore, cannot have been "forced" by a requirement to accommodate more neurons. The scenario is suggested by Eq. 6 where the increased numbers of neurons in larger brains have their cell bodies further apart. Some of the extra space for the greater distance between neuronal cell bodies in larger brains is provided by the greater thickness of the cortex (see Eq. 7), but most of the space is provided by the "expansion" of surface area. Much of the space between neuronal cell bodies is taken up by the axodendritic arborizations (see Armstrong, this volume), and therefore the anatomical "reason" for the peculiar exponent in Eq. 3 may be to accommodate increased *arborizations* (see Zilles et al., this volume).

The few data on dendritic arborizations that I considered earlier (Jerison 1973, 1977) show that this measure is approximately the 1/3 power of brain size. This, then, is the fundamental point that can be derived from Eq. 3. Larger brained mammalian species do not have disproportionate numbers of neurons in their cerebral cortices (the numbers increase in proportion to gross brain size; cf. Eq. 5), but they do have *more* neurons. These neurons also have *more* extensive axodendritic fields, on the average, than the fewer neurons in smaller brains. The disproportionality in surface represented by the exponent .91 in Eq. 3 can be explained by this disproportionality in arborization. If we keep in mind the neurobiologist's homily, that as a first approximation everything is connected to everything else in the nervous system, Eq. 3 gives evidence for that old saw.

FISSURIZATION

The disproportionate increase in cortical surface area in larger brains must occur in some physical way. It could change the shape of the brain case by making it less spherical, for example. But this would hardly be an efficient biomechanical solution for accommodating larger brains. It would be disastrous, making larger mammals unusually long-headed, and would produce serious problems for balancing the skull on the skeleton. The solution actually "adopted" in evolution was fissurization. When fissurization is measured as described earlier, the data of Figure 2 are generated.

As expected, the amount of fissurization measured by the total length of gyri in the brain appears almost entirely determined by brain size. The correlation for the data in Figure 2, like that for Figure 1, is r = 0.995, an almost perfect correlation.

I do not think of the result shown in Figure 2 as fundamental; it is essentially a necessary consequence of the relationships shown in Figure 1 and Eq. 3. The result is important for the methodology of paleoneurology, as suspected by Elias and Schwartz (1971) when they first collected and

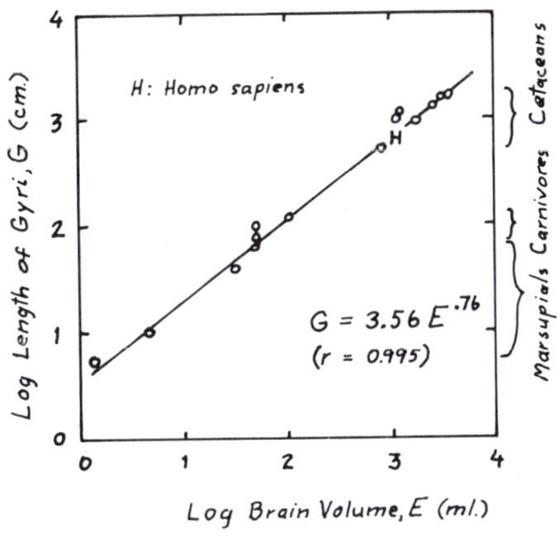

Fig. 2. Length of gyri as function of brain size in 15 species of mammals (data from Elias and Schwartz, 1971).

reported the data. It is important primarily for suggestions on how to interpret observed fissural patterns in brains and endocasts.

The result is, essentially, that the increase in gyri and sulci that can be seen in larger brains is almost entirely a result of the enlargement. An 800-ml hominid endocast, such as that of *Homo habilis* (KNM ER 1470), will inevitably have more gyri than a chimpanzee or australopithicine endocast of 400 ml or 450 ml. Functional interpretations of *new* gyri, such as a third frontal convolution (Broca's area), may, in fact, carry no more information than interpretations in terms of cerebral rubicons that are based on brain sizes (see also Falk, this volume; Galaburda and Pandya, this volume). The effect of brain size on fissurization is indicated in Figure 2, and the information in a measure of fissurization is equivalent to the information from size. If it can be shown that there have been changes in shape beyond those reflected by fissural pattern, e.g., changes in proportions of the several lobes of the brain, then arguments on functional roles of new gyri would have more force (see Gurche, this volume). If the new gyri were asymmetric in the two hemispheres, or if there were more new gyri in the frontal lobe than in the occipital lobe, an analysis of the shape of the "brain" and its fissural pattern, as revealed in fossil endocasts, could take on special significance. It could be interpreted legitimately as revealing functional reorganization of the brain in evolution, reorganization that is independent of changes in brain size.

REFERENCES

Baillarger, J., 1853. De l'étendue de la surface du cerveau et de ses rapports avec le développement de l'intelligence. Ann. Med. -Psych. (ser. 2), 5:1-9.

Brodmann, K., 1913. Neue Forschungsergebnisse der Grosshirnrinden-anatomie mit besonderer Berücksichtigung anthropologischer Fragen. Ver. 85 Vers. Deutsch. Naturf. und Aertze in Wien.

Eccles, J.C., 1979. The Human Mystery. Gifford Lectures 1977-1978. Springer-Verlag, New York.

Elias, H., and Schwartz, D., 1971. Cerebro-cortical surface areas, volumes, lengths of gyri and their interdependence in mammals, including man. Z. f. Säugetierkunde., 36:147-163.

Jerison, H.J., 1973. Evolution of the Brain and Intelligence, Academic Press, New York.

Jerison, H.J. 1977. The theory of encephalization. Ann. N.Y. Acad. Sci., 299:146-160.

Jerison, H.J., 1979. The evolution of diversity in brain size. In, Development and Evolution of Brain Size: Behavioral Implications, M. Hahn, C. Jensen, and B. Dudek, eds., Academic Press, New York, pp. 29-57.

Mountcastle, V.B., 1978. An organizing principle for cerebral function: the unit module and the distributed system. In, The Mindful Brain, G.M Edelman, and V.B. Mountcastle, eds., MIT Press, Cambridge, Mass.

Rockel, A.J., Hiorns, R.W., and Powell, T.P.S., 1980. The basic uniformity in structure of the neocortex. Brain, 103:221-244.

Scheibel, M.E., and Scheibel, A.B., 1970. Elementary processes in selected thalamic and cortical subsystems: the structural substrates. In, The Neurosciences Second Study Program, Rockefeller University Press, New York, pp. 443-457.

Szentagothai, J., 1978. The neuron network of the cerebral cortex: a functional interpretation. The Ferrier Lecture, 1977, Proceedings of the Royal Society, London. Series B 201:219-248.

Thompson, D'A.W., 1942. On Growth and Form, Cambridge University Press, Cambridge, England.

Towe, A.L., 1975. Notes on the hypothesis of columnar organization in somatosensory cerebral cortex. Brain Behav. Evol., 11:16-47.

ENCEPHALIZATION AND OBSTETRICS IN PRIMATES WITH PARTICULAR REFERENCE TO HUMAN EVOLUTION

Walter Leutenegger*

The existence of differences in pelvic morphology between Plio-Pleistocene hominids and modern humans is well documented (for a recent review see McHenry and Temerin, 1979). Pelvic remodeling during the evolution of the genus *Homo* has elicited two different interpretations: (1) it has been viewed as a reflection of increased efficiency in bipedalism (McHenry, 1975a; Zihlman, 1978), and (2) it has been interpreted in terms of rapid encephalization that would have generated selection pressure to increase birth canal dimensions (Lovejoy et al., 1973; Lovejoy, 1974, 1975, 1978). When the second hypothesis is scrutinized, the fossil evidence unambiguously indicates that the increase in brain size during human evolution was extraordinary both in magnitude and in rate (Pilbeam and Gould, 1974; Passingham, 1975; Sacher, 1975). This picture of encephalization from *H. habilis* to *H. erectus* to *H. sapiens* is based on estimated cranial capacities of a static series of adults. Obstetrical constraints and thus selection pressures on the pelvis are not, however, dependent on adult cranial dimensions but on those of the fetal cranium at term relative to birth canal dimensions. Moreover, evidence of differences between primate species in the growth rate of the brain and/or in the length of development during ontogeny (Schultz, 1941, 1956; Sacher and Staffeldt, 1974; Passingham, 1975) suggests that encephalization of adults cannot be expected to be proportionate to that of the neonate or to that of individuals at any particular developmental stage.

To clarify the question on the relationship between encephalization and obstetrics, I examine in this chapter the factors that affect the spatial relationship between neonatal cranial dimensions and the dimensions of the female lesser pelvis. Because reliable developmental data are scanty for

*From the Department of Anthropology, University of Wisconsin, Madison, Wisconsin 53706

prosimian primates and nonexistent for fossil hominids, comparisons can be made only for modern humans and other living anthropoid primates.

NEONATAL SIZE AND ENCEPHALIZATION

Individual fitness of mother and offspring in eutherian mammals is directly affected by the outcome of parturition. Successful delivery mainly depends on an adequate dimensional relationship between the cranium of the fetus at term and the lesser pelvis of the mother. This relationship is dependent on two main factors: (1) the size of the fetus at term relative to the size of the mother, and (2) the degree of encephalization at birth (Leutenegger, 1970a,b, 1972, 1974).

Among eutherian mammals, including primates, neonatal and maternal weight are allometrically related through the equation

$$y = bx^a$$

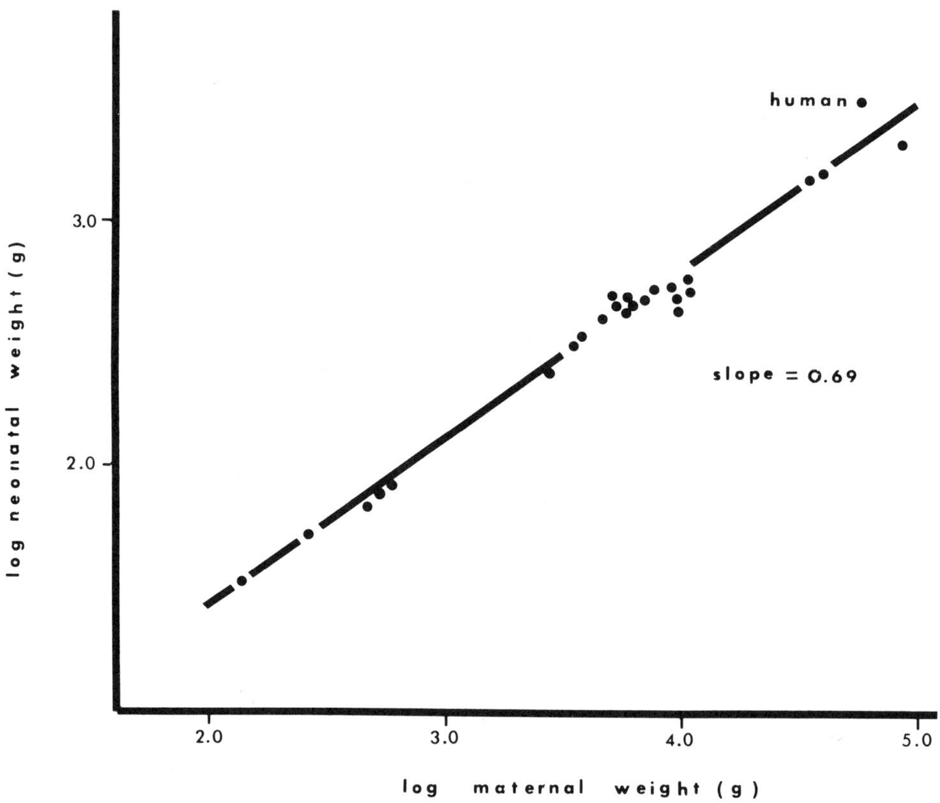

Fig. 1. Regression line and dispersion for neonatal weight on maternal weight for 25 anthropoid species. Note the strongly positive deviation for humans. Data from Leutenegger (1979).

where y is the neonatal weight (weight of the whole litter if there is more than one offspring), b the initial size constant (y intercept), x the maternal weight, and a the exponent of allometry (slope) (Leutenegger, 1973, 1976; Sacher, this volume). Figure 1 shows that for 25 anthropoid species, neonatal weight scales strongly (correlation coefficient r = 0.99) at a 0.69 power of maternal weight. The value of the exponent of allometry below 1.0 indicates that larger species have relatively smaller neonates than smaller species and vice versa. Interestingly, the data of human neonates display the highest positive deviation from the curve, with an actual average weight of 3,300 g being substantially above a predicted weight of 2,200 g.

Figure 2 shows that for 18 anthropoid species neonatal brain weight and body weight are also allometrically related. The two variables are correlated to a similarly high degree (r = 0.96). In contrast, however, neonatal brain weight scales at the much higher power of body weight (a = 0.94). The exponent of allometry just slightly below 1.0 indicates that

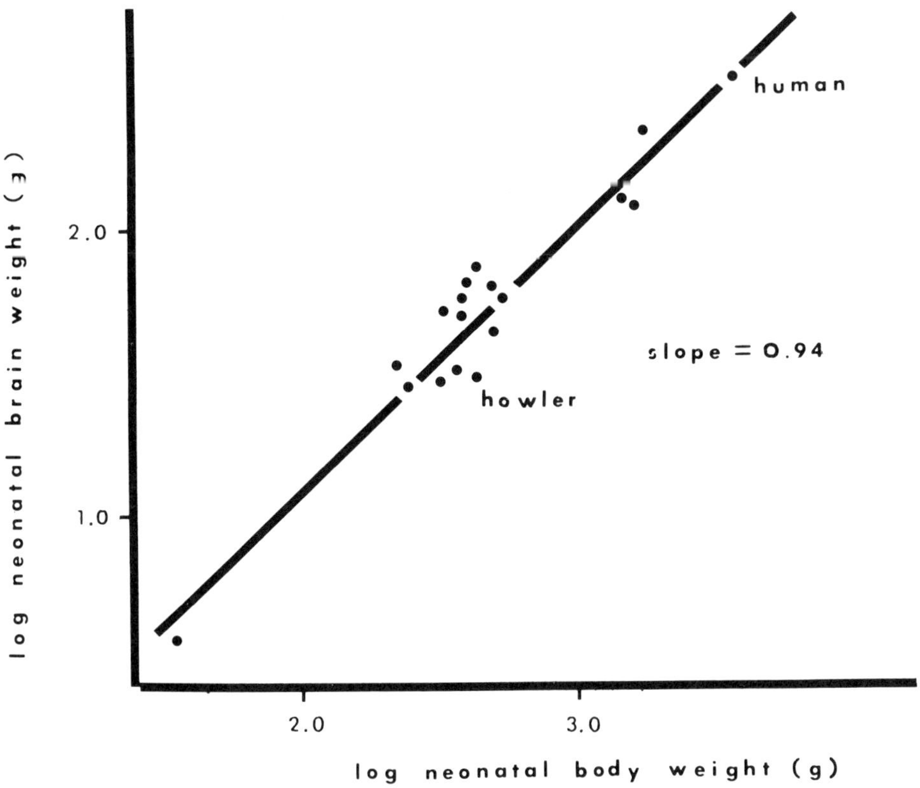

Fig. 2. Regression line and dispersion for neonatal brain weight on neonatal body weight for 18 anthropoid species. Note the highly negative deviation for the howler monkey and the lack of deviation for humans. Data from Leutenegger (1970a,b) and Sacher and Staffeldt (1974).

neonates of larger species tend to have only slightly smaller brains relative to their bodies than those of smaller species. In addition, the data of human neonates deviate only slightly and negatively from the curve, i.e., actual and predicted brain weight are virtually identical. This fact suggests that, in contrast to later developmental stages, encephalization at birth is not higher in humans than in nonhuman primates. Noteworthy also is the highly negative deviation of the data of neonate howler monkeys, with an actual brain weight of only 30.2 g compared with a predicted weight of 50.2 g.

Although neonatal brain weight scales nearly geometrically to body weight, the strongly negative allometric relationship between neonatal and maternal weight predicts neonatal brain size and consequently neonatal cranial dimensions of larger primates to be smaller relative to birth canal dimensions than those of smaller primates. This prediction is tested below by an analysis of the cranial dimensions of neonates in their relationship to female lesser pelvis dimensions.

OBSTETRICS

A most critical phase during parturition is the passage of the fetal cranium through the lesser pelvis (Fig. 3). In all nonhuman primates the

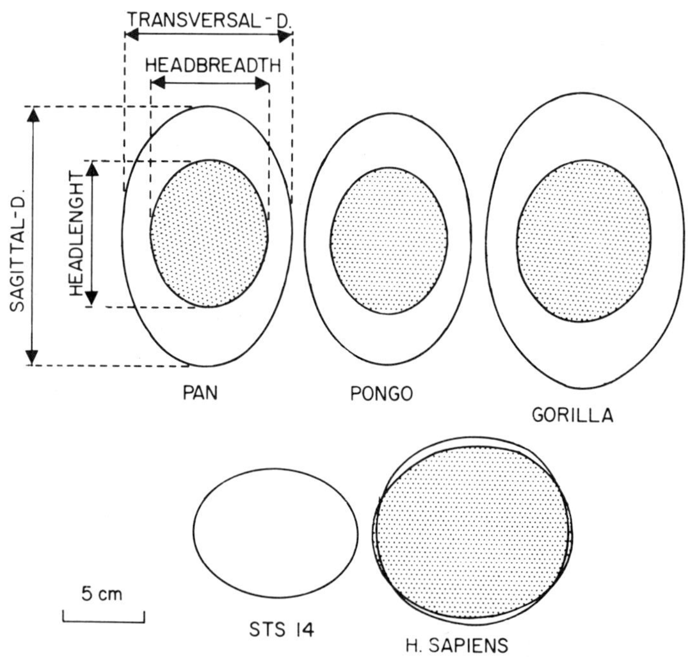

Fig. 3. Schematic representation of the relationship between neonatal cranial dimensions and female pelvic inlet dimensions in pongids, *Australopithecus africanus* (STS 14), and humans.

sagittal (conjugate) diameter exceeds the transverse diameter by a substantial amount. Because the sagittal dimensions of the fetal cranium, such as cranial length and height, also exceed transverse dimensions, such as cranial breadth, the cranium enters the pelvic inlet with the sagittal planes of both aligned. In the sagittal plane, cranial length is not as vitally important as cranial height (projective distance between vertex and gnathion) because during parturition the head can be tilted into a dorsoflexed position, so that the face, which in the sagittal direction is shorter than the neurocranium, may enter the pelvic inlet first. Furthermore, because in nonhuman primates the sagittal diameters of both pelvic inlet and outlet exceed the corresponding transverse diameters, the cranium does not rotate during passage through the lesser pelvis.

Table 1 shows that the relationship between corresponding dimensions of the neonatal cranium and female pelvic inlet varies considerably among anthropoid primates. As predicted, smaller species tend to produce neonates with relatively larger cranial dimensions than do larger species. The relatively largest neonatal crania occur in the common marmoset, a callitrichid, and the squirrel monkey, a small cebid. In both species all

TABLE 1. Relationship Between Neonatal Cranial Dimensions and Female Pelvic Inlet Dimensions

Species	No. of females	No. of neonates	Cranial length[a]	Cranial height[a]	Cranial breadth[b]
Callithrix jacchus	15	4	141.5	103.0	104.5
Saimiri sciureus	7	3	136.0	108.9	121.1
Cebus capucinus	4	2	116.0	101.5	118.2
Alouatta villosa	2	1	68.6	61.4	80.8
Lagothrix lagothricha	4	1	74.8	64.6	108.4
Ateles geoffroyi	7	8	74.0	67.4	96.0
Macaca mulatta	41	28	98.0	c	99.6
Nasalis larvatus	15	1	89.5	c	94.6
Hylobates lar	87	6	81.6	c	94.3
Pongo pygmaeus	26	4	56.2	c	73.1
Pan troglodytes	29	9	55.5	c	72.4
Gorilla gorilla	10	4	55.3	c	64.4
Homo sapiens	10	10	87.4[b]	c	101.8[a]

Data from Schultz (1949) and Leutenegger (1970a,b).

[a]% of sagittal diameter of female pelvic inlet

[b]% of transverse diameter of female pelvic inlet

[c]Data for cranial height of catarrhine neonates are not available. Since the sagittal diameter of the pelvic inlet exceeds cranial length, which in turn exceeds cranial height, the sagittal diameter of the pelvic inlet necessarily exceeds cranial height.

neonatal cranial dimensions are on the average larger than the corresponding female pelvic inlet dimensions. Cranial length, for example, exceeds the sagittal diameter of the pelvic inlet by more than a third, representing the most marked discrepancy. In the largest cebids, represented by the wooly monkey and the spider monkey, neonatal cranial dimensions are relatively smaller overall. Nonetheless, cranial breadth slightly exceeds (wooly monkey) or approximates (spider monkey) the transverse diameter of the pelvic inlet. By contrast, in another large cebid, the howler monkey, pelvic inlet dimensions are strikingly larger than corresponding neonatal cranial dimensions. In the rhesus monkey, proboscis monkey, and gibbon, representatives of cercopithecids and hylobatids, respectively, neonatal cranial dimensions approach the corresponding pelvic inlet dimensions, however, without exceeding them. Finally, in the pongids, the largest nonhuman primates, pelvic inlet dimensions greatly exceed, as predicted, those of the neonatal cranium (Fig. 3).

In contrast to the pelvic anatomy among nonhuman primates, the sagittal diameter of the pelvic inlet in humans is shorter than the transverse diameter. During the entry of the fetal cranium into the lesser pelvis, the sagittal plane of the cranium is therefore aligned with the transverse plane of the pelvic inlet. Because, however, the sagittal diameter of the pelvic outlet exceeds the transverse diameter, the fetal cranium must rotate 90 degrees when passing through the lesser pelvis. The spatial relationship between neonatal cranium and female pelvic inlet in humans is comparable to that in large cebids, cercopithecids and hylobatids, i.e., there is only a slight discrepancy between corresponding cranial and pelvic dimensions (Fig. 3).

Variation among primates in dimensional relationships between neonatal cranium and female pelvic inlet closely corresponds to variation in the degree of obstetrical difficulties. As just described, among callitrichids, cebids (except the howler monkey), cercopithecids, hylobatids and humans, one or another of the neonatal cranial dimensions exceeds or at least closely approximates the corresponding dimensions of the pelvic inlet. Although detailed observations are lacking for most species, successful delivery in these primates seems possible only by dilation of pelvic ligaments shortly before and during parturition or temporary deformation of the fetal cranium along its open sutures, or both. Despite these alterations, however, labor can be strenuous and delivery difficult as reported for callitrichids (Christen, 1974; Phillips, 1976), squirrel monkeys (Bowden et al., 1967; Hopf, 1967), chacma baboons (Gillman and Gilbert, 1946), rhesus macaques (Schultz, 1969; Brandt and Mitchell, 1971), and humans (e.g., Brandt and Mitchell, 1971). Furthermore, abortions, stillbirths, and miscarriages are especially frequent in those species with particularly large neonatal crania. The total frequency of such mishaps in, for example, the squirrel monkey amounts to slightly more than 50% of all recorded births, leaving less than half of the neonates surviving and developing into viable animals (Goss et al., 1968). An only slightly lower frequency of abnormal deliveries and nonviable offspring has been reported for the common marmoset (Phillips,

1976). In howler monkeys and pongids, on the other hand, cranial dimensions of neonates are much smaller than corresponding dimensions of the female pelvic inlet. Although parturition behavior in the howler monkey has yet to be extensively studied, delivery in pongids has been reported to be relatively quick and easy, and labor often hard to detect (Brandt and Mitchell, 1971).

DISCUSSION

The prediction that smaller species tend to produce neonates with proportionately larger cranial dimensions relative to those of the female pelvic inlet than do larger species has been fairly fully confirmed in the preceding analysis. As a correlate of this finding, obstetrical constraints also seem to be size-dependent, i.e., smaller species tend to have more obstetrical difficulties than larger ones. There are two exceptions, however, worth a closer inspection: howler monkeys and humans.

The howler monkey produces neonates whose cranial dimensions relative to female pelvic inlet dimensions are considerably smaller than predicted. As shown earlier, the disproportionate smallness of the neonatal cranium is solely because of an exceptionally small brain at birth and not because of an unusually low birth weight. The low degree of encephalization is also characteristic of subsequent developmental stages. At the adult stage, for example, the howler monkey is among the most lowly encephalized anthropoid primates measured by either the "index of progression" (Bauchot and Stephan, 1969) or the "constant of cephalization" (Hemmer, 1971). Gould (1975) interpreted the small brain of adult howler monkeys in terms of phyletic giantism, i.e., the decrease in encephalization is viewed as a result of downward intraspecific scaling. Because the neonate brain is disproportionately small and because no other structural or physiological features indicate phyletic giantism, this explanation does not seem entirely satisfactory. As an alternative, I suggest that the size of the howler monkey brain may have been limited by spatial constraints imposed by a structure situated sub-basally to the cranium, i.e., the vocal apparatus. The spatial requirements of the tremendously enlarged vocal apparatus have resulted in extreme alterations in the topography of both viscerocranium and neurocranium (Biegert, 1957). The changes in the neurocranium include an extreme backward relocation of foramen magnum and occipital condyles, and a complete extension of the cranial base. Because the cranial base in the howler monkey is even more extended than in, for example, the treeshrew with its tiny and primitive brain (Biegert, 1963), the exceptionally small brain of the howler monkey may be best explained in this context.

In contrast, humans produce neonates whose cranial dimensions relative to those of the female pelvic inlet are considerably larger than predicted. As indicated earlier, the relatively large cranium of human neonates is not the result of a higher degree of encephalization at birth but that of a disproportionately large overall neonatal size. Based on a comparison of human and macaque prenatal growth, Olivier and Pineau

(1958) interpreted the exceptionally large size of human neonates as the product of an extension of rapid embryonic growth well beyond the time of cessation in other primates. Using more recent and more extensive data on birth weights and gestation periods (Leutenegger, 1979; Doyle, 1979) and applying the formula of Huggett and Widdas (1951), I reinvestigated fetal growth rates for a large number of anthropoid species. Contrasting the results of Portmann (1941, 1945), I found considerable differences between species (Leutenegger, in preparation). Callitrichids, cebids, cercopithecids, and hylobatids have fetal growth rates that fall between 0.3 and 0.5 cm/day. Pongids have higher rates, ranging from 0.52 to 0.65 cm/day. Finally, humans have the highest rate with 0.69 cm/day. These figures translate into fetal growth rates for weight that are higher in humans than, for example, in chimpanzees and gorillas, by 20 and 60%, respectively. This difference indicates that the disproportionately large size of human neonates is due not only to the failure of fetal growth rates to decelerate as term approaches (Gould, 1977), but also to fetal growth rates that are simply higher than those in other primates. Even more striking than the differences in fetal growth rates are differences in neonatal brain growth advancement. The brain of human neonates weighs only about 25% of that of adults, whereas nonhuman primates have advancement factors of at least 35%. For human neonates, a similarly high advancement factor would correspond to a brain weight of at least 400 g, obviously much too large for successful parturition (Passingham, 1975; Sacher, 1975; Gould, 1977). Clearly then, the decrease in neonatal brain growth advancement and the postponement of three quarters of brain growth in humans to postnatal periods should be viewed in terms of obstetrical limitations. That is, to pass through the pelvic canal, a human neonate must be born at a more immature (more altrical) stage. The most severe constraint is the exceptionally short sagittal diameter of the pelvic inlet. A pelvic inlet that is wider than it is deep is a uniquely hominid feature. This pelvic configuration is a by-product of a shortening of the distance between sacroiliac and hip joints, which evolved as an adaptation to habitual erect posture and bipedalism in early hominids (Biegert and Maurer, 1972; Zuckerman et al., 1973; Leutenegger, 1974; 1977). There are strong indications that this and other major modifications in pelvic structure had been established before substantial brain expansion and thus before the emergence of the genus *Homo* (McHenry, 1975b,c). Minor pelvic alterations during the evolution of the genus *Homo*, such as a secondary increase in the sagittal diameter to produce a more cylindrical lesser pelvis, have been interpreted in terms of increased encephalization (Lovejoy, 1979). The results of this study indicate, however, that increased encephalization in humans is almost exclusively the result of increased postnatal brain growth. In particular, the remodeling of the human pelvis should be more appropriately viewed as reflecting selection against too low a neonate brain growth advancement and hence too high a degree of altricity.

Acknowledgments: I thank Henry McHenry, Susan Larson, and the editors of this volume for their comments. This research was supported in part by grants from the Wenner-Gren Foundation for Anthropological Research, Inc. and the Research Committee, University of Wisconsin-Madison.

REFERENCES

Bauchot, R., and Stephan, H., 1969. Encéphalisation et niveau évolutif chez les simiens. Mammalia, 30:235-275.

Biegert, J., 1957. Der Formwandel des Primatenschädels und seine Beziehungen zur ontogenetischen Entwicklung und den phylogenetischen Spezialisationen der Kopforgane. Morph. Jb., 98:77-199.

Biegert, J., 1963. The evaluation of characteristics of the skull, hands, and feet for primate taxonomy. In, Classification and Human Evolution, S. L. Washburn, ed., Aldine, Chicago, pp. 116-145.

Biegert, J., and Maurer, R., 1972. Rumpfskelettlänge, Allometrien und Körperproportionen bei catarrhinen Primaten. Folia Primatol., 17:142-156.

Bowden, D., Winter, P., and Ploog, D., 1967. Pregnancy and delivery behavior in the squirrel monkey (*Saimiri sciureus*) and other primates. Folia Primatol., 5:1-42.

Brandt, E.M., and Mitchell, G., 1971. Parturition in primates: behavior related to birth. In, Primate Behavior, Developments in Field and Laboratory Research, vol. 2, L.A. Rosenblum, ed., Academic Press, New York, pp. 177-223.

Christen, A., 1974. Fortpflanzungsbiologie und Verhalten bei *Cebuella pygmaea* und *Tamarin tamarin* (Primates, Platyrrhina, Callithricidae). Z. Tierpsychol., Suppl., 14:1-80.

Doyle, G.A., 1979. Development of behavior in prosimians with special reference to the lesser bushbaby, *Galago senegalensis moholi*. In, The Study of Prosimian Behavior, G.A. Doyle and R.D. Martin, eds., Academic Press, London, pp. 157-206.

Gillman, J. and Gilbert, C., 1946. The reproductive cycle of the chacma baboon (*Papio ursinus*) with special reference to the problem of menstrual irregularities as assessed by the behavior of the sex skin. S. Afr. J. Med. Sci. Biol., Suppl., 11:1-54.

Goss, C.M., Popejoy, L.T. II, Fusiler, J.L., and Smith, T.M., 1968. Observations on the relationship between embryological development, time of conception, and gestation. In, The Squirrel Monkey, L.A. Rosenblum and R.W. Cooper, eds., Academic Press, New York, pp. 171-191.

Gould, S.J., 1975. Allometry in primates, with emphasis on scaling and the evolution of the brain. Contrib. Primatol., 5:244-292.

Gould, S.J., 1977. Ontogeny and Phylogeny, Belknap Press, Cambridge, Mass.

Hemmer, H., 1971. Beitrag zur Erfassung der progressiven Cephalisation bei Primaten. In, Proc. 3rd Int. Congr. Primatol, vol. 1, J. Biegert and W. Leutenegger, eds., Karger, Basel, pp. 99-107.

Hopf, S., 1967. Notes on pregnancy, delivery and infant survival in captive squirrel monkeys. Primates, 8:323-332.

Huggett, A.St G., and Widdas, W.F., 1951. The relationship between mammalian foetal weight and conception age. J. Physiol. Lond., 114:306-317.

Leutenegger, W., 1970a. Das Becken der rezenten Primaten. Morph. Jb., 115:1-101.

Leutenegger, W., 1970b. Beziehungen zwischen der Neugeborenengrösse und dem Sexualdimorphismus am Becken bei simischen Primaten. Folia Primatol., 12:224-235.

Leutenegger, W., 1972. Newborn size and pelvic dimensions in *Australopithecus*. Nature, 240:568-569.

Leutenegger, W., 1973. Maternal-fetal weight relationships in primates. Folia Primatol., 20:280-293.

Leutenegger, W., 1974. Functional aspects of pelvic morphology in simian primates. J. Hum. Evol., 3:207-222.

Leutenegger, W., 1976. Allometry of neonatal size in eutherian mammals. Nature, 263:229-230.

Leutenegger, W., 1977. A functional interpretation of the sacrum of *Australopithecus africanus*. S. Afr. J. Sci., 73:308-310.

Leutenegger, W., 1979. Evolution of litter size in primates. Am. Nat., 114:525-531.

Lovejoy, C.O., 1974. The gait of australopithecines. Yrbk. Phys. Anthrop., 17:147-161.

Lovejoy, C.O., 1975. Biomechanical perspectives on the lower limb of early hominids. In, Primate Functional Morphology and Evolution, R.H. Tuttle, ed., Mouton, The Hague, pp. 291-326.

Lovejoy, C.O., 1978. A biomechanical review of the locomotor diversity of early hominids, in: Early Hominids of Africa, C.J. Jolly, ed., Duckworth, London, pp. 403-429.

Lovejoy, C.O., 1979. A reconstruction of the pelvis of AL-288 (Hadar Formation, Ethiopia). Paper presented at 48th Annual Meeting of the Am. Ass. Phys. Anthrop., San Francisco, Calif.

Lovejoy, C.O., Heiple, K.G., and Burstein, A.H., 1973. The gait of Australopithecus. Am. J. Phys. Anthrop., 38:757-780.

McHenry, H.M., 1975a. Biomechanical interpretation of the early hominid hip. J. Hum. Evol., 4:343-356.

McHenry, H.M., 1975b. Fossils and the mosaic nature of human evolution. Science, 190:425-431.

McHenry, H.M., 1975c. Fossil hominid body weight and brain size. Nature, 254:686-688.

McHenry, H.M. and L.A. Temerin, 1979. The evolution of hominid bipedalism; evidence from the fossil record. Yrbk. Phys. Anthrop., 22:105-131.

Olivier, G. and H. Pineau, 1958. Croissance prénatale comparée du macaque et de l'homme. C. R. Acad. Sci. Paris, 246:1292-1293.

Passingham, R.E., 1975. Changes in the size and organisation of the brain in man and his ancestors. Brain Behav. Evol., 11:73-90.

Phillips, I.R., 1976. The reproductive potential of the common cotton-eared marmoset (*Callithrix jacchus*) in captivity. J. Med. Primatol., 5:49-55.

Pilbeam, D. and Gould, S.J., 1974. Size and scaling in human evolution. Science, 186:892-901.

Portmann, A., 1941. Die Tragzeiten der Primaten und die Dauer der Schwangerschaft beim Menschen: ein Problem der vergleichenden Biologie. Rev. Suisse Zool., 48:511-518.

Portmann, A., 1945. Die Ontogenese des Menschen als Problem der Evolutionsforschung. Verh. Schweiz. Naturforsch. Ges., 125:44-53.
Sacher, G.A., 1975. Maturation and longevity in relation to cranial capacity in hominid evolution. In, Primate Functional Morphology and Evolution, R. H. Tuttle, ed., Mouton, The Hague, pp. 417-441.
Sacher, G.A., and Staffeldt, E.F., 1974. Relation of gestation time to brain weight for placental mammals: implications for the theory of vertebrate growth. Am. Nat., 108:593-615.
Schultz, A.H., 1941. The relative size of the cranial capacity in primates. Am. J. Phys. Anthrop., 28:273-287.
Schultz, AH, 1949. Sex differences in the pelves of primates. Am. J. Phys. Anthrop., 7:401-424.
Schultz, A.H., 1956. Postembryonic age changes, in: Primatologia, vol. 1, H. Hofer, A.H. Schultz and D. Starck, eds., Karger, Basel, pp. 887-964.
Schultz, A.H., 1969. The Life of Primates, Universe Books, New York.
Zihlman, A., 1978. Interpretations of early hominid locomotion. In, Early Hominids in Africa, C. J. Jolly, ed., Duckworth, London, pp. 361-377.
Zuckerman, S., Ashton, E.H., Flinn, R.M., Oxnard, C.E., and Spence, T.F., 1973. Some locomotor features of the pelvic girdle in primates. Symp. Zool. Soc. Lond., 33:71-165.

THE ROLE OF BRAIN MATURATION IN THE EVOLUTION OF THE PRIMATES

George A. Sacher*

In 1959 I showed a close allometric relationship between the maximum lifespans for captive extant mammalian species and their adult brain weights and body weights (Sacher, 1959). The same quantitative relationship was later found to hold within individual taxa as diverse as the Rodentia (Sacher, unpublished), the Odontoceti (Sacher, 1981), and, in particular, the Haplorhini (Sacher, 1975). The multivariate regression for the Haplorhini, based on data from 43 species, yielded the following relationship of lifespan (L, in years) to brain weight (E, in grams) and body weight (S, in grams):

$$\log L = (0.65 \pm 0.11) \log E - (0.20 \pm 0.08) \log S + 0.87 \pm 0.11 \quad (1)$$

This relationship has a squared multiple correlation of 0.74, so that three-fourths of the longevity variance in this group of primates is accounted for by their brain weights and body weights.

The relationship of L to E and S in Eq. 1 is not significantly different from the formula for the mammals as a whole, based on 273 species (Sacher, 1975); it may therefore be concluded that the great longevity of the primates, as compared with other mammalian orders, is due not to a unique evolutionary adaptation by the primates, but instead to the uniform operation of a general mammalian relationship of species lifespan to brain size. Moreover, because the L (E, S) function holds uniformly among recent orders that diverged in the late Cretaceous (Kielan-Jaworowska et al., 1979), this basic relationship may be assumed to hold throughout the evolutionary history of the mammals. Acceptance of this assumption enables estimation of longevities of extinct mammalian species if there is adequate information in the fossil record about their cranial capacities and body sizes. The fossil record of well-preserved primate postcranial skeletons is too scant to

*From the Division of Biological and Medical Research, Argonne National Laboratory, Argonne, Illinois 60439

support an estimation of evolutionary trends in primate encephalization before the Pliocene (Radinsky, 1977). The relative abundance of remains of the Hominidae, however, has permitted tentative inferences about the increase in longevity during hominid evolution from the late Pliocene to recent time, *pari passu* with the threefold increase of hominid brain size (Sacher, 1975; Cutler, 1976).

Inasmuch as the long lifespans of primates (Jones, 1968) are closely linked to their high degree of encephalization, it became pertinent to inquire into the factors the made the increase of brain size in the primates both necessary and possible. Sacher and Staffeldt (1974, hereafter referred to as SS74) showed that mammalian gestation times are governed more strongly by the amount of brain growth reached at birth than by any other factor. This finding, together with the demonstration of a negative association of fetal brain to litter size, provided a basis for the conclusion that the increase of lifespan with brain size in mammals is necessary because the reduction of reproductive rate concomitant with increased brain size can only be paid back by a prolongation of reproductive span and lifespan (SS74).

This paper shows that the developmental schedule for the primate brain differs in a major respect from that for other mammalian orders. It is proposed that this changed development schedule occurred early in primate evolution, in response to certain unique features of the aboreal niche that the early progenitors of the modern stepsirhine and haplorhine primates sought to exploit.

THE RELATIONSHIP OF BRAIN WEIGHT TO BODY WEIGHT DURING FETAL DEVELOPMENT AND AT BIRTH FOR MAMMALS

Some important implications of the SS74 data on neonatal brain weight and body weight were not discussed. I present here an examination of those additional aspects of that data set. Figure 1 is a plot of the relation of neonatal brain weight to neonatal body weight for 91 species of mammals from 19 orders (SS74). The following features are noteworthy.

1. The data points clearly separate into two groups. Uppermost is a band of points representing primate species, together with--although at much greater body weights--data points for elephants and toothed whales. Lying below these is a band of points for a larger group of species drawn entirely from the other mammalian orders. The two clusters show comparatively little scatter and virtually no overlap, despite the heterogeneous provenance of the data, and despite the fact that the neonatal status is heterogeneous in itself, ranging as it does from the extremely altricial birth status seen in the marsupials, myomorph rodents, and ursids to the highly precocial birth condition of the ungulates and monkeys.

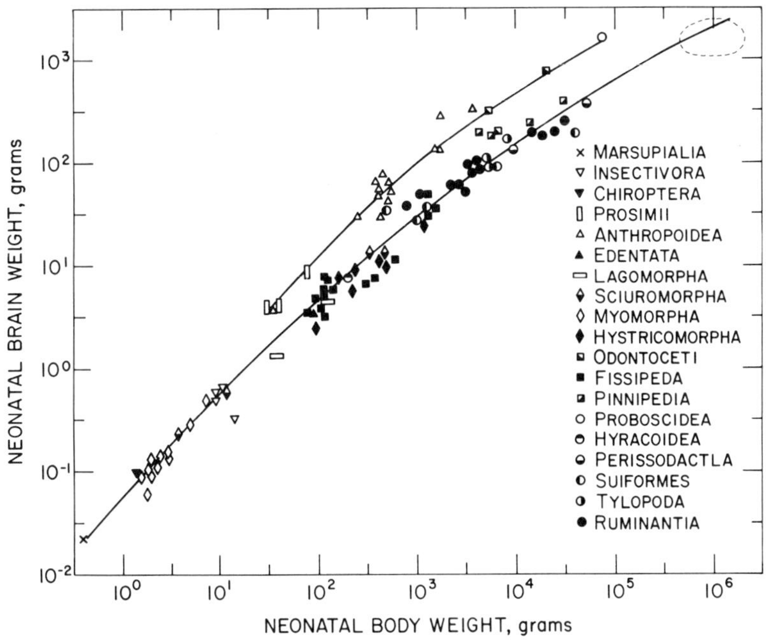

Fig. 1. Relations of brain weight to body weight for newborn mammals, plotted on double logarithmic scales. Data from Sacher and Staffeldt 1974. The dashed line ellipse in the upper right is an unpublished rough estimate for the Greenland right whale. The solid curved lines are fitted to the two sets of data by least squares, as described in the text. Regression coefficients are given in Table 1.

2. Both data clusters show a well defined convex-upward relation of neonatal brain weight to body weight, as shown by the two second-degree curves that are fitted to the respective data sets by least squares (Table 1). The larger, nonprimate, data set shows an approximately isometric (slope equal to unity) relation of log brain weight to log body weight up to a body weight of about 100 g, and then a gradual decrease of slope as body weight increases further. No systematic divergences can be seen among the orders that make up the nonprimate group. The primate trend, which seems to be shared by the toothed whales, shows the same convex-upward pattern, and lies above and approximately parallel to the non-primate trend. There is approximate isometry of brain weight to body weight for primate body weights up to about 1 kg.

The two distributions can be characterized by the mean ratio of neonatal brain weight to body weight within the body weight range for which there is isometry of neonatal brain weight and body weight. The mean ratio

TABLE 1. Statistical Evaluation: Neonatal Brain Weight vs. Neonatal Body Weight[a]

		Non-Primates	Primates
Intercept (logarithmic units)	a_0	-1.2790 ± 0.0470	-1.3920 ± 0.2310
First power term	a_1	1.0610 ± 0.0450	1.4690 ± 0.1520
Second power term	a_2	-0.0486 ± 0.0094	-0.1122 ± 0.0240
Standard error of estimate	SE_{est}	0.1313	0.1192
Correlation ratio	R^2	0.9870	0.9680
Sample size	n	74	25

[a]Coefficients of least squares quadratic regression of logarithm of neonatal brain weight, E_n, on logarithm of neonatal body weight, S_n, with data from Sacher and Staffeldt (1974).

for 17 primate species with body weights less than 1 kg (SS74) is 11.8%; the primates can therefore be said to follow the *12% fetal brain growth trajectory*. A sample of 10 species from the nonprimate group that have body weights less than 100 g gave a mean neonatal brain:body weight ratio of 5.7%; they can be said to belong to the *6% trajectory*.

All major subgroups of strepsirhine and haplorhine primates are represented on the 12% trajectory. Table 2 gives the data for three strepsirhine species, from (SS74), and a new data point for *Tarsius*. These 4 species average 11.6%. On the other hand, the tupaiids adhere to the 6% distribution, as do the other insectivore families.

As noted above, the neonatal brain-to-body weight relationship contains a major element of heterogeneity because of the extremely wide range of developmental advancement at birth. A numerical measure for brain growth advancement used by Portmann (1962) and in SS74 is the ratio of neonatal brain weight to adult brain weight. (Portmann defined the "Vermehrungsfaktor" (factor of increase) as adult brain weight divided by neonatal brain weight, the reciprocal of the "advancement factor" defined in SS74.) The species tabulated in SS74 have brain growth advancements ranging from 0.0013 for a marsupial and 0.02 for two bear species to 0.6 or higher for some arboreal primates. The level of behavioral performance at birth and the pattern of postnatal brood care corresponds in general with the anatomical advancement of brain growth. The nidicolous species, with altricial young, also show a low level of brain growth advancement at birth,

TABLE 2. Neonatal Brain Weights and Body Weights for Primates of Three Infraorders

Species[a]	Brain Weight (g)	Body Weight (g)	Ratio %
Galago crassicaudatus (Lorisiformes; Strepsirhini)	4.0	40.0	10.0
Nycticebus coucang (Lorisiformes; Strepsirhini)	4.0	30.5	13.1
Lemur catta (Lemuriformes, Strepsirhini)	8.8	78.0	11.3
Tarsius spectrum (Tarsiiformes; Haplorhini)	3.4	28.5	11.9

[a]Data for first three species are from SS74. Data for *Tarsius* are from a personal communication with L.A. Griner, San Diego Zoo, 1973.

whereas nidifugous species have precocial young with high levels of neonatal growth advancement (Portmann, 1962).

Despite the more than hundred-fold range of advancement values for brain weight (and for body weight as well), the two distribution bands, for primates and for nonprimates, respectively, show comparatively little scatter around their respective mean trends, and no evidence of stratification of altricial and precocial species.

The explanation for the ruggedness of the neonatal brain-to-body weight relationship can be found in Figure 2, in which data are plotted for the fetal and early postnatal brain weight and body weight growth of 6 species from 4 different orders of mammals (Dobbing and Sands, 1973, and personal communication--human; Hubbert et al., 1972 and personal communication--cattle; Kerr et al., 1969--macaque; R. D. Phemister, personal communication--beagle dog; Frahm, 1973--hamster; Sykora et al., 1965--rat). The least squares quadratic equations for the interspecific neonatal brain weight to body weight relationships of the primate (12%) and nonprimate (6%) groups (Table 1) are also drawn on the chart, and one sees that the ontogenetic growth trajectory for each of the 6 species conforms closely to the quadratic between-species trend defined by the neonatal brain-to-body weight data for the appropriate 6% or 12% group. The agreement is especially good for the primates, where the fetal growth data for *H. sapiens* and *M. mulatta* lie very close to each other and to the interspecific line. The fetal growth data for *H. sapiens* (Dobbing and Sands, 1973) extend back to early fetal life (fetal body weight of 1 gram) and show a remarkably close adherence to 12% growth isometry over more than a hundred-fold range of fetal body weight.

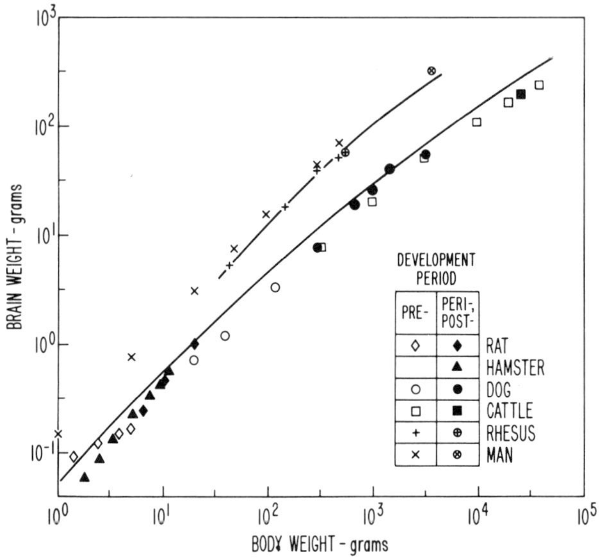

Fig. 2. The relation of brain weight to body weight during fetal and early postnatal growth, for six species of mammals from four orders. Each symbol identifies an ontogenetic series from one experiment. Curved lines redrawn from Figure 1 and Table 1.

Therefore, the ruggedness of the interspecific relation of brain weight and body weight arises from the fact that the between-species relationship of neonatal brain weight to body weight closely tracks the early ontogenetic growth trends for mammals. The implication of this finding is that the primate and nonprimate mammals respectively follow a single quantitative curvilinear allometric relation of brain weight to body weight throughout fetal life and, in some altricial species, well beyond birth.

The uniform mammalian mass growth pattern during fetal development is superseded in the later stages of development by taxon-specific ontogenetic growth trajectories that branch off from the general fetal growth function and continue on a low slope--about 0.10 to 0.25--until adult brain weights and body weights are attained. Adult brain and body weights for each taxon lie on or near to an allometric trend with a slope in the range from 1/3 to 2/3 (Jerison, 1973), as shown in Figure 3, which gives the adult brain-to-body allometry for 13 orders or suborders, all drawn from the nonprimate, 6%, brain-to-body growth trajectory. The adult trends are taxonomically stratified, have lower slopes than the fetal-to-neonatal trend, and always lie below the fetal-to-neonatal trend, which is represented by the solid curved line redrawn from Figure 1. Figure 7 (p. 109) should also be consulted.

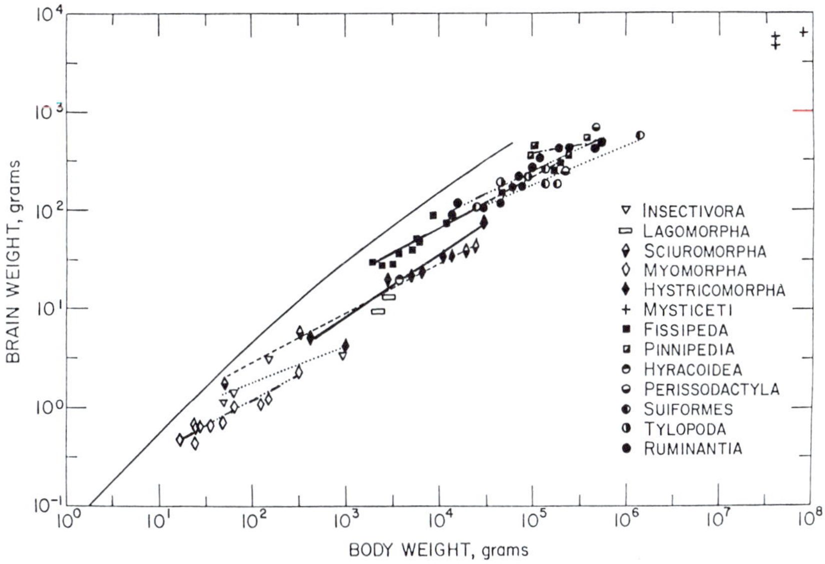

Fig. 3. Brain-to-body weight allometry for adult mammals, all from the 6% neonatal brain-to-body allometric trend. The curved line is the fitted least-squares neonatal allometric trend for these orders (from Fig. 1) and is also the approximate common fetal growth trajectory for members of these orders (cf. Fig. 2).

One superficially plausible explanation for the evolution of the 12% trajectory should be disposed of before we go further. The 12% developmental pattern did not evolve because of the evolution of large brains in the primates. The transition to the 12% trend occurred at or before the onset of the second primate radiation, when the ancestors of the modern primates were not notably big-brained (Radinsky, 1977; see Gurche, this volume). Therefore, the emergence of the 12% trajectory had an initial cause unrelated to large adult brain size.

Although the data on primate brain growth allometry were available to me as early as 1974 (SS74), I was unable until recently to find a key to their explanation. The needed clue was provided by Leutenegger (1979), who showed that when the Haplorhini and the Strepsirhini are compared with respect to the total litter weight at birth, the Haplorhini consistently produce about twice the weight of young for comparable maternal weight, as shown in Figure 4 (from Leutenegger, 1979).

Following up on Leutenegger's observation, I compared the available data on the perinatal reproductive investment of insectivore species (from SS74) with Leutenegger's data on strepsirhine reproductive investment. The outcome is seen in Figure 5. The maternal investments of the four insectivore species lie above the trend line for the strepsirhines, and below that for the haplorhines. *Tupaia glis*, the insectivore species that is closest to the Strepsirhini in terms of immunodiffusion systematics (Dene et al., 1976), is a typical insectivore in regard to maternal investment, just as it is in regard to fetal brain-to-body weight ratio, as noted above (see also Bauchot, this volume).

From Figures 5 and 6 (p. 108) and the above discussion, one can infer that at some time between the radiation of the earliest insectivore-like primates in the late Cretaceous and the second radiation of the strepsirhine-like "primates of modern aspect" (Simons, cited by Cartmill, 1974) in the mid- to late Paleocene (Kielan-Jaworowska et al., 1979; Szalay and Delson, 1979), an extraordinary evolutionary event took place. The schedule of primate fetal development was modified by reducing by half the amount of non-neural somatic tissue associated with a given amount of neural tissue throughout the greater part of fetal life.

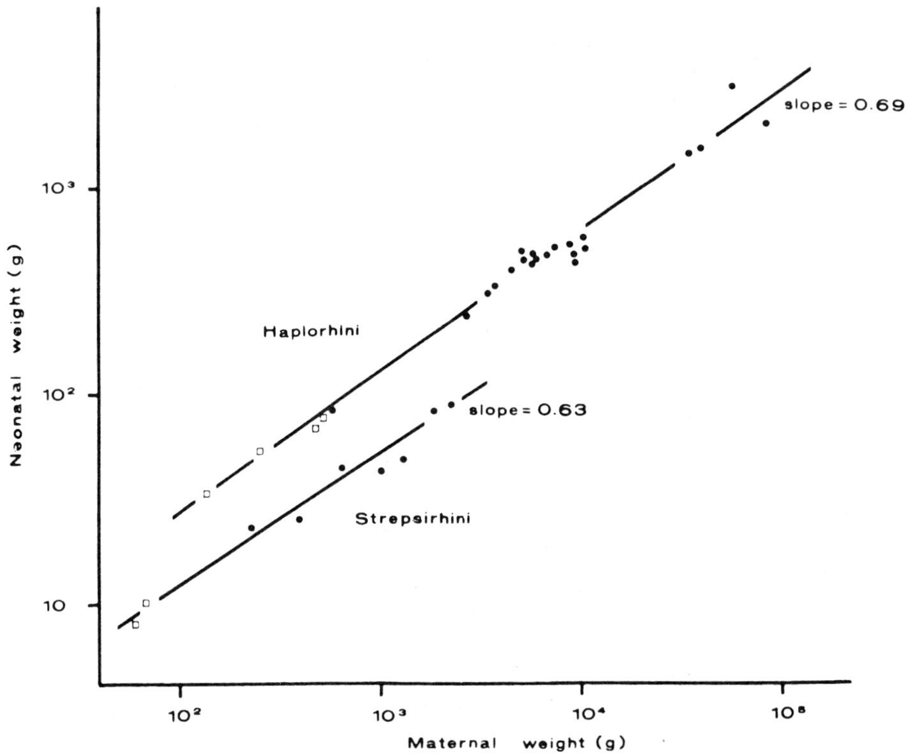

Fig. 4. The maternal investment for primates, estimated by the total litter weight for each species. The data form two discrete trends, the higher for the suborder Haplorhini and the lower for the suborder Strepsirhini. The lines are fitted by least squares as calculated by Leutenegger. (Leutenegger, W., 1979. Am. Natur., vol. 114. Courtesy of the University of Chicago Press.)

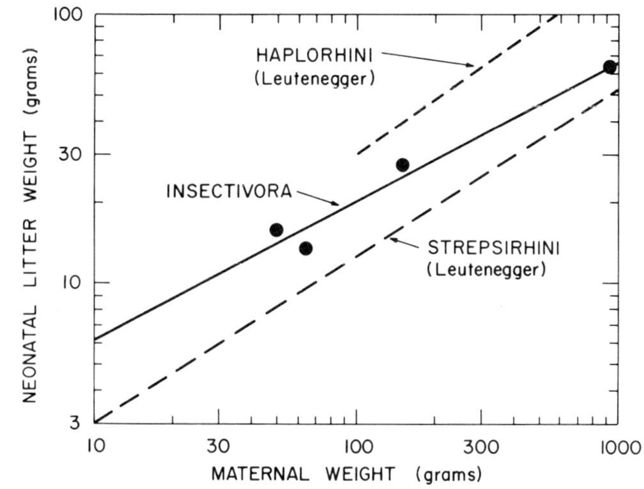

Fig. 5. The maternal investment for four insectivore species (SS74). The trend lines for the Haplorhini and Strepsirhini are from the least-squares fits by Leutenegger (Fig. 4). The four insectivore data points are, from left to right: *Elephantulus myurus*, *E. Intufi*, *T. glis*, and *Erinaceus europaeus*.

THE TIME OF ORIGIN OF THE 12% TRAJECTORY

Kay and Cartmill (1977) give evidence that *Palaechthon*, a plesiadapiform primate, shows no specific aboreal adaptations, and its facial morphology suggests that it was a predominantly terrestrial insect eater, guided largely by tactile, auditory, and olfactory sensation in pursuit of prey. This is in contrast to the specialization of Eocene strepsirhines toward a niche defined by arboreal existence and visually directed manual prehension of prey (Cartmill, 1974; Charles-Dominique, 1977). The differences between plesiadapiforms and strepsirhines in feeding patterns, and the absence of any clear distinctions between plesiadapiforms and the erinaceid insectivorans (Kay and Cartmill, 1977) raises serious doubt that the early plesiadapiform primates had developed an ecological adaptation sufficiently distinctive to account for the appearance of the 12% fetal brain growth trajectory early in the initial Cretaceous primate radiation. The hypothesis adopted here, therefore, is that the 12% trend evolved in conjunction with the second primate radiation in the mid- to late Paleocene.

FACTORS THAT FAVORED SELECTION FOR THE 12% TRAJECTORY

The second radiation of the primates was based on a greatly increased commitment to arboreal life, but the distinctive new strepsirhine characteristic was the development of visually guided grasping of arthropod prey, either at rest or at the beginning of flight (Cartmill, 1974). This foraging strategy was feasible for the early strepsirhines, which typically had small body sizes (Radinsky, 1977; Szalay and Delson, 1979); Charles-Dominique (1977) reported that the stomach contents of *Galago demidovii* (body weight 65 g) consisted of almost 70% arthropod matter. He also found, however, that the absolute amount of arthropod material in the stomach contents was almost constant (range 1.2 to 3.4 g) for 5 species of lorises and galagos with body weights ranging from 65 to 1000 g. This finding suggests that the stalking-and-grasping strategy imposes an upper limit on the rate of prey capture that would be restrictive even for the small ancestral strepsirhines, especially under the metabolic demands of late pregnancy and lactation.

The new eye-hand coordinations were associated with an increase of the degree of neocorticalization, and with an increase of the encephalization quotient (EQ) beyond the insectivore stage (Jerison, 1979; Radinsky, 1977). On the evidence from living strepsirhines, the newborns of the ancestral form appear to have been more precocial in their behavioral development than was the case for the contemporary insectivores (Charles-Dominique, 1977; SS74). Greater precociality at birth, however, entails an increase of neonatal brain size, and this requires both a longer gestation time and increase of metabolic burden on the dam. At the same time, the increase of neocorticalization is associated with an increased importance of learned behavior, requiring a longer period of postnatal dependence.

The ancestral strepsirhines were therefore in a critical situation. They had discovered an attractive new niche, but were constrained in their ability to exploit it by the limitations on the rate of intake of the insect prey on which they were still strongly dependent, and by the simultaneous demand for an increased maternal investment in more precocial, larger-brained young.

My hypothesis is that both problems abated when the common ancestor of the strepsirhines evolved a change in the sequence of fetal development whereby brain growth shifted from the 6% to the 12% trajectory, and the perinatal maternal investment was simultaneously reduced by half. This shift was presumably accomplished early in fetal development, perhaps in an early stage of organogenesis, by retarding non-neural cell proliferation for the equivalent of one-cell doubling. The ontogenic brain-to-body growth allometry data for human fetuses in Figure 2 shows that growth is isometric throughout a large part of fetal life. This means that brain tissue and all the body tissues grow at the same rate at like gestation age, so that once the 50% retardation of non-neural tissue growth is effected in early fetal life, the subsequent growth rates of all tissues are the same on the 12% trajectory as they are in the 6% trajectory.

The degree of precociality of behavior at birth is primarily a function of the ratio of neonatal brain weight to adult brain weight (Portmann, 1962; SS74). The maximum maternal investment for a given degree of precociality of behavior, then, would be cut roughly in half by the shift to the 12%

trajectory, thus affording some relief to the burden of reproduction in a resource-limited ecological niche.

The analysis in SS74 supports the conclusion that the shift of trajectory was accomplished by a retardation of non-neural tissue growth, and also provides cogent evidence that the logically possible alternative, advancement of neural tissue growth by a factor of 2, was not an option available to the ancestral strepsirhines. The analysis in SS74 also brings out the fact that the brains of all mammalian species come close to obeying a single law of absolute mass growth throughout fetal life. This finding was made the basis for the *minimax hypothesis* of mammalian development, that: (a) neural tissue is the most slowly growing of mammalian tissues; (b) growth of neural tissue in all species proceeds at the maximum rate possible, which is a general mammalian parameter with little species-specific influences; and (c) during the initial, or *neurotrophic*, stage of development, in which growth of neural and non-neural tissue is isometric, this isometry is maintained by means of an inhibitory control by neural tissue on the rate of proliferation in the potentially rapidly growing non-neural tissues (SS74).

Theoretically, the change of developmental sequence could be effected by means of a single mutational event. The real problem is not how it could have happened, but rather why it did not happen repeatedly. Why do most mammalian taxa adhere to the 6% pattern? A complete explanation is not possible now, and there may be several explanations applicable to different cases. For example, the 12% trajectory would evidently have no advantage for altricial species because the gross disproportion between brain development and muscular development at birth would increase the vulnerability of altricial young.

THE 12% TREND AS A PREADAPTATION FOR HAPLORHINE BRAIN EVOLUTION

The original, restrictive form of the 12% trend, in which overall reproductive investment is cut in half, is retained by all extant strepsirhine families (Leutenegger, 1979). The restriction to a primarily insect diet was progressively lifted by the larger, more progressive strepsirhines, as Charles-Dominique (1977) showed for the West African nocturnal Lorisinae and Galaginae. This means that it would have been possible for the large strepsirhines to discard the characteristic strepsirhine reduction of reproductive investment. This step was not taken by any strepsirhine group, but it was taken at the threshold of the transition to the Haplorhini, for the higher maternal investment that characterizes the Haplorhini is possessed by *Tarsius* as well as by all species in the Anthropoidea (Table 2, SS74). When the higher level of maternal investment was fully assumed by the Anthropoidea, it became possible for them to produce highly precocial "normal-sized" young with the neonatal brain weight doubled.

The sequence of changes in brain-to-body weight relationships in the course of primate evolution is reviewed schematically in Figure 6. The upper box is based on the data in SS74 and Figure 1, the second box summarizes Leutenegger's data on maternal investment in the Haplorhini and the Strepsirhini (Fig. 4), together with my data on reproductive effort in the insectivores (Fig. 5), and the third box draws on the data on brain advancement at birth from SS74.

The lowermost box in Figure 6 schematizes the advances in the adult encephalization quotient, EQ, that took place during primate evolution. This increase, and especially the increase of EQ from the Strepsirhini to the Haplorhini, can be shown to be a consequence of the adoption of the 12% growth trajectory. Figure 7 brings together diagrammatically some of the features of fetal brain and body growth and of adult brain-to-body weight relationships that were significant for the evolutionary increase in the size of the primate brain.

A mammal on the 6% growth trend follows a brain-to-body growth trajectory schematized in Figure 7 by the two-branched growth function

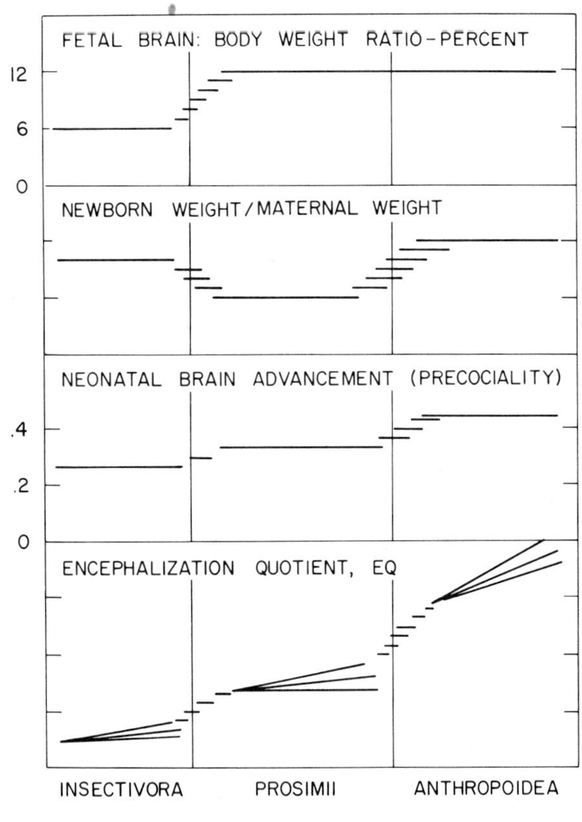

Fig. 6. Diagrammatic representation of the changes in fetal and adult brain weight and body weight in the course of evolution from the insectivores to the higher primates. See text.

A'B'C', in which B' is the point of transition from the phase of active brain growth, characterized by mitotic increase in neuronal numbers and by isometry of brain and body weight, to the final, somatotropic growth phase devoted to the maturation of neural processes and the passive growth of the perikarya and axons to match body growth. C' is the stationary adult brain-to-body relation, shown as lying on a line of slope 2/3, as is typically but not invariably the case for the interspecific allometry of adult brain weight versus body weight (Sacher, 1975; see Martin, this volume; Radinsky, this volume).

The first benefit of the transition to the 12% trajectory is shown by the two vertical dashed lines in Figure 7, the one on the right representing the original birth epoch of the forms that follow the nonprimate 6% developmental sequence and the line on the left representing the birth status of the ancestral strepsirhine species after the jump to the 12% trajectory. The growth trajectory is now ABB'C'. The offspring born at point B on the 12% trajectory are behaviorally about as mature as those born at the corresponding brain size on the 6% trajectory, and enter the world at almost the same gestation age but at a lower metabolic cost. However, there is

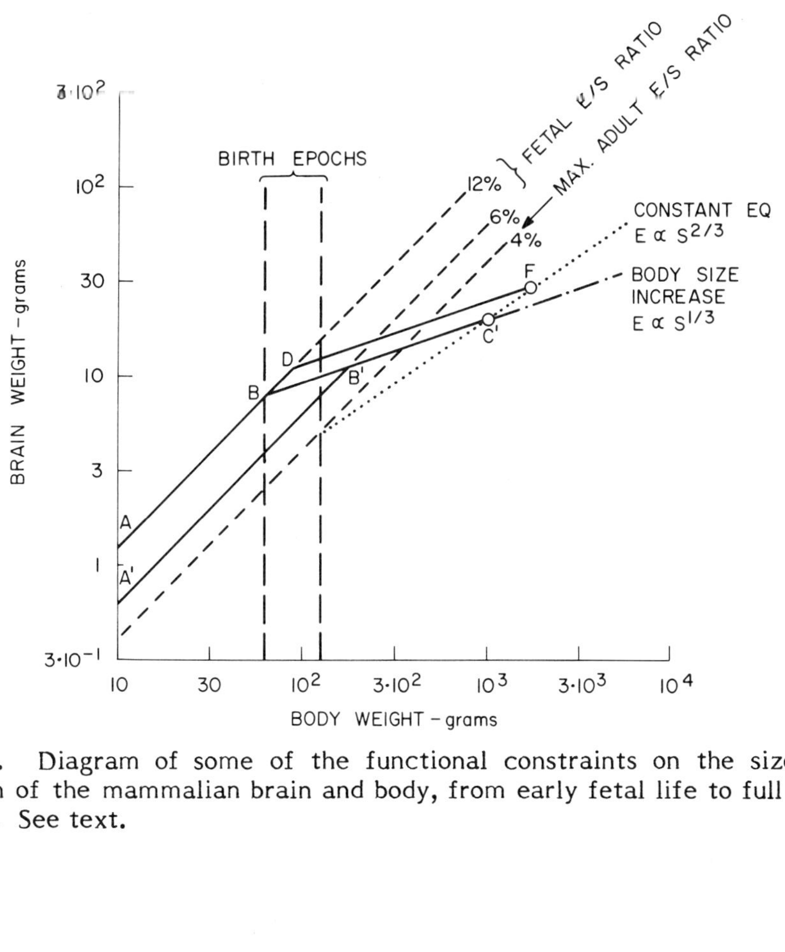

Fig. 7. Diagram of some of the functional constraints on the size and growth of the mammalian brain and body, from early fetal life to full adult status. See text.

another important consequence. The highest brain-to-body weight ratio attained by any adult mammal is 4% (Sacher, 1975). This limiting 4% adult ratio is achieved, but not exceeded, by a number of mammalian families, including the smaller members of several strepsirhine and haplorhine families (Sacher, 1975). In the case of the less highly encephalized mammalian taxa, e.g., the myomorph rodents, the small gap between the 6% trajectory of fetal brain growth and the 4% brain-to-body weights ratio of the fully adult animal would occasion no difficulty for postnatal behavioral development, because birth is altricial. A highly encephalized and highly precocial species, however, would have serious difficulty in developing a satisfactory schedule of postnatal behavioral maturation if its fetal brain growth followed the 6% trajectory. The pinnipeds, which are carnivores with precocial birth, also may feel the difficulty of the 6% trajectory, for the otariid seals show a tendency toward brain-to-body ratios for the newborn that are higher than the mean 6% trend (Fig. 1), and may possibly be evolving toward the 12% trajectory.

THE SIGNIFICANCE OF THE CHANGE IN GROWTH TRAJECTORY FOR THE THEORY OF HETEROCHRONIC DEVELOPMENT

These findings underline the importance of the concept of the heterochrony, as developed by Gould (1977), and yet they do not fit easily into his typological terminology. The shift to a higher brain growth trajectory is a case of retardation, in Gould's sense, but so far as can be ascertained it has no effect on morphology or on the age of sexual maturity; therefore Gould's other rubrics of heterochronic growth cannot be applied. Nevertheless, it is important for the basic theory of differentiation that we have found in mammals an instance of a heterochronic "mutation" that has its expression very early in the differentiation process, and that was of profound importance for the cladogenesis of the modern primates, while yet having no direct expression in adult structure or in the schedule for sexual maturation.

SUMMARY

All extant primate species conform to a 12% ratio of brain weight to body weight at birth and throughout a considerable part of fetal life, in contrast to a 6% trajectory for the nonprimate taxa. This growth pattern probably evolved at the beginning of the strepsirhine radiation in mid- to late Paleocene, initially as a means to reduce maternal reproductive investment, in adaptation to a requirement for precocial birth in a strictly arboreal niche, coupled with a relatively restricted insect food resource. When a wider range of plant food resources became available, and body size increased, the Haplorhini, including *Tarsius*, retained the 12% trajectory while returning to a high level of maternal investment. This was an essential preadaptation for their subsequent rapid evolution toward large brain size, which would not have been possible within the restrictions imposed by the 6% brain growth trajectory.

Acknowledgments: I thank J. Altman, W. Leutenegger, L. Radinsky, and R. Tuttle, who read and commented on a draft of this paper. I am grateful to J. Dobbing, R.D. Phemister and W. T. Hubbert, who kindly made unpublished data available to me. This work was supported by the U.S. Department of Energy under contract No. W-31-109-ENG-38.

REFERENCES

Cartmill, M., 1974. Rethinking primate origins. Science, 184:436-443.

Charles-Dominique, P., 1977. Ecology and behavior of nocturnal primates. In, Prosimians of Equatorial West Africa, R.D. Martin, trans., Columbia University Press, New York.

Cutler, R.G., 1976. Evolution of longevity in mammals. I. Primates. J. Hum. Evol., 6:189-204.

Dene, H., Goodman, M., Prychodko, W., and Moore, G. W., 1976. Immunodiffusion systematics of the primates. III. The Strepsirhini. Folia Primatol., 25:35-61.

Dobbing, J., and Sands, J. 1973. Quantitative growth and development of the human brain. Arch. Dis. Childhood, 48:757-767.

Frahm, H., 1973. Metrische Untersuchungen an den Organen von Hamstern der Gattung *Phodopus*, *Mesocricetus*, and *Cricetus*. Zool. Jb. Anat. Bd., 90:55-159.

Gould, S.J., 1977. Ontogeny and Phylogeny. Harvard University Press, Cambridge, Mass.

Hubbert, W.T., Stalheim, O.H.V., and Booth G.D., 1972. Changes in organ weights and fluid volumes during growth of the bovine fetus. Growth, 36:217-233.

Jerison, H.J., 1973. Evolution of the Brain and Intelligence, Academic Press, New York, pp. 482

Jerison, H.J., 1979. Brain, body, and encephalization in early primates. J. Hum. Evol., 8:615-635.

Jones, M.L., 1968. Longevity of primates in captivity. International Zoo Yearbook, 8:183-192. Additional comments as of 25 April 1979 (private communication).

Kay, R.F., and Cartmill, M., 1977. Cranial morphology and adaptation of *Palaechthon naciamenti* and other Paromomyidae (Plesiadapoidea, Primates) with a description of a new genus and species. J. Hum. Evol., 6:19-53.

Kerr, G.R., Kennan, A.L., Waisman, H.A., and Allen, J.R., 1969. Growth and development of the fetal rhesus monkey. I. Physical Growth. Growth, 33:201-213.

Kielan-Jaworowska, Z., Bown, T.M., and Lillegraven, J.A., 1979. Eutheria. In, Mesozoic Mammals: The First Two-Thirds of Mammalian History, J.A. Lillegraven, Kielan-Jaworowska, and W.A. Clemens, eds., University of California Press, Berkeley, pp. 182-191.

Leutenegger, W., 1979. Evolution of litter size in primates. Am. Natur., 114:525-531.

Portmann, A., 1962. Cerebralisation und Ontogenese. Medizinische Grundlagenforsch., 4:1-62.

Radinsky, L., 1977. Early primate brains: Fact and fiction. J. Hum. Evol., 6:79-86.

Sacher, G.A., 1959. Relation of lifespan to brain weight and body weight in mammals. In, Ciba Foundation Colloquia on Ageing, Vol. 5, The Lifespan of Animals, G.E.W. Wolstenholme and M. O'Connor, eds., Churchill, London, pp. 115-141.

Sacher, G.A., 1970. Allometric and factorial analysis of brain structure in insectivores and primates. In, Advances in Primatology, C.R. Noback and W. Montagna, eds., Appleton-Century-Crofts, New York, pp. 245-287.

Sacher, G.A., 1975. Maturation and longevity in relation to cranial capacity in hominid evolution. In, Primate Functional Morphology and Evolution, R. H. Tuttle, ed., Mouton, the Hague, pp. 417-441.

Sacher, G.A., 1981. The constitutional basis of longevity in the Cetacea: Do the whales and terrestrial mammals obey the same laws? In, International Whale Commission, Special Issue 3, Age Determination of Toothed Cetaceans and Sirenians, W.F. Perrin and A.C. Myrick, Jr., eds., Cambridge, England, in press.

Sacher, G.A., and Staffeldt, E., 1974. Relation of gestation time to brain weight for placental mammals. Implications for the theory of vertebrate growth. Am. Natur., 108:593-615.

Sykora, I., Wildt, S., and Hradil, F., 1965. Veränderungen des Körper- und Organgewichtes der Wistarratte während der postnatalen Entwicklung. Zeitschr. Versuchstierkunde, 7:23-34.

Szalay, F.S., and Delson, E., 1979. Evolutionary History of the Primates, Academic Press, New York, pp. 550.

THE DEVELOPMENT OF THE PRIMATE PULVINAR

Marilee P. Ogren*

When the nervous systems of different species are compared, homologous structures can nonetheless exhibit differences that are related to their evolutionary histories. For example, there are dramatic differences in the morphology of the rabbit and monkey retinae that correspond to the requirements of their different environments, yet there is little doubt that they are homologous. The retina of the rabbit contains a horizontally oriented stripe of densely packed ganglion cells and high visual acuity that monitors the horizon for the essentially stationary, laterally placed eyes. The retina of the monkey contains a fovea that is a small centrally located area of much higher ganglion cell density and visual acuity, that can be directed via eye movements to scan the environment and to fixate binocularly. In some ways the ontogenetic development of the brain seems also to reflect its phylogenetic history. For instance, at early gestational ages the brains of widely divergent species are virtually indistinguishable, as if they pass through developmental stages that reflect vestiges of their common ancestors.[1] In this communication, a comparison is made between the prenatal development of the pulvinar nucleus in the rhesus monkey (*Macaca mulatta*) and the human in an attempt to provide insight into its phylogenetic development.

---

[1]For a comprehensive consideration of the relationship of ontogeny to phylogeny, the reader is referred to Gould (1977).

*From the Section of Neuroanatomy, Yale University School of Medicine, New Haven, Connecticut 06510

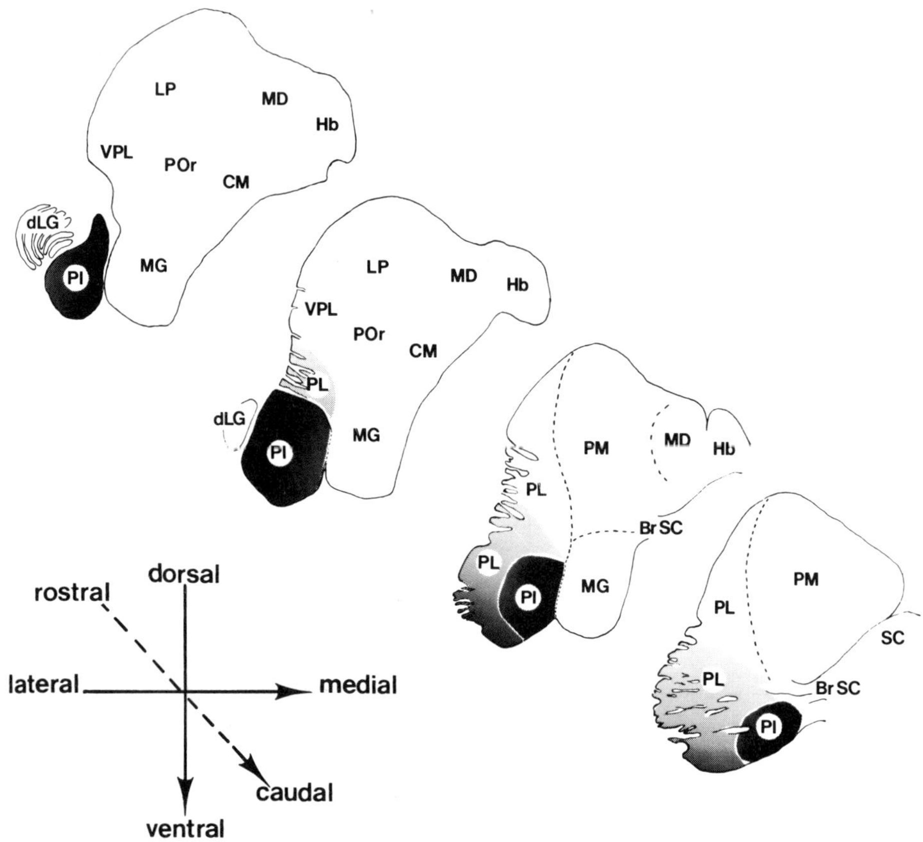

Fig. 1. Coronal sections through the dorsal thalamus of one hemisphere of the adult monkey are shown in four rostro-caudal planes. The most rostral section is on the upper left. The boundaries of three pulvinar subdivisions have been previously determined by cytoarchitectonics and connectivity. On the basis of recent neuroanatomical findings, lateral PL is further divided into two portions: PLα (shaded area) and PLβ (nonshaded area). Neighboring dorsal thalamic nuclei are also indicated. Abbreviations: BrSC, brachium of the superior colliculus; CM, centrum medianum; dLG, dorsal lateral geniculate; Hb, habenula; LP, lateral posterior nucleus; MD, medial dorsal nucleus, MG, medial geniculate nucleus; POr, oral pulvinar; PI, inferior pulvinar; PL, lateral pulvinar; PM, medial pulvinar; SC, superior colliculus; VPL, ventroposterior lateral nucleus. (From Ogren, M.P., and Hendrickson, A., 1979. J. Comp. Neurol., vol. 188. Courtesy of Wistar Press.)

HISTORICAL PERSPECTIVE

Pulvinar Morphology and Connections

Compared with those of other mammals, the primate pulvinar is a conspicuously large diencephalic nucleus relative to the other thalamic nuclei. In absolute terms, the pulvinar nucleus reaches its greatest proportions in apes and man (Armstrong, 1979, 1981, this volume; Bergquist, 1954; Harting et al., 1972; Kuhlenbeck, 1954; Walker, 1938). Generally, the pulvinar in the adult primate occupies over one third of the dorsal thalamus at the posterior pole of the diencephalon. It can be divided into three major subdivisions on the basis of cytoarchitecture and connectivity (Fig. 1).

Classically the pulvinar has been considered to be a sensory association nucleus. Unlike the sensory relay nuclei of the thalamus, early degeneration studies in the rhesus monkey suggested that the pulvinar received no input from ascending sensory pathways originating in the periphery. Its inputs were thought to come from association cortex and neighboring sensory thalamic nuclei, and its projections were considered to be exclusively to areas of association cortex, rather than to primary sensory cortices that are recepients of inputs from sensory relay nuclei (Le Gros Clark and Northfield, 1937; Papez, 1939; Rose and Woolsey, 1949; Walker, 1938).

Recent studies of pulvinar connectivity in the rhesus monkey, as well as other primate species, have used more sensitive methods for tracing neural connectivity, such as silver degeneration, autoradiography and horseradish peroxidase histochemistry. They have confirmed and extended earlier findings concerning pulvinar connections with association cortex, revealing reciprocal connections between various pulvinar subdivisions and association areas of the occipital, temporal, parietal, and frontal cortex (Benevento and Davis, 1977; Burton and Jones, 1976; Kasdon and Jacobson, 1978; Ogren et al., in preparation; Trojanowski and Jacobson, 1976). Intrathalamic connections, however, have not been amply demonstrated. Although the above findings concerning pulvinar-cortical relationships confirm the classical definition of the pulvinar as an association nucleus, other recent studies have provoked questions about the validity of other aspects of this definition (Diamond, 1979). Such studies have shown that the inferior subdivision of the rhesus monkey pulvinar receives indirect retinal input via the superficial layers of the superior colliculus (Benevento and Fallon, 1975; Partlow et al., 1977). In addition, the inferior and adjacent portions of the lateral pulvinar have been shown to have reciprocal connections with primary visual cortex (Benevento et al., 1975; Benevento and Rezak, 1976; Ogren and Hendrickson, 1976, 1977) as well as with extrastriate visual areas.

Evolutionary Development of the Pulvinar

It is generally agreed that the expansion of the pulvinar during phylogenetic evolution may be related to the elaboration of the areas of neocortex with which it is connected because comparative studies show that the size of these cortical areas and the size of the pulvinar increase together (for reviews, see: Allman, 1977; Chalupa, 1977; Diamond, 1979; Harting et al., 1972; Rose and Woolsey, 1949; Walker, 1938). Recently, Armstrong (1979, 1981, this volume) showed in a quantitative comparative study of several hominoids that the size of the pulvinar keeps pace with total brain size, such that the ratios of both the volume of the pulvinar and the numbers of pulvinar neurons to the total brain weight in apes can be used to predict pulvinar volume and numbers of neurons in the larger human brain. Sensory relay nuclei of the thalamus, on the other hand, do not show a positive correlation, implying a more conservative development for these structures. Several authors suggested that during evolution, selection pressures favoring visual predation, eye-hand coordination, sensory integration, and visual abstraction, influenced the development of the primate visual system in a variety of ways, including the growth of the pulvinar (Allman, 1977, this volume; Armstrong, 1979, 1981; Masterton et al., 1974; Walker, 1938). Although the large size of the pulvinar in primates is generally accepted to be a relatively recent phylogenetic development, its specific functional significance currently remains obscure.

Prenatal Development of the Pulvinar

On the basis of classical landmarks that apply to all mammalian brains, the primate diencephalon can be recognized as a distinct zone in the forebrain at early embryonic ages, even though few of its cells have yet been generated (Kuhlenbeck, 1954). When the lateral surface of the embryo is viewed, the diencephalon lies between the telencephalic vesicles, and the incipient mesencephalon (see inserts, Fig. 5, p. 121). Its rostral border is marked by the teleodiencephalic fissure and its caudal border by the pineal gland. In cross-section, the immature diencephalon is divided into four regions by shallow depressions in the wall of the third ventricle, which were first observed in the amphibian by Herrick (1910) (Kuhlenbeck, 1954). These four diencephalic regions give rise to the nuclei of the epithalamus, dorsal thalamus, ventral thalamus, and hypothalamus. The region of the dorsal thalamus (enclosed by brackets in Fig. 5, p. 121), is delineated by the dorsal and medial sulci in the ventricular wall and gives rise to the sensory, motor, limbic, and association nuclei of the thalamus (as defined classically), including the pulvinar.

Neurons destined for these thalamic nuclei are generated in the germinal matrix lining the third ventricle, which consists of the ventricular and subventricular zones (Figs. 2A and 5A, p. 121). The immature neurons migrate laterally along radial glial fibers (Levitt and Rakic, 1980) toward the lateral surface of the dorsal thalamus, where they form the anlage of the dorsal thalamic nuclei. At these early ages, the brains of different mammalian species are virtually indistinguishable.

Until recently, the prenatal development of the pulvinar had been studied only in the human. These studies were primarily based on examination of the Nissl-stained histological sections of the diencephalon. Several authors noted that the pulvinar is among the last thalamic nuclei in which neurons mature morphologically, and that its size appears to increase dramatically during the second half of gestation (Cooper, 1950; Gilbert, 1934; Rakic and Sidman, 1969; Yakovlev, 1969). Furthermore, Rakic and Sidman (1969) noted that most of the growth of the pulvinar occurs after cell proliferation in the diencephalon has ceased. On the basis of their Nissl-stained and Golgi-impregnated material from human fetuses they suggested that although some, or perhaps most (Rakic, 1974), of the pulvinar

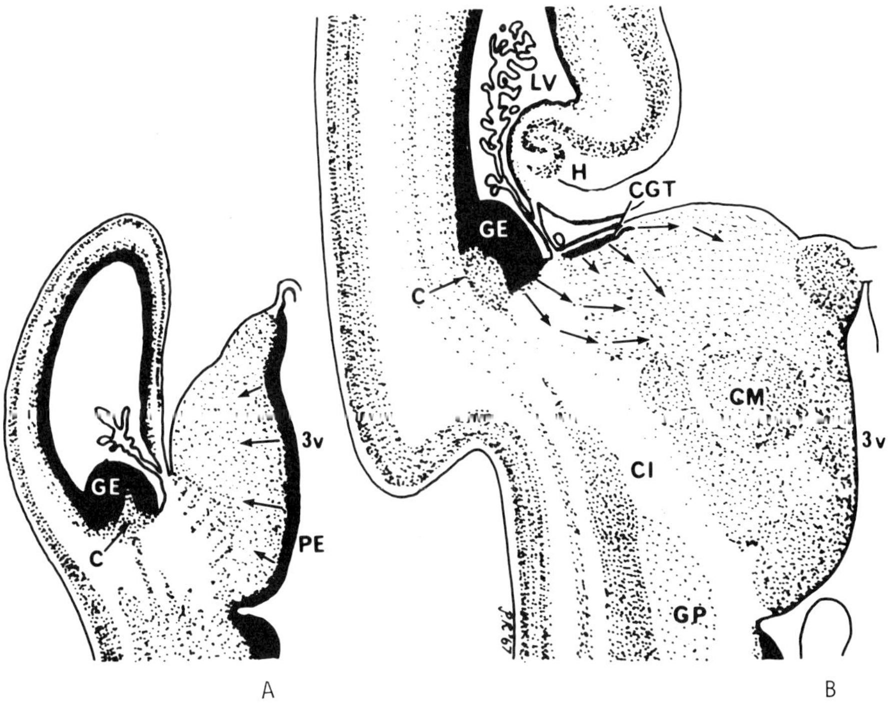

Fig. 2. Camera lucida drawings of two horizontal sections through the fetal human brain showing two modes of pulvinar neuronal genesis. During the first quarter of gestation (A) the proliferative epithelium near the third ventricle (3V) of the diencephalon is mitotically active, and contributes immature cells to the developing dorsal thalamus. Arrows indicate the route of their migration. During the second half of gestation (B) the proliferative epithelium near the third ventricle is depleted of proliferative cells, whereas the ganglionic eminence (GE) in the telencephalon continues to generate new cells. Some cells from the ganglionic eminence migrate through a transient structure, the corpus gangliothalamicus (CGT), to the posterior thalamus. Arrows indicate their route of migration. Abbreviation: 3V, third ventricle; C, caudate nucleus; CGT, corpus gangliothalamicus; CI, internal capsule; CM, centrum medianum; GE, gaglionic eminence; GP, globus pallidus; H, hippocampus; LV, lateral ventricle. (From Rakic, P., and Sidman, R.L., 1969. Z. Anat. Entwickl. -Gesch., vol. 129. Courtesy of Springer Press.)

neurons are generated in the germinal matrix of the diencephalon early in gestation (Fig. 2A), a second late-forming population of pulvinar cells may be generated in the telencephalon (Fig. 2B). These late-forming cells appear to migrate from the proliferative epithelium in the telencephalon, the ganglionic eminence, to the posterior thalamus via a transient structure, the corpus gangliothalamicus. This migration occurs during the last four to six months of gestation. These authors suggested that the late-forming population of cells derived from the telencephalon may contribute to the expansion of the human pulvinar during late stages of gestation, and in phylogeny.

A recent study of pulvinar ontogenesis in the rhesus monkey (Ogren and Rakic, 1981) was initiated to determine whether the large size of the pulvinar in the monkey may be related to a late-developing population of cells derived from the telencephalon as described for man (Rakic and Sidman, 1969), and to further our understanding of the development of the primate pulvinar by comparing its prenatal development in monkey and man. Tritiated thymidine autoradiography was used to determine the time and place of cell origin in the monkey. Nissl-stained material from monkey fetuses was also examined to make direct comparisons with the earlier findings in the human concerning the tempo and characteristics of morphological maturation of the pulvinar during prenatal development.

TIME AND PLACE OF PULVINAR CELL ORIGIN IN THE MONKEY

Pulvinar Neurons Are Generated Early in Gestation

The time of pulvinar cell origin was determined by exposing prenatal monkeys to a short pulse of tritiated (^{3}H) thymidine on different days of the 165-day gestation period. This method is particularly useful for analysis of neuronal genesis because it permanently labels neurons that are generated, i.e., that undergo their last mitotic division, soon after the time of injection. Prenatal monkeys can be exposed to short pulses of 3H-thymidine by intravenous injection of the mother at selected embryonic days. All cells that are in the S phase of the cell cycle (DNA synthesis) at the time of the injection incorporate the ^{3}H-thymidine. Those cells that do not divide again are heavily labeled, containing large quantities of radioactive label in their nuclei, and are considered to have been generated on the day of injection. Cells that continue to divide after the injection dilute the label and are not considered to be generated on that day. This procedure and the method of timing pregnancies has been described by Rakic (1973). Standard autoradiographic procedures are used to detect labeled cells in histological sections (Cowan et al., 1972); an example of autoradiographically labeled pulvinar neurons is shown in Figure 3.

Twenty-one monkeys were exposed to ^{3}H-thymidine on embryonic (E) days ranging between E25 and E140, and sacrificed about two months postnatally, when the brain appears to be morphologically mature. Autoradiograms of histological sections through the diencephalon of these

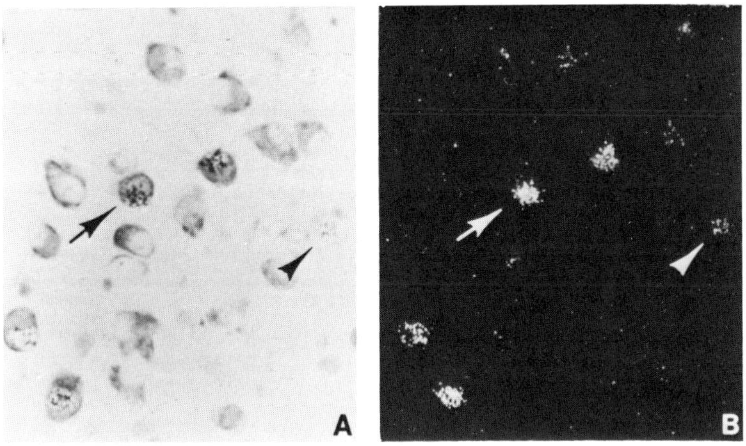

Fig. 3 Photomicrographs of autoradiographically labeled neurons in the mature pulvinar nucleus. (A) Light field illumination shows Nissl-stained cell bodies with nuclei overlaid by silver grains as a result of having incorporated ^{3}H-thymidine during pulvinar neuron genesis. Heavily labeled cells (arrow) had undergone their last mitotic division at the time of injection. Lightly labeled cells (arrow head), overlaid by half or less silver grains, underwent further mitotic division after injection. (B) Darkfield illumination of the same section in (A).

monkeys showed that pulvinar neurons in the rhesus monkey are generated early, during the first quarter of gestation in a relatively short period of just ten days. As shown in Figure 4, only the animals that had been injected at E36, E38, E40, E43, and E45 have radioactively labeled pulvinar neurons. An analysis of the spatio-temporal gradients of proliferation, derived from plots of the positions of labeled neurons with respect to the three pulvinar subdivisions, indicates that although there is a great deal of overlap, the three subdivisions are generated sequentially. Rostral portions including the inferior and adjacent portion of the lateral pulvinar are generated first, and the caudal parts of the nucleus including the remainder of lateral pulvinar and the medial subdivision are generated last. Nonetheless, all cells destined for the pulvinar nucleus are generated by E45, therefore subsequent morphological changes during the remainder of the 165-day gestation are not a result of the addition of new neurons.

Of note is the fact that pulvinar genesis occurs simultaneously with neuronal genesis in the sensory relay nuclei, and neighboring association nuclei. The dorsal lateral geniculate (Rakic, 1977), medial geniculate (Rakic and Bump, unpublished), and the ventrobasal nuclei, as well as the lateral posterior nucleus and the medial dorsal nucleus (Ogren, unpublished), are all generated during the same general period.

The early genesis of the monkey pulvinar was unexpected because it is in contrast to the general notion that the pulvinar develops late in gestation,

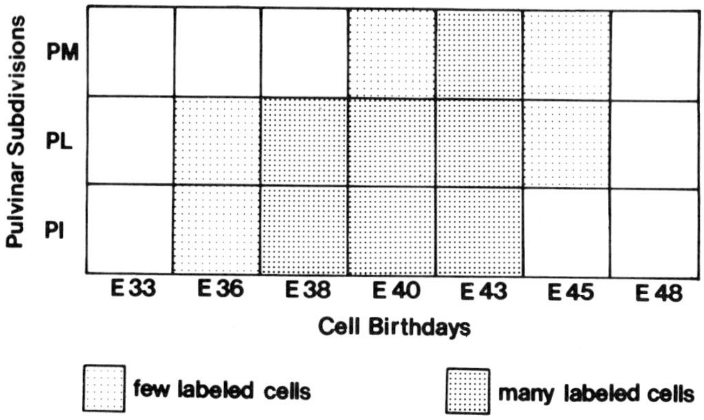

Fig. 4 Schematic representation of the time of cell origin for neurons in the three pulvinar subdivisions. The presence of a small number (less than 10 per $1mm^2$) of heavily labeled cells is represented by light shading; the presence of many (greater than 10 per $1mm^2$) heavily labeled cells is represented by dark shading. The age of injection for each embryonic (E) day is indicated in a row at the bottom of the figure. Abbreviations: PI; inferior pulvinar, PL; lateral pulvinar, PM; medial pulvinar.

and to the findings by Rakic and Sidman (1969), which indicate that in the human a second population of pulvinar neurons is derived from the telencephalon during the second half of gestation. Analysis of fetal material (see below), however, shows that the diencephalon is mitotically active throughout the period of pulvinar genesis, suggesting that in the rhesus monkey it is derived entirely from the diencephalon.

Pulvinar Neurons Are Generated in the Diencephalon

A second group of animals was exposed ^{3}H-thymidine during the time of pulvinar cell genesis (determined by the first set of animals as occurring between E36 and E45) and the fetuses were sacrificed and examined at short intervals after injection. In this way, neurons can be examined in successive stages of genesis and migration to the site of the prospective pulvinar nucleus. These data indicate that pulvinar neurons in the rhesus monkey are indeed derived from the diencephalic proliferative zones, as illustrated in Figure 5. The photomicrographs are taken from autoradiograms of two monkey fetuses that were exposed to ^{3}H-thymidine during the peak of pulvinar cell genesis (E40 and E41) and resemble the diencephalon of the human fetus at 9-10 weeks after conception (Fig. 2A). One animal was sacrificed one hour after injection (Fig. 5A, E41-E41.1) and one was sacrificed one week after injection (Fig. 5B, E40-E47). Radioactive labeling is at first (one hour after injection) confined to the diencephalic precurser cells in the proliferative zone lining the third ventricle. One week later, the band of radioactively labeled cells is located at the lateral aspect of the

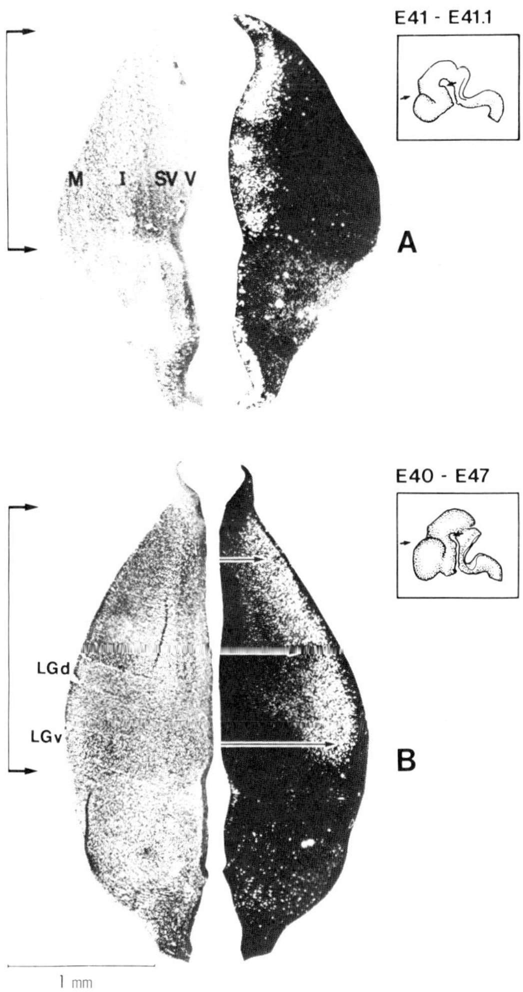

Fig. 5 Light field illumination (left) and darkfield illumination (right) photomicrographs of coronal sections cut through the caudal diencephalon (see inserts for orientation) in E41- and E47-day-old monkey fetuses. The animal shown in (A) was exposed to ^{3}H-thymidine one hour before sacrifice (E41-E41.1), the other shown in (B) was injected one week before sacrifice (E40-E47). Left hemispheres photographed in light field illumination show cytoarchitecture, whereas right hemispheres photographed in darkfield illumination show the distribution of heavily labeled cells (bright spots on dark backgrounds). Arrows in the left margin bracket the dorsal thalamus, and arrows through the dorsal thalamus in (B) show the route of cell migration. Abbreviations: M, mantle layer; I, intermediate mantle layer; SV, subventricular zone; V, ventricular zone; LGd, anlage of dorsal lateral geniculate nucleus; LGv, anlage of ventral lateral geniculate nucleus.

dorsal thalamus, the region in which the pulvinar becomes identifiable at later fetal ages (see below). This lateral migration occurs in three days or less (Ogren and Rakic, 1981). Furthermore, by E45 all pulvinar neurons have been generated (as shown by analysis of the first set of animals and in Fig. 4), and it is at about this time that the germinal matrix in the dorsal thalamus of the monkey narrows and is exhausted of proliferating cells (Fig. 5B). These observations also show that the dorsal thalamus as a whole is the last portion of the diencephalon to cease mitosis in these monkeys, as is the case in the human.

Together, these data indicate that the monkey pulvinar is not only generated earlier than expected from previous work in the human, and derived entirely from diencephalic precurser cells, but it is generated simultaneously with neighboring dorsal thalamic nuclei during a short time, even though the pulvinar in the rhesus monkey is a much larger nucleus and contains many more neurons than do its neighbors. This would imply that the number of neurons contributing to the size of the monkey pulvinar is determined by the size of the pool of precurser cells in the diencephalon, and/or by the rate of their proliferative cycle, and not by the addition of new neurons from the telencephalon late in gestation.

POSTMITOTIC DEVELOPMENT OF THE MONKEY PULVINAR: A COMPARISON BETWEEN MONKEY AND MAN

Pulvinar neuronal genesis is completed in the rhesus monkey in a short period between E36 and E45, during the first quarter of gestation, and the lateral migration of pulvinar neurons from the diencephalic proliferative zones to the incipient pulvinar nucleus occurs within three days after neuronal genesis ceases (see above and Ogren and Rakic, 1981). This mode of cellular genesis and lateral migration has also been described for human pulvinar neurons at early gestational ages (Fig. 2A; Rakic and Sidman, 1969; Rakic, 1974), and for the monkey dorsal lateral geniculate nucleus (Rakic, 1977).

During the second quarter of gestation, the pulvinar remains as a homogeneous mass of cells in the rather amorphous thalamus in both species of primates. During this period, by about E60 (Fig. 6A) in the monkey and by about 13 weeks after conception in the human, the pulvinar nucleus as a whole becomes delineated from neighboring dorsal thalamic nuclei by the appearance of the medullary laminae. These thin acellular regions separate the pulvinar and the dorsal lateral geniculate from the ventral thalamus and from the ventrobasal nuclei. A less distinctive acellular region separates the pulvinar from the lateral geniculate (Fig. 6A). Fiber connections that comprise the thalamo-cortical radiations are few and cytoarchitectonic differentiation is not yet apparent within the pulvinar or in the other thalamic nuclei at these early ages.

The period around the midpoint of gestation, which is about E80 in the monkey (Fig. 6B) and just over four months in the human (18 weeks after

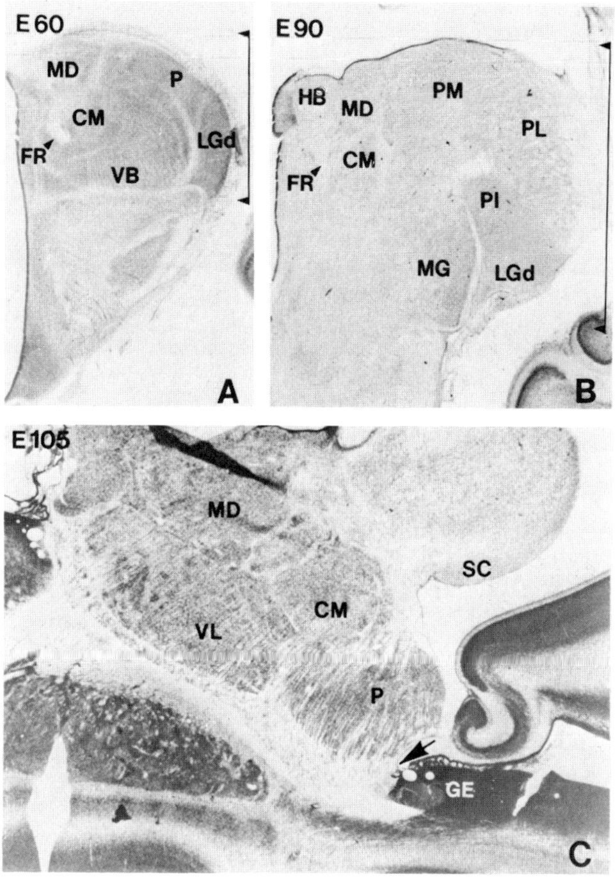

Fig. 6 Photomicrographs of Nissl-stained sections through fetal monkey brains. (A) Coronal section through one diencephalic hemisphere in a monkey fetus sacrificed at E60. Arrows (right) bracket the dorsal thalamus. Note cell sparse zones that delineate the pulvinar from the anlage of adjacent dorsal thalamic nuclei. (B) Coronal section through one hemisphere of the dorsal thalamus of monkey fetus sacrificed at E90. At this age, individual pulvinar subdivisions can be distinguished by cytoarchitectonic criteria although inferior pulvinar has not yet settled into its final position and is not easily distinguished from the dorsal lateral geniculate. (C) Horizontal section through the diencephalon of a monkey fetus sacrificed at E105, when pulvinar subdivisions assume adult-like dispositions. Note the absence of the corpus gangliothalamicus, which is clearly visible (position indicated by arrow) in horizontal section of human fetus at comparable ages. Abbreviations: CM, centrum medianum; FR, fasciculous retroflexus; GE, ganglionic eminence; HB, habenula, LGd, dorsal lateral geniculate nucleus; MG, medial geniculate nucleus; MD, medial dorsal nucleus; P, pulvinar nucleus; PI, inferior pulvinar; PL, lateral pulvinar; PM, medial pulvinar; SC, superior colliculus; VB, ventrobasal nuclei; VL, ventrolateral nucleus.

conception), marks the beginning of cytoarchitectonic differentiation of the pulvinar. In the monkey, midgestation is also the time when afferents from the striate cortex first appear in the pulvinar; their arrival begins by E72 (Shatz and Rakic, 1981). Other fiber systems that do not terminate in the pulvinar but pass through it *en route* to the superior colliculus, are the retino-tectal fibers, which are apparent by E64 (Rakic, 1976), and the cortico-tectal fibers, which are present at E78 (Shatz and Rakic, 1981). The further accumulation and myelination of these fibers, along with the addition of cortico-pulvinar fibers from extrastriate cortex, the tecto-pulvinar fibers and pulvinar efferents, eventually results in the characteristic fasciculated appearance of the pulvinar (Fig. 6B and C).

By the end of the third quarter of gestation (E124 in the monkey and 24 weeks after conception in the human), the efferent and afferent fibers have contributed substantially to the overall volume of the nucleus and probably play some organizational role in its cytoarchitectonic differentiation. Pulvinar neurons in the monkey become cytologically mature by E124 (Ogren and Rakic, 1981), and by the seventh to eighth fetal month in the human (Yakovlev, 1969). This is the time at which the cytoarchitectonic criteria that are applied to the adult animal can be used reliably to differentiate the pulvinar subdivisions in both species.

It is during the second half of gestation in the human, between the fourth and sixth months after conception, that Rakic and Sidman (1969) noted the presence of the corpus gangliothalamicus, a transient structure that serves as a conduit for cells streaming from the ganglionic eminence of the telencephalon toward the posterior thalamus. The unlikely possibility that this late-forming population of cells is present in the monkey but fails to become labeled by ^{3}H-thymidine autoradiography was tested by examining Nissl-stained histological sections from the diencephalon of 11 coronally sectioned, and 2 horizontally sectioned, monkey fetuses between E45 (the end of pulvinar genesis as determined by ^{3}H-thymidine autoradiography) and birth for the presence of the corpus gangliothalamicus. This structure was never observed in the rhesus monkey thalamus in the position were it is typically found in the human, even though the ganglionic eminence is apparent in the monkey fetus during late gestational ages (compare Fig. 3B and 6C).

Nonetheless, the tempo and characteristics of morphological maturation of the pulvinar, as described above, appear to proceed at the same rate in both human and monkey during the last half of gestation. This implies that the late-forming population of telencephalically derived pulvinar neurons in humans contributes less to the overall volume and morphology of the human pulvinar than previously expected. It is puzzling then that so many authors (Cooper, 1950; Gilbert, 1934; Kuhlenbeck, 1951; Rakic and Sidman, 1969; Yakovlev, 1969) report that the growth of the human pulvinar has such a protracted time course. We tested the possibility that the influx of afferent fibers, their myelination and the concomitant growth of pulvinar neuronal dendrites and axons result in an increase in neuropil area and expansion of the monkey pulvinar late in gestation. Our morphometric analysis of neuropil area in the rhesus monkey (Ogren and Rakic, 1981) shows that as early as E58 the neuropil occupies 65% of the area in the pulvinar, exceeding that in the dorsal lateral geniculate (51%) by 14%. The neuropil area, however, increases steadily in both nuclei, such that the difference between neuropil area in the two nuclei has increased only by an additional 5% by the second postnatal month (pulvinar = 83%, dorsal lateral geniculate = 64%). Although these data indicate that, indeed, more area is devoted to neuropil in the pulvinar than in the lateral geniculate, it should be emphasized that the difference between the two nuclei is already expressed early in gestation, and increases only moderately during the second half of gestation. Therefore, either the 5% increase in pulvinar neuropil area over that in the dorsal lateral geniculate represents a sufficiently large volume to result in the appearance of disproportionate growth, or the late gestational growth of the pulvinar is illusory.

An explanation for this apparent late growth might be found by focusing on developmental events that take place in the dorsal thalamus as a whole. During the course of this investigation, it became apparent that many of the developmental events in the pulvinar nucleus occur simultaneously with those in neighboring dorsal thalamic nuclei. For example, the dorsal thalamus is the last diencephalic region to cease mitosis, and neurons destined for the pulvinar, dorsal lateral geniculate, medial geniculate, ventrobasal nuclei, lateral posterior nucleus, and medial dorsal nucleus are generated within the same period during the first quarter of gestation. Yakovlev (1969) has shown that cytological maturation within these same nuclei in the human occurs at about the same rate, just as we have shown for the pulvinar and dorsal lateral geniculate nucleus of the monkey (Ogren and Rakic, 1981). Moreover, cytoarchitectonic differentiation of this group of thalamic nuclei appears to occur simultaneously, and to be associated with the accumulation of afferent fibers during the second half of gestation. Tectogenic changes in the dorsal thalamus during the second half of gestation, such as the greater than 90° rotation and ventral displacement of the dorsal lateral geniculate, have been previously attributed to the late expansion of the pulvinar (Cooper, 1950; Gilbert, 1934; Kuhlenbeck, 1951; Rakic and Sidman, 1969; Yakovlev, 1969). In view of the present observations, and the simultaneous development of the dorsal thalamic nuclei, however, the tectogenic changes during the second half of gestation appear to result more from a complicated interplay of events. The apparent growth of the pulvinar, rather than being the source of these changes, may be in part a result of them, as it is displaced dorsally and caudally during the second half of gestation (Fig. 6A-C).

In conclusion, these findings indicate that the size of the pulvinar in the monkey, and probably to a large degree in man, is governed by the number of cells generated in the diencephalic proliferative zone and to the relatively large area devoted to neuropil, both of which are established early in gestation. Growth of the pulvinar late in gestation appears to have been overestimated in previous studies. This suggests that the telencephalically derived neurons in the human pulvinar may contribute less to the overall size of the human pulvinar than previously expected. From a neurobiological standpoint, the ontogenetic mechanisms governing the genesis of the large number of pulvinar neurons from diencephalic precursors deserve further investigation. For comparative neurology, the dual source of pulvinar neurons in man should draw attention to whether this is a singular quality of the human pulvinar, or applies to other hominids as well.

SPECULATION

If the human pulvinar contains even a small number of neurons for which there is no counterpart in the monkey, one might imagine that this is a recently evolved species difference related to some function unique to man. One obvious example of such a function is language. Experiments in awake humans undergoing neurological surgery have implicated the pulvinar in language performance. Electrical stimulation of the left (but not right) pulvinar impairs short-term verbal memory (Ojemann, 1974). This result, however, is complicated by the nearness of the site of stimulation to other thalamic nuclei associated with primary sensory and motor pathways to the cortex that may also be related to linguistic functions, and deserves further investigation.

Allometric studies of the phylogenetic development of the pulvinar and neighboring thalamic nuclei (Armstrong 1979, 1981, this volume) have provided additional relevant data. The number of neurons in the pulvinar nucleus keeps pace with total brain weight in phylogeny, but the numbers in its sensory relay nuclei do not show this positive correlation. Moreover, these data indicate that the medial and lateral pulvinar subdivisions have a greater growth rate than the inferior pulvinar, suggesting that their development is the most progressive of the pulvinar subdivisions. Perhaps these two subdivisions are the recipients of the telencephalically derived neurons reported for the human pulvinar (Rakic and Sidman 1969). Studies of pulvinar connectivity would suggest that these regions are the most likely to be connected to cortical language areas in humans. Alternatively, it is possible that the medial dorsal nucleus (a neighbor of the medial pulvinar) is the recipient of the additional popultion of the telencephalically derived neurons in the human, as this nucleus shows an even greater growth rate than the pulvinar (Armstrong, this volume). At present, however, the fate and nature of such neurons remains uncertain.

Acknowledgment: Supported by USPHS grant EY02593 and fellowship EY05348.

REFERENCES

Allman, J., 1977. Evolution of the visual system in the early primates. In, Progress in Psychobiology and Physiological Psychology, J. Sprague, ed., Academic Press, New York, pp. 1-34.

Armstrong, E., 1979. A quantitative comparison of the hominoid thalamus. I. Specific sensory relay nuclei. Am. J. Phys. Anthrop., 51:365-382.

Armstrong, E., 1981. A quantitative comparison of the hominoid thalamus. IV. Posterior assoication nuclei - the pulvinar and lateral posterior nucleus. Am. J. Phys. Anthrop., 55:369-383.

Benevento, L.A., and Davis, B., 1977. Topographical projections of the prestriate cortex to the pulvinar nuclei in the macaque monkey: An autoradiographic study. Exp. Brain Res., 30:405-424.

Benevento, L.A., and Fallon, J., 1975. The ascending projections of the superior colliculus in the rhesus monkey (*Macaca mulatta*). J. Comp. Neurol., 160:339-362.

Benevento, L.A., and Rezak, M., 1976. The cortical projections of the inferior pulvinar and adjacent lateral pulvinar in the rhesus monkey (*Macaca mulatta*): An autoradiographic study. Brain Res., 108:1-24.

Benevento, L.A., Rezak, M., and Bos, V., 1975. Extrageniculate projections to layers VI and 1 striate cortex (area 17) in the rhesus monkey (*Macaca mulatta*). Brain Res., 96:51-55.

Bergquist, H., 1954. Ontogenesis of diencephalic nuclei in vertebrates: A comparative study. Kgl. Fysiogr. Sallsk, Lunk, Handl. N. F., 65:1-34.

Burton, H., and Jones, E.G., 1976. The posterior thalamic region and its cortical projection in New World monkeys. J. Comp. Neurol., 168:249-302.

Chalupa, L.M., 1977. A review of cat and monkey studies implicating the pulvinar in visual function. Behavioral Bio., 20:149-167.

Cooper, E.R.A., 1950. The development of the thalamus. Acta Anat., 9:201-226.

Cowan, W.M., Gottlieb, D.I., Hendrickson, A.E., Price, J.L., and Woolsey, T.A., 1972. The autoradiographic demonstration of axonal connections in the central nervous system. Brain Res., 37:21-51.

Diamond, I.T., 1979. The subdvision of neocortex: A proposal to revise the traditional view of sensory, motor, and association areas. Pro. Psychobiol. Physiol. Psychol., 8:1-43.

Gilbert, M.S., 1934. The early development of the human diencephalon. J. Comp. Neurol., 62:81-151.

Gould, S.J., 1977. Ontogeny and phylogeny, The Belknap Press of Harvard University Press, Cambridge, Massachusetts.

Harting, J.K., Hall, W.C., and Diamond, I.T., 1972. Evolution of the pulvinar. Brain Behav. Biol., 6:424-452.

Kasdon, D.L., and Jacobson, S., 1978. The thalamic afferents to the inferior parietal lobule of the rhesus monkey. J. Comp. Neurol. 177:685-706.

Kuhlenbeck, H., 1951. The derivations of the thalamus dorsalis and epithalamus in the human brain. Their relation to cortical and other centers. The Military Surgeon, 198:205-256.

Kuhlenbeck, H., 1954. The human diencephalon. A summary of development, structure, function and pathology. Confin. Neurol. Suppl., 14:1-230.

Le Gros Clark, W.E., and Northfield, D.W.C., 1937. The cortical projections of the pulvinar in the macaque monkey. Brain, 126-142.

Levitt, P., and Rakic, P., 1980. Immunoperoxidase localization of glial fibrillary acidic protein in radial glial cells and astrocytes of the developing rhesus monkey brain. J. Comp. Neurol., 193:815-840.

Masterton, B., Skeen, L.C., and Robards, M.J., 1974. Origins of anthropoid intelligence. Brain Behav. Evol., 10:322-353.

Ogren, M.P., and Hendrickson, A., 1976. Pathways between striate cortex and subcortical regions in *Macaca mulatta* and *Saimiri sciureus*: evidence for a reciprocal pulvinar connection. Exp. Neurol., 53:780-800.

Ogren, M.P., and Hendrickson, A.E., 1977. The distribution of pulvinar terminals in areas 17 and 18 of the monkey. Brain Res., 137:343-350.

Ogren, M.P., and Hendrickson, A.E., 1979. The structural organization of inferior and lateral subdivisions of the *Macaca* monkey pulvinar. J. Comp. Neurol., 188:147-178.

Ogren, M.P., and Rakic, P., 1981. The prenatal development of the pulvinar in the monkey: ^{3}H-thymidine autoradiographic and morphometric analyses. Anat. Embryol., 162:1-20.

Ojemann, G.A., 1974. Speech and short term verbal memory alterations evoked from stimulation in pulvinar. In, The Pulvinar-LP Complex, I. S. Cooper, M. Riklan, and P. Rakic, eds., Charles C Thomas, Springfield, Illinois, pp. 173-184.

Papez, J.W., 1939. Connections of the pulvinar. Arch. Neurol. Psychiatr., 41:277-289.

Partlow, G.D., Colonnier, M., and Sazbo, J., 1977. Thalamic projections of the superior colliculus in the rhesus monkey *Macaca mulatta*: a light and electron microscopic study. J. Comp. Neurol., 171:285-318.

Rakic, P., 1974. Embryonic development of the pulvinar-LP complex in man. In, The Pulvinar-LP Complex, I.S. Cooper, M. Riklan, and P. Rakic, eds., Charles C Thomas, Springfield, Illinois, pp. 3-35.

Rakic, P., 1973. Kinetics of proliferation and latency between final division and onset of differentiation of the cerebellar, stellate and basket neurons. J. Comp. Neurol., 147:523-546.

Rakic, P., 1977. Genesis of the dorsal lateral geniculate nucleus in the rhesus monkey: Site and time of origin, kinetics of proliferation, routes of migration and pattern of distribution of neurons. J. Comp. Neurol., 176:23-52.

Rakic, P., and Sidman, R.L., 1969. Telencephalic origin of pulvinar neurons in the fetal human brain. Z. Anat. Entwickl. -Gesch., 129:53-82.

Rezak, M., and Benevento, L.A., 1979. A comparison of the organization of the projections of the dorsal lateral geniculate nucleus, the inferior pulvinar and adjacent lateral pulvinar to primary visual cortex (area 17) in the macaque monkey. Brain Rcv., 167:19-40.

Rose, V.E., and Woolsey, C.N., 1949. Organization of the mammalian thalamus and its relationship to the cerebral cortex. Electroenceph. Clin. Neurophys., 1:391-405.

Shatz, C., and Rakic, P., 1981. The genesis of efferent connections from the visual cortex of the fetal rhesus monkey. J. Comp. Neurol., 196:187-308.

Trojanowski, J.Q., and Jacobson, S., 1976. Areal and laminar distribution of some pulvinal cortical efference in rhesus monkey. J. Comp. Neurol., 169:371-393.

Walker, A.E., 1938. The Primate Thalamus, University of Chicago Press, Chicago.

Yakovlev, P.I., 1969. The development of the nuclei of the dorsal thalamus and of the cerebral cortex; morphogenic and tectogenic correlation. In, Modern Neurobiology, S. Locke, ed., Little, Brown and Company, Boston.

MOSAIC EVOLUTION IN THE PRIMATE BRAIN:

DIFFERENCES AND SIMILARITIES IN THE HOMINOID THALAMUS

Este Armstrong*

QUANTITATIVE CYTOARCHITECTURE

Development of evolutionary hypotheses about the brain or central nervous system (CNS) in any particular taxon depends on solid comparative neurobiological data. Differences among closely related species are likely to be quantitative, and in comparative neuroanatomy a histological technique amenable for quantitative comparisons is a Nissl stain that colors all neuronal perikarya and glial nuclei. Nissl-stained serial sections throughout the brain permit identification of discrete populations of neurons according to their cytoarchitecture, i.e., the arrangement, orientation, density, and coloring features of the nerve cells. Measurements on and within architecturally distinct regions called nuclei or laminae can then be made (Armstrong, 1979; Bauchot, this volume; Zilles et al., this volume; Galaburda and Pandya, this volume).

Measurements from Nissl-stained material can provide size estimates of tissue volume, numbers of neurons or glia, sizes of neuronal perikarya, neuronal density and/or neuropil volume. One of the more popular methods used in comparative neurobiology is to measure a region's tissue volume. After correction for shrinkage, volumetric data can be relatively rapidly accumulated and compared. Bauchot (e.g., 1963, 1979, this volume) and Stephan and Andy and their colleagues (e.g., Stephan and Andy, 1964; Stephan et al., 1970) have analyzed many regions in this way.

A given volume of nervous tissue, however, is composed of many variables: cells, capillaries, extracellular space, and neuropil (axons and dendrites). A change in one or several of these variables cannot be

*From the Department of Anatomy, Louisiana State University Medical Center, New Orleans, Louisiana 70112

discerned from volumetric measurements. Different combinations of CNS features may produce equivalent volumes. One refinement of quantitative volumetric work that avoids some of this ambiguity is to measure neuropil (Zilles et al., this volume), whereas a second approach, which I have chosen, is to analyze numbers of neurons.

Studying neuronal numbers has several advantages. First, volume and neuronal density in a delimited region will change in a reciprocal manner during tissue shrinkage, so specimens prepared at different laboratories may be compared (Armstrong, 1979). Subcortical regions, where all three spatial dimensions can be bounded, are particularly well suited for such work. Second, neurons are basic units of the brain. Indeed, investigators have used volumes to derive estimates of numbers of neurons (Jerison, 1973). While it is appreciated that larger brains have less neuronal density, estimating numbers of neurons from volumetric figures assumes a precise correlation between nerve cell densities and brain weight. This assumption is based on three different studies of neuronal density in cortical regions (Bok, 1959; Shariff, 1953; Tower, 1954), but the values of the slopes of these three studies differ from the relationships found in two other investigations of the same areas (Mayer, 1912; Chow et al., 1950). Slopes representing morphometric correlations frequently alter depending on which phylogenetic or taxonomic groups are used (Martin, this volume; Holloway and Post, this volume). Whether this occurs in the neuronal-density-to-brain-weight relationship is not known. Indeed, primates may have a higher neuronal density per brain weight than do other mammals (Haug, 1972). Third, density variability among different cortical regions and in subcortical nuclei needs study. A similar number of neurons in a vertical slab of cortex has been found in nonvisual cortex of 3 mammalian orders, suggesting that the total number of cortical neurons may covary with surface area, but also that this regularity may change in specialized regions, such as the primate visual cortex (Rockel et al., 1980). Direct estimation of numbers of neurons bypasses the problems of shrinkage artifacts and the as yet untested assumptions of a precise correlation between brain size and neuronal density.

Whether the variable studied is tissue volume, neuropil volume, or numbers of neurons, the major premise of such work is that the bigger the nucleus is, the more important it must be for the animal. The analyses and interpretations change, however, depending on whether one compares absolute quantities or sizes relative to either the total brain size, some CNS subregion, or body weight.

All of the above comparisons have validity. The kinds of comparisons to be made will depend on the questions asked and the nature of the materials. Since the brain must remain integrated with the total body, quantitative data need to be analyzed in light of the size of the organism. That the human septum is as large as expected for a primate of our body size was overlooked for years until quantitative septal data were examined in relation to body weight (Stephan and Andy, 1964, 1974). This latter technique determines encephalization indices (EI), known more specifically as progression indices, isoponderal indices, or encephalization quotients, and body weights are of prime importance in establishing a given organism's or region's EI (Bauchot, this volume; Holloway and Post, this volume). If increases in body size have been favored (e.g., the gorilla), however, the EI values must be interpreted with care and not be assumed to represent the relative degree of brain development anymore than it is the relative degree of body development.

The brain and its subregions form a complex network of interconnected systems, and an investigation of changes among the different parts is necessary for an evolutionary perspective. It is particularly important to try to distinguish changes that are expected because of growth in the overall brain size from changes that are not size-related. Although the human cortex is the size one expects for a primate whose brain weight attains human dimensions (Passingham, 1973), it is not clear whether this relationship holds for all intracortical or noncortical regions. The major focus of this preliminary study was to examine a small region of the CNS that contains different populations of cells (nuclei) and to determine which nuclei and which features covary with increased brain size.

THALAMIC CYTOARCHITECTONIC STUDIES

Data collected from quantitative investigations of the dorsal thalamus form the basis of this study. The thalamus was chosen as the region of study because it is relatively small and is composed of cytoarchitectonically identifiable nuclei that are juxtaposed. The nuclei were identified according to their intrinsic cellular architecture. Some nuclei, such as the lateral geniculate body, are closely associated with one particular set of functions (vision), and have a limited number of connections, whereas other nuclei, such as the medial dorsal complex, contain subregions involved in diverse functions (such as olfaction, regulation of emotions, and acquiring and choosing among cognitive tasks) and connections. A correlation between cytoarchitecturally defined nuclei and a general set of functions can be made, but correlations with any one set of connections or physiologically or chemically defined units cannot be considered exact.

Volumes of the nuclei were generated from areal measurements and neuronal nucleoli were counted in known volumes systematically throughout the cytoarchitecturally defined thalamic regions. This information produced estimates of numbers of neurons. The sizes of neuronal perikarya were also measured. Because the amount of shrinkage is not known, only patterns of nerve cell size distributions that distinguish nuclei within a specimen as well as between specimens are considered.

The data are considered preliminary in that they come from a limited number of specimens, 2 gibbons, 1 chimpanzee, 2 gorillas, and 1 human for numbers of neurons or 2 humans in considering volumes. Additional specimens, 6 chimpanzees (4 with known brain weights), 3 gorillas, and 3 gibbons, were used in the analysis of the anterior nucleus. This nucleus has been analyzed in 11 humans. Details of the procedures are published elsewhere (Armstrong, 1979, 1980a).

Nuclear volume was studied as a function of the summation of the other thalamic nuclear areas plus fiber tracts. The number of neurons were analyzed as a function of brain and body weights. The latter relationships were analyzed somewhat differently. The correlation between numbers of neurons and brain weight is slightly stronger when untransformed data are used, but the correlations between numbers of neurons and body weight are stronger when logarithmically transformed body weights are used. The differences may reflect differences in the magnitude of changes between brain and body weights. Tissue volumes correlate better with brain weight when all the data are logarithmically transformed.

Linear regressions were used in the analyses. The correlation of the volumetric data of nuclei to the remaining thalamus is so strong that no important differences between the results of linear regression and other systems of analyses (e.g., major axes, reduced major axes) are expected (Martin, this volume). Because the variance of the numbers of neurons is much greater than the variance of brain or body weights, linear regressions were chosen to analyze those relationship. The small number of extant hominoids necessitates caution in interpretations of the slopes.

An increase both in numbers of specimens per taxonomic group and in numbers of species will help future analyses. An increased sample size will enhance comparisons among hominoids, although generating slopes with few points will remain problematical. Increasing the numbers of specimens will involve collecting and analyzing nonhominoid primate data. For the predictive slopes to be meaningful, they cannot just be generated from homologous structures, but the structures should be similar in connections, chemistry, and physiology. Among anthropoids and/or primates many, but not all, thalamic or other brain regions are sufficiently similar to permit easy interpretations of quantitative differences. One of the major challenges for the future will be the synthesis of qualitative alterations and quantitative shifts.

This study examines the relative composition of the thalamus among extant hominoids and the differences between the human sizes and those of the other hominoids. A major question asked is, as the thalamus and/or brain expanded, did all thalamic nuclei expand equally? If not, what are the regions that appear particularly large or small for the human thalamus or brain? To answer these questions, I analyzed both hominoid slopes (slopes determined from ape and human data) and ape slopes.

The data are presented on a continuum of quantitative divergence for the human values, going from relatively conservative to divergent states. A conservative condition is one that indicates little or no absolute differences in numbers of neurons among extant hominoids. Conservative structures are more likely to have retained the same configuration of afferents, efferents and integration, although system studies (e.g., Allman, this volume; Galaburda and Pandya, this volume) are necessary to validate and refine such statements. Allometrically expected nuclear regions are those in which the observed human values are close to what one would expect for an ape whose brain attained human dimensions. Divergent structures are those regions in which humans have more neurons than expected for an ape brain of human size. These regions are likely to be especially important in supporting human-specific behavior.

Evolutionary interpretations based on features observed in Nissl-stained material are limited to tissue volumes, numbers of neurons, neuropil spacing, and neuronal perikaryal sizes. Other studies are required to define the functionally important components of a region, i.e., its connections, physiology, and chemistry. Even where correlations between the latter features and cytoarchitectonic regions are known and changes in number of neurons are documented, one cannot tell whether there are more of all the neurons, so that the distribution of connections has not changed, or whether the increase was in only a particular subpopulation, e.g., those nerve cells projecting to area A increased in number, but not those projecting to area B. An intrinsic population of interneurons could have increased in number or the additions of nerve cells may have been restricted to a particular neurochemical type. All of these variations would produce extremely different capabilities for the organisms and remain undetected from analyses of Nissl-stained material alone.

Moreover, a conservation of numbers of neurons does not imply that nothing has changed; neurochemical and other structural-functional characteristics could have altered. It is possible for source A, for example, to have increased its input to an area under study. This difference may not be reflected in numbers of neurons, but, for example, in changes in numbers or preferential positioning of dendrites, spines, and other receptor sites. Likewise, selection may change the pattern and number of axonal collaterals without affecting nerve cell numbers. Numbers of neurons provide important comparative information, but the analyses will raise additional questions because cell number is but one aspect of a complex organ.

CONSERVATIVE NUCLEI

Of the 12 major nuclei examined for numbers of neurons, three contained equal numbers of nerve cells among the great apes and human specimens (Table 1). Thus by the criterion of numbers of neurons they are considered conservative. One, the parafascicular complex, is part of an allometrically expected complex and is discussed in that section. Two are sensory relay nuclei that share common physiological features among living hominoids (Farrer and Young, 1970; Farrer and Prim, 1965; Grether, 1941, 1942; Riesen, 1970; Elder, 1934; Stebbins, 1971). They are the visual relay nucleus, the dorsal lateral geniculate body (LGB), and the auditory relay

TABLE 1. Estimated Number of Neurons: Conserative Nuclei

Specimen	Thalamic Region		
	LGB	MGBp	PF
Hylo.-s	1,131,000	378,000	75,000
Hylo.-h	1,333,000	413,000	109,000
Hylo. lar	--	407,000	--
Gorilla g.	1,884,000	707,000	515,000
Pan t.	2,172,000	913,000	380,000
Homo s.	2,065,000	870,000	371,000

Abbreviations: *Hylo.-s* and *Hylo.-h* are the two sides from one gibbon brain, the species of which is not known, and *Hylo. lar* is *Hylobates lar.*

center, the medial geniculate body, pars parvicellularis (MGB). The observed similarities suggest that these nuclei have not changed their structural features from the time hominids separated from the great apes. The gibbon's sensory nuclei contain fewer neurons, but the distance separating the lesser apes from the other hominoids are less in these nuclei than in the remainder of the thalamus, suggesting that selection pressures favored the stabilization of numbers of neurons within LGB and MGB. Despite this similarity, LGB and MGB have differences that merit separate discussions.

The LGB has been termed a progressive structure among primates because of its greater relative size in anthropoids than prosimians (Stephan, 1969), yet it does not seem to have diverged as much as other thalamic nuclei. Indeed, in absolute terms, the volumes of these hominoid LGB's are more alike than are the other thalamic nuclei. The similarity is also indicated by the slope (α) being much less than 1 when nuclear volume is regressed on the volume of the rest of the thalamus (hominoid $\alpha = 0.57$, $r = 0.96$ $p < .05$). Similarly a low slope is derived from Hopf's (1965) *Lemur catta*, *Macaca cynomolgus*, *Pongo pygmaeus*, and *Homo* data ($\alpha = 0.45$, $r = 0.95$ $p < .05$). In addition, encephalization indices of LGB volumes are lower than nonsensory thalamic regions and they are lower in hominoids than in monkeys (Bauchot, this volume), indicating that other regions have changed more in hominoids. Neither the volumetric data of this study nor Hopf's predict the human condition (Armstrong, 1979; Hopf, 1965; Passingham, 1979).

When the number of neurons is regressed on brain weight, the correlation between the number of LGB neurons and brain weight is not significant for these hominoids ($r = 0.57$ $p > .10$). Larger hominoid brains have not yet been demonstrated to have more LGB neurons than smaller ones. Similarly the growth in numbers of neurons per body weight is lower in LGB than in any other thalamic nucleus studied and the correlation is not significant ($\alpha = .14$ $r = .86$ $p > .05$). Apparently, among hominoids neither brain weight nor body weight can accurately predict the numbers of LGB neurons (Fig. 1). Whether a statistically significant correlation will be found with retinal size and/or numbers of ganglion cells remains to be determined.

The nerve cell sizes also have a similar distribution among these extant hominoids. The differences in the relative frequencies of particular group sizes are as yet inseparable from individual variation and shrinkage artifact (Fig. 2). Similarities of the shape of the great ape and human patterns support the concept of the conservative nature of LGB.

The chief target of LGB, the striate cortex, may be a relatively conservative cortical structure among hominoids. Volumetric dimensions of the human striate cortex are predictable from primate data when the striate cortex is studied as a function of body weight (Passingham, 1973), but like LGB, the hominoid striate cortex has incremented in size less than other cortical regions (Shariff, 1953; Passingham, 1973; Holloway, 1968, 1979). Unlike LGB, the human striate cortex may contain more neurons than does that of the chimpanzee (Shariff, 1953).

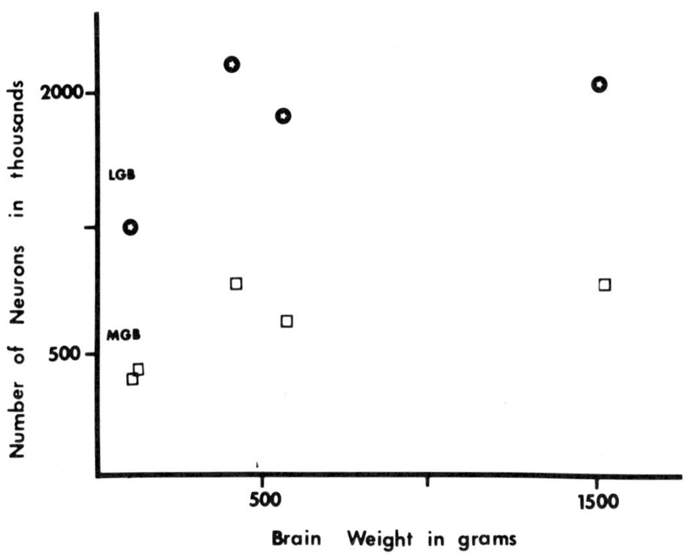

Fig. 1. Relationship of the number of neurons and brain weight in conservative nuclei, the medial (MGB) and lateral (LGB) geniculate bodies. The relationship is not statistically significant among either the apes or all hominoids.

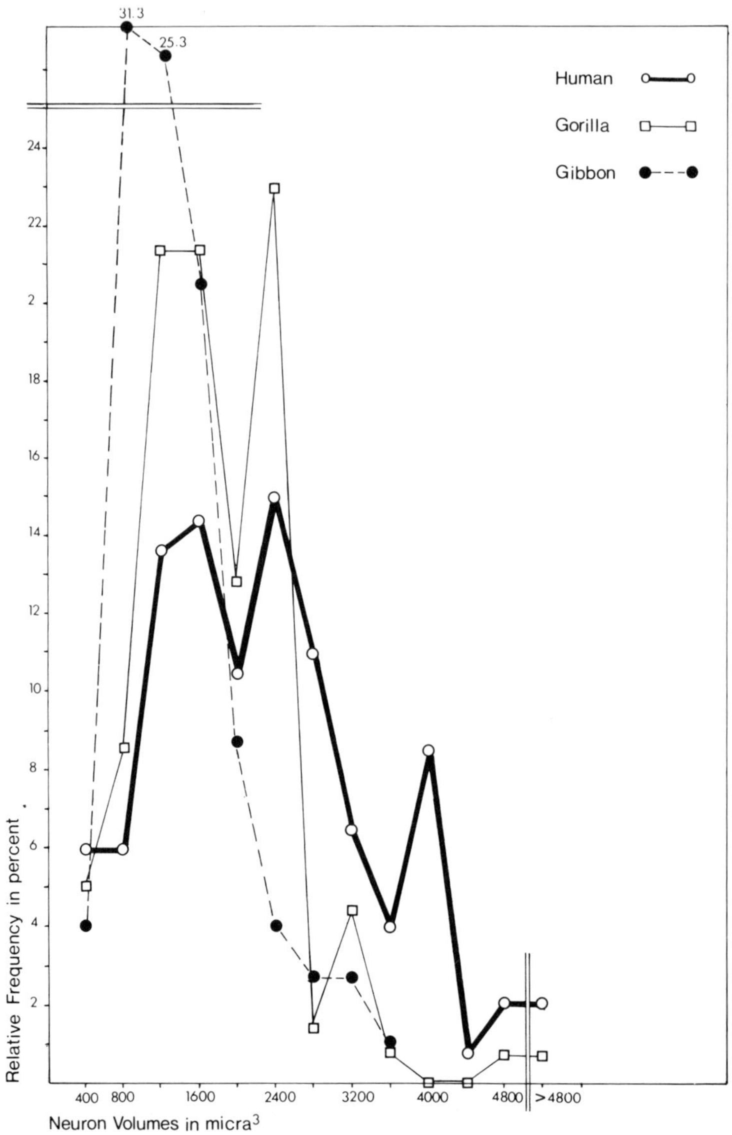

Fig. 2. Frequency distribution of neurons according to the sizes of their perikarya in LGB pars parvicellularis. The differences in the relative frequencies of the small cell sizes between the human and great ape specimens (here represented by the gorilla) most likely represent individual variation. The gibbons typically lacked the larger cells in all the conservative nuclei.

Certain Old World and New World monkeys have relatively enlarged visual structures. Shariff (1953), for example, found that a *Cercopithecus* had approximately the same number of neurons in the striate cortex as did his human brain. An index of progression based on *Lepilemur* shows *Homo* to have relatively less visual cortex and LGB volume per body weight than do several anthropoids (Stephan, 1969). An LGB encephalization index based on a slope of 0.667 also shows the largest relative development in nonhominoid haplorhines (Bauchot, this volume). *M. mulatta*, with a brain size similar to that of gibbons, may have an absolutely larger LGB (Armstrong, 1979) and a more complexly macrolaminated structure than any hominoid (Le Gros Clark, 1941). The relatively large volumetric sizes of ceropithecoid visual structures (Stephan, 1969; Shariff, 1953) and cell counts in *M. mulatta* (Chow et al., 1950; Sullivan et al., 1958; Vyshinskaya, 1961) suggest that the number of neurons in the cercopithecoid LGB may equal that found in great apes and humans (Chow, 1951; Shariff, 1953).

My suspicion about the above quantitative similarities between the macaque monkey, great apes, and humans is that it represents parallelism. That is, the common ancestor of apes and macaques, and presumably all Old World monkeys, most likely had fewer LGB neurons than found in either extant group and the numbers increased separately in both groups. It is unlikely that the common ancestor had already attained the same number of LGB neurons because gibbons shared a more recent common ancestry with the other hominoids than did Old World monkeys, yet they have fewer neurons. Presumably the common hominoid ancestor also had fewer neurons than observed in extant great apes or humans. Second, the greater elaboration of macroscopically defined LGB lamination patterns in *M. mulatta* than in humans (Le Gros Clark, 1941) or other hominoids (Armstrong, 1979) suggests that some aspect in the LGB organization differs. Given that gibbons have a much simpler macroscopic LGB organization than the other hominoids (Armstrong, 1979; Kanagasuntheram and Wong, 1968), the LGB in the common ancestor of hominoids probably contained a simple lamination pattern, unlike that found in Old World monkeys. The increase in numbers of neurons occurred separately in Old World monkeys and hominoids and as the neurons were added they were organized into slightly different patterns.

On the other hand, one may speculate that the similar number of LGB neurons in great apes and humans more likely represents a stabilization or conservation of an ancestral condition. The common ancestor to humans and great apes was undoubtedly smaller and had an absolutely smaller brain than any extant great ape. After controlling for shrinkage artifact, a nuclear volume containing the same number of LGB neurons as found in modern great apes can be placed within the dimensions of a smaller brain. This produces a larger relative size of the visual center, one that is observed in the modern monkey range according to Bauchot's data (this volume). Thus it is not unreasonable to predict that the common ancestor of great apes and man contained the same number of LGB neurons as found among its extant descendants.

The auditory relay nucleus, MGB, also appears conservative among great apes and humans (Armstrong, 1979). Because the auditory and vocalization systems probably evolved together in primates, specializations associated with human speech detection may yet be found in the heterogeneous MGB.

Volumetrically, MGB is larger in the larger hominoid brains. In two independent studies a similar ratio of the increase in MGB size between gibbons, chimpanzees, and humans was found to be about 1:3:6 (Armstrong, 1979; Blinkov and Zvorykin, 1950). Although the volumetric increases are less than those of the rest of the thalamus, they are closer to being isometric with the latter than is LGB (all hominid MGB $\alpha = 0.88$, $r = 0.98$ $p < 0.05$). Similarly, Hopf's (1965) data correlating volume to the rest of the thalamus yielded a higher slope in MGB than in LGB (MGB $\alpha = 0.77$, $r = 0.99$ $p < 0.01$). The encephalization indices among Bauchot's specimens (this volume) indicate that hominoids have relatively less MGB volume per body weight than do other primates. This decline, however, is not precipitate as with LGB, but on a continuum, suggesting a linkage between the two factors. The addition of MGB neurons in these hominoids correlates significantly with body weights ($r = .80$ $p < .05$), but no significant correlation has yet been found with brain weight (Fig. 1). Whether the numbers of MGB neurons covary with cochlear size and/or numbers of hair cells remains to be determined.

The apparent stability of numbers of neurons among great apes and humans and its significance deserves further study. Volumetric data, numbers of neurons in the inferior colliculus, and assumed parallels with the visual system led to postulations that the human and macaque MGB contain a similar number of cells (Chow, 1951; Blinkov and Glezer, 1968). Although an equivalent number of neurons in LGB is likely in these taxa given the macaque's high LGB EI (Bauchot, this volume) and relative volume (Hopf, 1965), it is less likely in MGB, unless there have been drastic shifts in nerve cell densities.

Although the LGB and MGB both appear conservative with respect to numbers of neurons among these hominoids, they maintain different relationships to the thalamus, brain, and body, and thus their evolution may have been different. This is consistent with the idea that CNS regions do not respond uniformly to a single evolutionary force, but that different parts respond to different selection pressures. Maintenance of an integrated organization places some limits on evolutionary changes, but not confinement to a single pathway.

ALLOMETRICALLY EXPECTED NUCLEI

Most of the human thalamic nuclei have more neurons than do those of apes, and in many instances the numbers correlate significantly with body and/or brain weights. The nuclei that show either a moderate increase in

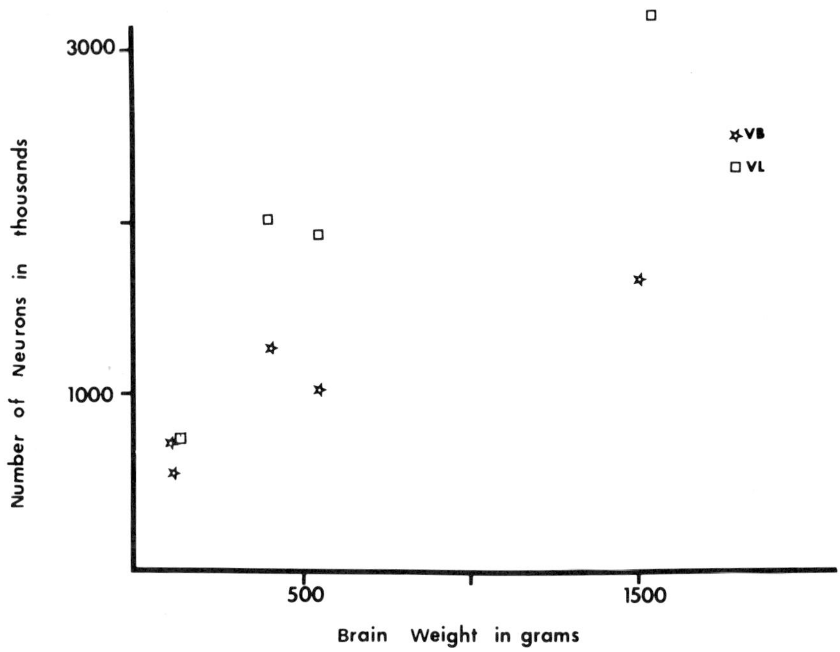

Fig. 3. Relationship of the number of neurons and brain weights in the allometrically expected nuclei, ventrolateral (VL) and ventrobasal (VB). The relationship of the number of neurons with brain weight among the apes suggest that the human complexes have fewer neurons than expected; however, the variates are not correlated significantly. The number of neurons do correlate significantly with brain weight when human values are added to the computations.

numbers of neurons between the great apes and human specimens or a substantial increment that covaries with brain and/or body weights are classified here as allometrically expected nuclei (Fig. 3).

The ventrobasal complex (VB) is a sensory relay center, but unlike the auditory or visual relay centers, it has about 1½ times as many neurons in people as in great apes (Table 2). The increase in numbers of neurons among all hominoids is significantly correlated with brain weight ($r = 0.92$ $p < .05$). At the same time, the slope derived from the ape data seems to predict more human VB neurons than are observed, although the small sample size precludes statistical significance. Volumetrically VB increases in size at a rate only a little less than the rest of the thalamus (all hominoid $\alpha = .86$, $r = .98$ $p < .05$) and it parallels MGB in this regard.

Although the gain in neurons in VB is dissimilar to that of the other sensory relay nuclei, it is comparable to that observed in the *motor* thalamic

TABLE 2. Estimated Number of Neurons[a]: Allometrically Expected Nuclei

Specimen	Thalamic Region VB	VL	CM	MD	PU
Hylo.-s	587	740	128	827	1950
Hylo.-h	458	789	145	705	1950
Hylo lar	770	--	223	1030	1890
Gorilla g.	1070	1940	590	2350	4560
Pan t.	1226	2064	410	2500	5450
Homo s.	1690	3306	895	7570	10110

[a]Numbers are in thousands

Abbreviations: See Table 1

relay region, the ventrolateral complex (VL). The number of VL neurons is also significantly correlated with brain weight (all hominoid VL, pars caudalis and medialis $r = .91$, $p < .05$; VL pars oralis $r = .94$, $p < .05$). As with VB, ape data seem to predict more human VL neurons than were found. In gibbons the number of VL motor neurons represents about 1/3 (34%) of all the sensory neurons (VB+LGB+MGB), in great apes motor neurons are about half as numerous as sensory neurons (gorilla 53%, chimpanzee 48%), whereas in the human thalamus motor neurons are about 2/3 (71%) as numerous as all thalamic sensory nerve cells. The ratios show that bigger hominoid brains carry more information per input load toward the motor cortex and its portion of the final common pathway, the corticospinal tract. Changes in the volume of VL keep pace with those observed in the rest of the thalamus (Armstrong, 1980b).

VL neurons are closely related to the motor cortex and corticospinal tract. The hominoid motor cortex (Brodmann areas 4 and 6) may have enlarged relatively more than VL, and the corticospinal tract less than VL (Armstrong, 1980b). An expansion of motor regions not directly tied to the corticospinal tract may permit greater flexibility in the integration of motor programs, and thereby foster skill and dexterity (Phillips and Porter, 1977).

Another moderately derived region, the centro-medianum-parafascicular complex (CM-Pf), is apparently involved in a wide range of functions, including the modulation of pain, participation in cortical recruiting, augmenting, and contingent negative variation waves, and in the integration of motor activities through its widespread connections (McGuinness and Krauthamer, 1980; Morison and Dempsy, 1942; Tsubokawa and Moriyasu, 1978).

The complex has been studied as a whole in primates by several investigators. Volumetric analyses indicate that CM-Pf has expanded at approximately the same rate as the rest of the thalamus (this study $\alpha = 1.0$; $r = .89$; $p < .05$; Hopf's primates (1965) $\alpha = 1.1$, $r = 0.998$ $p < .01$; and Kurepina's primates (1938) $\alpha = 1.0$; $r = 0.997$ $p < .05$).

CM and Pf may, however, be under different selection pressures. Niimi and coworkers (1960; Niimi and Kuwahara, 1973) found greater a divergence in CM than Pf among different mammals. Bauchot detected more CM neurons than Pf neurons in prosimians, but an equal number in the two nuclei in insectivores (Bauchot, 1963). Hominoids also differ in the changes in CM and Pf.

Pf has a relatively low volumetric accretion that is not significantly correlated with changes in brain or body weights. The human Pf has fewer than half as many nerve cells as CM, the gibbon Pf has about 2/3 as many neurons as CM, but among the great apes Pf has about 90% as many. Because Pf has an equivalent number of neurons in great apes and humans, the shifts in the ratio probably stem from CM's expansion during hominid evolution. Pf may be stabilized like LGB and MGB. Interestingly, Pf appears quantitatively derived in gorillas (Armstrong, 1976, and in preparation).

The centromedianum nucleus expanded volumetrically at a faster rate than the rest of the thalamus, particularly among apes (ape $\alpha = 1.4$, $r = 0.99$, $p < .01$; all hominoid $\alpha = 1.1$, $r = 0.98$ $p < .01$). CM manifests a substantial increase in numbers of neurons (Table 2) when humans are compared with great apes, and this number covaries with brain weight ($r = .96$ $p < .01$). In this region, ape data predict the observed human numbers.

ALLOMETRICALLY EXPECTED NUCLEI WITH CYTOARCHITECTURAL DIFFERENCES

The above hominoid nuclei, conservative and derived, have a similar distribution of the different sizes of their constituent neuronal soma. In this study shrinkage could not be measured; therefore only differences that affect some, but not all, of the nuclei can be interpreted as representing biological aspects. Gibbon nuclei frequently lack the largest neurons, but whether this fact reflects differences in the histological preparations, plane of sectioning, or biology is not clear. Small differences in the relative frequencies are evident among the other hominids (Fig. 2), but artifact and individual variation can account for these differences. Overall, the similarities suggest that as the neuronal pools of the above nuclei either stabilized or expanded there was no selection for different proportions of nerve cell sizes.

In two allometrically expected regions, however, a common profile of perikaryal sizes is not maintained among these hominoids. They are the pulvinar-lateral posterior (Pu-LP) and the medial dorsal (MD) complexes. Both are prominent in primate thalami including man. They have reciprocal connections with cortical "association" areas, Pu-LP with the posterior association cortex and MD to the prefrontal association cortex. MD also has important limbic connections.

The Pu-LP complex is involved in diverse functions including the integration of sensory information, especially vision (e.g., Chalupa, 1977; Ogren, this volume). It can switch information between sensory modalities, a function that may be necessary, but is not sufficient for cross-model transfer (Armstrong, 1976, 1981). Additionally Pu may influence "higher cognitive" processes, particularly language (Ojemann, 1974).

In this study the Pu-LP complex as a whole has enlarged volumetrically at a slightly greater pace than has the rest of the thalamus (ape and all hominoid $\alpha = 1.2$, $r = 0.99$ $p < .01$). Likewise Hopf (1965) observed that Pu volumes diverged a little bit more than the rest of the thalamus among his primates (all primates $\alpha = 1.2$, $r = 0.99$ $p < .01$). The volumetric rate of increase, however, is less than that of the whole brain (ape $\alpha = .83$, $r = .99$ $p < .05$; all hominoid α 0.86, $r = 0.99$ $p < .01$). The above data imply that other parts of the brain enlarged more than Pu-LP among hominoids.

The Pu-LP complex contains more neurons than any other thalamic complex in hominoids (Armstrong, 1976, 1981) and is probably the largest thalamic nucleus in all primates (Harting et al., 1972). The evolutionary accretion of Pu-LP neurons per brain weight is moderate among these extant hominoids (Fig. 4). The rate of adding neurons is less than that found in VL and CM, for example, and the numbers of human Pu-LP neurons are predictable from ape values with the pulvinar data providing the predictability.

Although the human Pu-LP appears to have as may neurons as one would expect in an ape whose brain attained human dimensions, the measurements also indicate that not all parts of the complex underwent equal modifications. The human LP has fewer neurons than expected, a subregion of Pu, the pars inferior, shows a conservative volumetric growth rate, and, interestingly, the pattern of relative frequencies of nerve cell sizes suggests two populations of cells in the human Pu-LP complex (Fig. 5), but not in those of the apes (Armstrong, 1976 and 1981).

Analyses of Pu-LP's embryological development has led to suggestions that *some* human Pu-LP neurons may develop from a telencephalic rather than a diencephalic anlage (Rakic and Sidman, 1969; Rakic, 1974; Ogren, this volume). The additional telencephalic neurons are thought to be small interneurons. To date the data imply that although people have as many pulvinar neurons as expected for an enlarged ape brain, the increase was brought about through a differential expansion of the neuronal pool. Such an increase would change the organizational arrangements within Pu and could produce disparate functional capabilities. The correlation between a

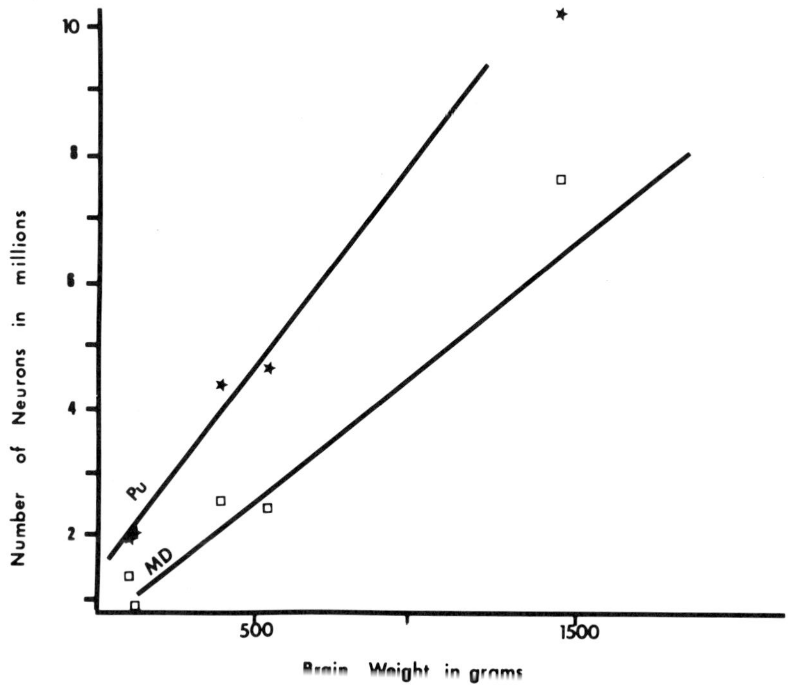

Fig. 4. Relationship of the number of neurons and brain weight in two allometrically expected regions, the pulvinar (Pu) and medial dorsal (MD). In these complexes the numbers of neurons correlate significantly with brain weight among apes and the ape data predict the observed human values. The slope was determined using the least-squares technique.

bimodal adult neuronal population and a dual embryological origin, however, must be more firmly established. In the medial dorsal (MD) complex and other limbic nuclei (see below), both the great apes and humans have bimodal distributions of neuronal perikarya and no dual embryological origin of neurons have been demonstrated for these regions. Detailed morphometric analyses in ontogenetic series are needed.

The neurons comprising the limbic and association complex, MD, are extremely heterogeneous in appearance, size, connections and functions. Although the primate MD is frequently divided into two parts, pars magnocellular and parvocellular (e.g., Olszewski, 1952; Walker, 1938), and the human MD into more than six parts (Hassler, 1959; Namba et al., 1958), the cytoarchitectural divisions among hominoids are difficult to localize consistently (e.g., Heiner 1960; Kanagasuntheram and Wong, 1968; Van Buren and Borke, 1972). Consequently in this study I analyzed MD as one unit. Important reciprocal connections with the prefrontal cortex exist

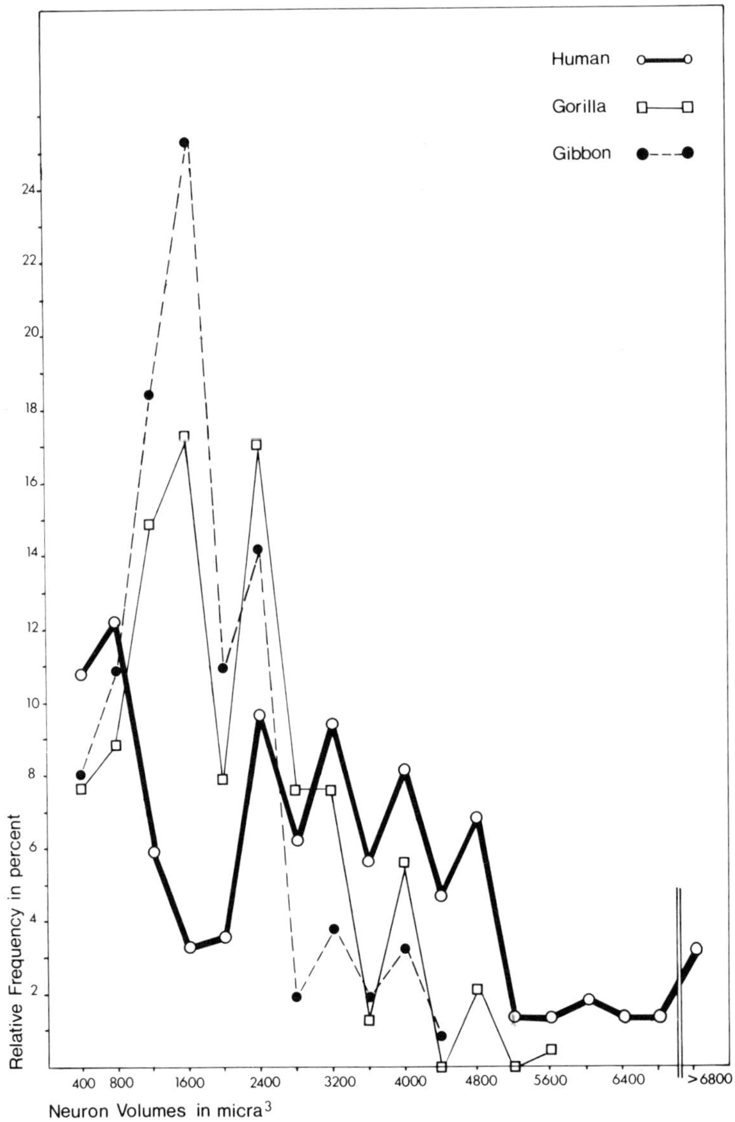

Fig. 5. Frequency distribution according to the relative frequency of different sized perikarya in the pulvinar. The pronounced human bimodal distribution was also observed in the lateral posterior nucleus. (Courtesy of Allan R. Liss, from Armstrong, 1981).

along with many other less massive ones (Kievit and Kuypers, 1977; Van Buren and Borke, 1972).

Among the hominoids of this study and Hopf's (1965) primates, the volumes of MD enlarged like the rest of the thalamus (ape $\alpha = 1.12$, $r = 0.98$, $p < .05$; Hopf's data $\alpha = 1.04$, $r = 0.999$, $p < .05$) and in both cases the addition of human data increased the slope of the line (this study $\alpha = 1.32$, $r = 0.99$, $p < .01$; Hopf's $\alpha = 1.11$, $r = 0.99$ $p < .01$). The encephalization index of the human MD is higher than those of other thalamic regions, except for the dorsal tier group (Bauchot, this volume). During primate evolution MD increased in volume and during hominid evolution selection for increasing MD's size was particularly strong.

The number of MD neurons is highly correlated with brain weight (all hominoid $r = .99$, $p < .001$) and the scaling of ape data (ape $r = .95$ $p < .05$) show the predicted human number of neurons to be close to, but lower than, the number calculated from the observations. Future data may show the numbers of neurons or some other component of MD to be a derived characteristic. Until individual variation in the numbers of neurons is determined and/or the number of taxonomic groups is increased, the significance of the upward deviation of the human values will be unclear.

A similar distribution of neuronal perikarya sizes is found among humans and great apes (Fig. 6), but this distribution contrasts with that of the gibbons. The difference between the taxonomic groups suggests that a change in organization may have taken place after the great and lesser apes separated. The extreme complexity of MD demands a more detailed investigation.

Thalamic regions were classified in this paper as allometrically expected if the observed number of neurons in the human nucleus was increased and the larger number was considered to be either close to or less than the number predicted from ape data. Although the ape data do not predict the human estimates in all cases, the numbers of neurons in the allometrically expected nuclei are statistically significantly correlated with brain weight among all hominoids. The data do not imply, however, that the changes in number of neurons were the result of nonspecific selection pressures or pressure for only one variable, that of brain size. Three important kinds of data argue against such an interpretation. First, the existence of conservative nuclei shows that not all thalamic regions necessarily expand in numbers of neurons as the overall brain volume does. Second, the regions typified as allometrically expected are all nuclear complexes, i.e., they can be divided into smaller regions, each of which may be functionally distinct. Volumetric fluctuations within VL (Armstrong, 1980b) and Pu-PL (Armstrong, 1981) show that evolutionary changes are not constant in large complexes. When smaller, more circumscribed regions are analyzed, fluctuations around the expected values will likely be observed, with some subnuclear regions having conservative and others divergent characteristics. Third, the addition of nerve cells in Pu-LP seems not to

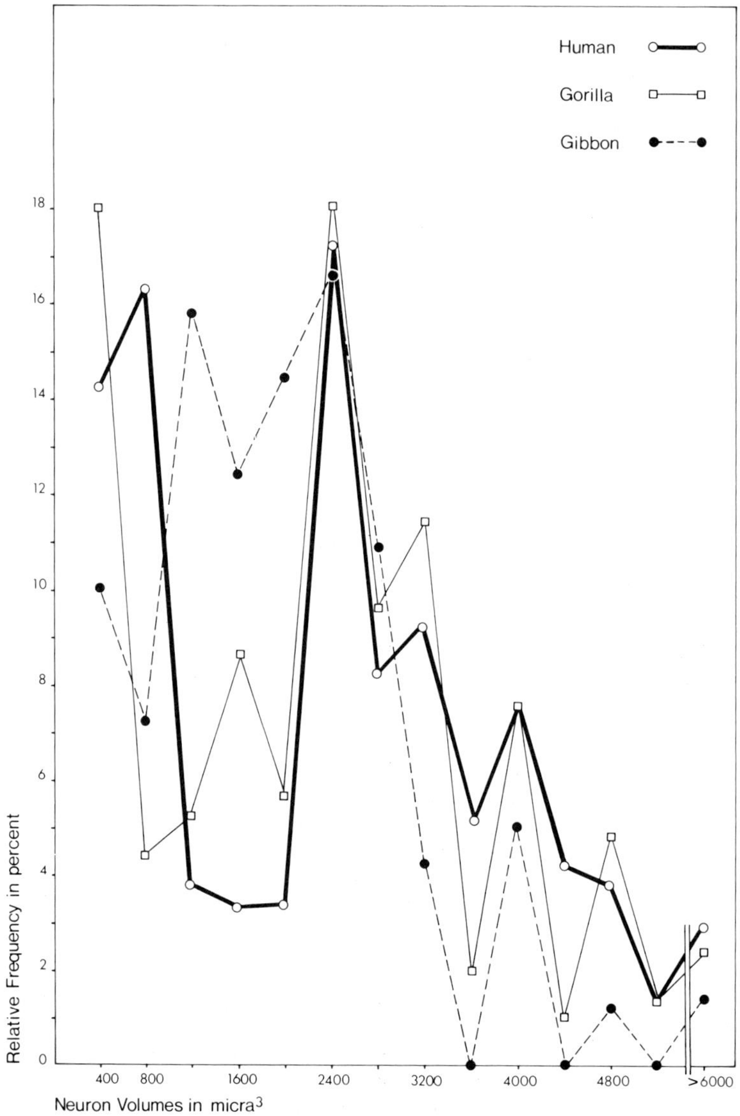

Fig. 6. Distribution of different-sized MD perikarya according to their relative frequencies. Both the gorilla and human MD complexes have a bimodal distribution that the gibbon does not share. In this complex the gibbon's range of perikaryal sizes equals that of the other hominoids.

have occurred through an increase in all neurons but in one or two particular subsets. Such differential shifts cannot be explained solely on the basis of the expansion of the total brain or else they would be observed in all enlarged nuclei. Thus the allometrically expected nuclei of humans are likely to have attained their numbers of neurons through different evolutionary processes.

DIVERGENT NUCLEI

The human limbic nuclei, anterior principal (AP) and lateral dorsal (LD), have many more nerve cells than do those of the other hominoids, both absolutely and as expected on the basis of body or brain size. Both nuclei are intimately related to the limbic system and specifically Papez' circuit (MacLean, 1973, this volume; Mikol et al., 1977; Papez, 1937). As components of this circuit, the nuclei are considered to be involved with emotion and memory, particularly in states of motivation, anxiety, and aggression (Mirsky et al., 1957; Riss et al., 1969; Victor et al., 1971; MacLean, 1973, this volume). The anterior thalamic nuclei have been studied more recently in nonprimate mammals, where they appear to modulate the animal's attention towards incentives and memory systems (Brady, 1960; Hirsh et al., 1979; McDonnell and Flynn, 1968; Thompson et al., 1979; Vinagradova, 1975).

In nonhuman primates, the limbic nuclei do not expand volumetrically as much as the rest of the thalamus (AP: α ape = 0.81, r = 0.97 $p < .05$; Hopf's primates α = 0.81, r = 0.95, n.s.; LD: α ape = 0.75, r = 0.97, $p < .05$; Hopf's primates α = 0.54, r = 0.95, n.s.).

In addition to this relative conservatism, AP and LD do not maintain a constant quantitative relationship among themselves. AP and LD were the same size in *Macaca cynomolgus* and *Lemur catta*, whereas in hominoids, AP is much larger than LD (Armstrong, 1976, 1980a; Hopf, 1965). In addition, except for *Tarsius*, nonhominoid primates have an internal cytoarchitectural distinction between anterior pars medialis and pars ventralis (In preparation). My observations agree with those who find no such distinction in hominoids (Andrew and Watkins, 1969; Feremutsch, 1963; Walker, 1938; Van Buren and Borke, 1972). Nuclei that differ in cytoarchitecture are also thought to differ in connectivity (Galaburda and Pandya, this volume; Zilles et al., this volume). The cytoarchitectural variations suggest a difference in AP's connections between hominoids and other primates.

Interesting volumetric shifts have also been observed (Armstrong, 1976, 1980a, 1981a; Hopf, 1965; Passingham, 1979). In absolute terms the limbic nuclei of *L. catta* are as big as those *M. cynomolgus* even though the thalamus and the brain of the latter are more than twice the size of the former (Hopf, 1965). Likewise Bauchot's data (this volume) showed *M. mulatta* to have a lower anterior nuclear EI than two lemurs, *L. fulvus* and *L. variegatus*. In my study, the smaller *Pan* brain and thalamus contained a larger LD nucleus than did the gorilla (Armstrong, 1980a). Despite these interesting reversals, overall the volumes of the anterior complex correlate with brain weights among primates (Bauchot, this volume).

The human limbic nuclei show a substantial increase in volume over what would be expected in an ape thalamus of human size. The slope representing the volumetric enlargement of the limbic nuclei as a function of the enlargement of the rest of the thalamus has a substantial upward deviation compared with the nonhuman primate slope (AP all hominoid α = 1.08, $r = 0.977$, $p < .01$; Hopf $\alpha = 0.97$, $r = 0.97$ $p < .05$; LD all hominoid α = 0.95, $r = 0.98$, $p < .01$; Hopf $\alpha = 0.72$, $r = 0.96$, $p < .05$).

The numbers of neurons in humans not only have a vastly expanded pool, but are more numerous than the ape data predict (Table 3). Although the human LD also has a large upward separation from the expected numbers of neurons, data for AP have been gathered from an enlarged set of specimens (3 gibbons, 6 chimpanzees, 3 gorillas, 11 people); the discussion will, therefore, focus on AP. The numbers of AP neurons have not been observed to covary with brain weight within any of these taxonomic groups. The numerical distribution of AP neurons in the human population is neither sexually dimorphic nor bilaterally asymmetric.

Use of the means of the different taxonomic groups (1 value each for gibbons, chimpanzees, and gorillas) suggests that the numbers of neurons covary with brain weight ($r = 0.95$), but the small numbers of specimens preclude statistical significance and taxonomic group predicts the numbers of neurons in the human AP better than brain weight does. For a more conservative estimate, the slope describing the ape relationship is extended to predict human values, and the expected number is much less (more than two standard deviations from the human mean) than the observed values

Table 3. Estimated Number of Neurons: Divergent Nuclei in Thalamic Region

	Nucleus Lateral Dorsal		Nucleus Anterior Principal		
Taxonomic Group	No. of Specimens	No. of Neurons	No of Specimens	No. of Neurons	(S.E.)[a]
Hylobates	1	84,000[b] 69,000[b]	3	258,000	(38,105)
Gorilla	1	153,000	3	384,000	(33,486)
Pan t.	1	149,000	6[c]	386,000	(30,619)
Homo s.	1	410,000	11	1,190,000	(55,024)

[a]S.E. = Standard error of the mean
[b]Measurements are from two sides of one gibbon brain
[c]Only 4 of the chimpanzees had recorded brain weights. The mean estimated number of neurons for these specimens is 409,000 (S.E. 39,460).

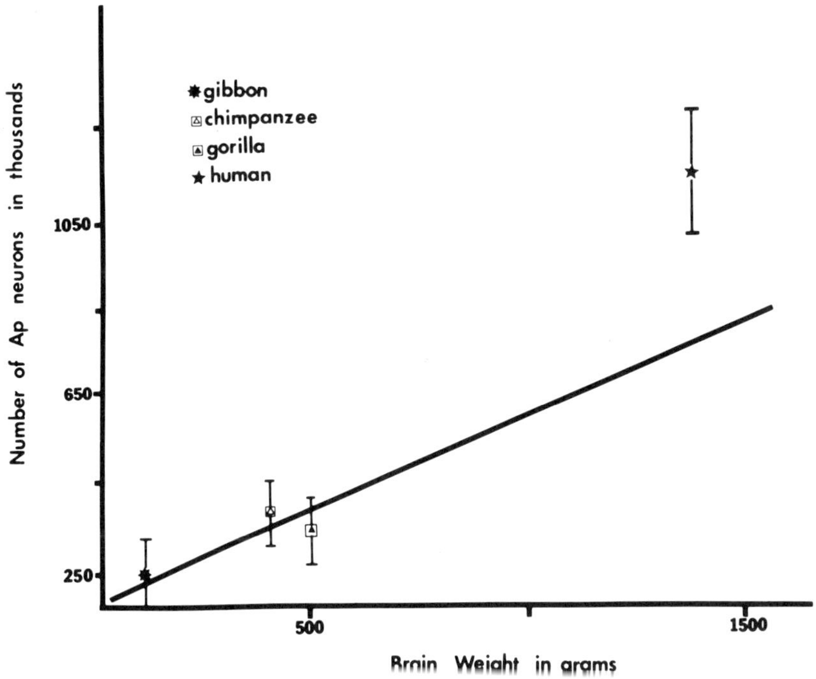

Fig. 7. Relationship of the mean number of neurons and brain weight for each taxonomic group in the anterior principal nucleus (AP). The ape data appear to predict fewer AP neurons than observed; however, the probability of the correlation of numbers of neurons with brain weight is greater than 0.05. The line was determined from the least-squares technique. The bars represent $\pm$ 1 standard deviation.

(Fig. 7). Thus during hominid evolution, neurons appear to have been preferentially increased in AP.

The connections of the anterior complex allow it to influence activities in both limbic and association regions. The primate AV and AM nuclei receive large inputs from the medial and lateral mamillary nuclei respectively and from the hippocampal subiculum (Krayniak et al., 1979). In turn they project to the cingulate gyrus (Vogt et al., 1979), subiculum (Amaral and Cowan, 1980) and, particularly AM, to the orbital frontal cortex (Kievit and Kuypers, 1977). In addition to the thalamic input, the posterior cingulate gyrus receives fibers from the inferior parietal association area (Vogt et al., 1979; Mesulam et al., 1977). The connections involve AP in a circuitry connected to both emotional (limbic) and higher cognitive (associative) centers.

Behavioral studies of the anterior nuclei in primates are still rudimentary, but these areas are thought to involve modulation of emotional states, particularly anxiety and aggression (Spiegel et al., 1957; Brady, 1960; Kuhlenbeck, 1954). Close connections with the posterior cingulate gyrus suggest a role in attention systems (Mesulam and Geschwind, 1978; Mesulam et al., 1977), as do findings in nonprimate studies (Hirsh et al., 1979).

The known connections and hypotheses of function permit a line of further speculations and questions. The increase in numbers of AP neurons occurs at a low, but consistent rate with increases in brain weight among nonhuman hominoids. The increase in neuronal numbers during human evolution may have been selected either to relay a relatively larger emotional or limbic component into the posterior cingulate gyrus and posterior association region or to modulate the integration of emotional and cognitive functions. Future behavioral and neuropathological studies will be necessary to determine which of these two hypotheses is the more correct. Although the correlation between function and anatomy is, at this time, most rudimentary, the difference in cytoarchitectural patterns between nonhominoid primates and hominoids and a quantitative shift between humans and other primates preclude an assumption that the neurological circuitries for emotional behaviors are the same. This in turn indicates that we must be careful when arguments are advanced concerning the conservative nature of behavior traits such as aggressive dominance systems, the scaling of emotional responses, and the degree and length of maternal care and socialization (Wilson, 1975).

On the other hand, the differences, particularly the increase in numbers of human AP neurons, support hypotheses based on behavioral data that implicate the importance of emotional factors in human evolution (Holloway, 1975; Reynolds, 1976; Lovejoy, 1981) and suggest further speculation about how emotional factors influence cognitive ones and vice versa. Language is recognized as a particularly human expression for our higher cognitive functions, but as a system of symbolization, it is also social (Singer, 1980) and as manifest by the babbling of even deaf babies, it may have a unique motivational system (Gould and Gould, 1981). Under severe emotional trauma, language may not develop (Curtiss, 1977). The limbic system, those neural circuits that support and modulate our emotional behavior, structure limits for our social and motivational states. The importance of limbic input into higher cognitive functions is not restricted to language, but can be found in all symbolic expressions, be they making tools or arranging social relations. A tool's appearance and the ways to make it and use it are culturally learned and require cognitive abilities; the ability to focus attention over sufficient time for its creation and maintenance requires tapping limbic sources.

Limbic functions are not only channeled into cognitive activities, but the latter also modulate our emotions. We share with higher primates intense maternal and sibling ties (Reynolds, 1976; Lovejoy, 1981). Additionally we have strong emotional ties to other relatives. These emotionally charged relationships have become integrated into cognitive systems, namely, kinship structures. A confluence of limbic and cognitive forces may also be shown in role-affect differences (Hsu, 1979). Human emotional reactions are modulated by our limbic systems and regulated by culturally structured, cognitive processes. All human emotional expressions and activities, ranging from lovemaking to killing, have culturally prescribed ways to accomplish them. When conflict arises between a person's desires and cognitive rules, the result is frequently anxiety or guilt. These qualities may denote the linkage of cognitive and limbic systems or as Freud stated (1930), the linkage of culture and aggressive motives. It is also these qualities of anxiety, agitation, and aggression that are said to be partially suppressed by lesions in AP (Spiegel et al., 1951). Such suppression could occur through a decrease of limbic impulses into the cognitive sphere or from interference with circuitry connecting the cognitive and emotion spheres. An important part of that integration may pass through the thalamic limbic nuclei, which may have enlarged preferentially during hominid evolution.

SUMMARY AND CONCLUSION

The large size of the human brain and its neocortex have been universally accepted as evidence of its increased contribution to human complexity. As other regions of the brain are increasingly studied and understood, it is natural that the same types of questions are now being asked about smaller neural entities. Which regions are larger than expected and what does this tell us about hominid evolution?

Most of this sort of research used volumetric data as an indicator of size and thus, by extension, importance. Although there are problems of shrinkage in the use of this technique, careful histological controls can overcome this artifact (Bauchot, this volume; Stephan et al., 1970). Volume, however, is a function of both numbers of neurons and glia, as well as their size, and numbers of processes, capillaries, and extracellular spacing. Controlling for these factors is more difficult. Some attention has been given to defining the constituent determinants of volume (Tower and Young, 1973), but too few regions of the brain have been analyzed to be certain of a constant relationship between volume and neuronal densities. The research presented here circumvents both the problem of shrinkage and uncertainty of the relationship of volume to neuronal density, by comparing estimated numbers of neurons in different thalamic regions. Further insight into the results of these comparisons of neuronal populations is given by data on regional volumes and sizes of the nerve cell bodies.

Not all parts of the thalamus changed at the same rate during hominid evolution. Two of the three sensory relay nuclei appeared unchanged in numbers of neurons when the data of great apes and humans are compared. Volumetric data support the conservatism of LGB somewhat more than that of MGB in that the increase of LGB volume per increment in brain weight is much lower than MGB's rate or those of any of the other thalamic nuclei. Perikaryal sizes also appear similar among these species. Comparison of these data with a larger primate series suggests that the smaller-brained common ancestor of humans and apes could have easily contained an LGB complex with the same volume as is found in modern apes by its encompassing a larger proportion of the thalamus and brain. The theoretical volumetric proportion for the common ancestor of great apes and man is observed in modern monkeys; by extension, the common ancestor could have had as many LGB neurons as found in extant great apes and man.

Most of the human thalamic regions contain more neurons than do their ape counterparts and in some regions the numbers correlate significantly with body and/or brain weights. Like LGB and MGB, the nuclei VB and VL are specific relay nuclei. The latter are both involved with the somatosensory-motor systems. They differ from the former relay nuclei in that the human nuclei contain more neurons than do those of the apes, but at the same time the human nuclei are composed of fewer neurons than the ape data predict for a brain of human size. Another complex, CM, was classified with VB and VL because the human complex contained more neurons than that of apes and no differences in soma sizes were observed. CM differs from VB and VL in that ape data predict the human condition. CM is not primarily concerned with the relay of specific sensory information to the cortex and the predictability of the human numbers of neurons on the basis of brain weight may reflect CM's integrative roles. CM is also a nuclear complex and smaller divisions may possibly be shown to have distinctive alterations in neuronal numbers. The similar distribution of the sizes of the nerve cell bodies in the allometrically expected nuclei among these hominoids suggest that their circuitries have not changed as much as Pu-LP and MD.

The regions, Pu-LP and MD, are both association regions and, like the above areas, they are both nuclear complexes. Interestingly, when these hominoid taxa are grouped on the basis of neuronal somata size, two different clusters are formed. The MD perikarya in humans and great apes manifest similar frequency patterns that differ from that of the gibbons, whereas in Pu-LP humans are separate from the other hominids. It is not clear what other anatomical or physiological variables correlate with differences in cellular architecture, although a reasonable hypothesis would be that changes in connectional patterns would do so. This reasoning suggests that some part of the limbic system changed either among hominoids or in hylobatids after their separation and that some of the cognitive regions changed in hominids after the pongid-hominid lines separated. Given the development of pair territories in gibbons, a rather unusual behavioral repertoire for a primate, one may hypothesize that major shifts would have occurred in their limbic structures. Refined studies on the anatomy and physiology of the limbic system may discern such correlates. On the other hand, differentiation in the association regions of hominids is to be expected in the further development of cognitive functions and symbolism. The evidence suggests a unique bimodal neuronal population in the human Pu-LP, a condition that could have arisen from a selection of small and large cells. A more detailed investigation of the histology of Pu-LP among hominoids is warranted. Whether the human Pu-LP small cells are derived from a different embryological anlage (Rakic and Sidman, 1969) than are the small cells in other primates needs further investigation (Ogren, this volume).

Finally, two nuclei appear divergent on the criterion of numbers of neurons. These two nuclei are limbic nuclei according to their connections with the cingulate gyrus, subiculum, and mamillary bodies. Although volumetric data suggest that these nuclei are the size one would expect for a primate whose brain attained human dimensions, estimates of numbers of neurons do not. The human limbic nuclei, AP and LD, have increased their pool of neurons more than would be estimated from brain size. While doing so they have maintained a profile of cell sizes similar to that of the apes, implying that changes in circuitry may not have been as important as increasing the number of cells. In other words, the way the limbic system in the anterior nucleus operates may be the same among hominoids, but this operation may be of much more importance for humans than apes. The increase in these limbic neurons were speculatively interpreted as supporting part of the human ability to symbolize and have a cultural way of life.

The above data show the human thalamus to be composed of neural substrates that, on the criterion of numbers of neurons, range from being conservative to divergent. The thalamus itself has enlarged, in part as a result of the additions of numbers of neurons in its different nuclei. Without separate analyses of its nuclei, however, the variations in degree of changes could not have been observed. Thus, the thalamus is too large a mass to indicate where important shifts have occurred. Until more quantitative comparative neurobiological data is collected, brain size is also too large an entity to provide information on changes that might be of interest in an analysis of the phylogenetic history of a particular taxon or of a particular behavioral complex or feature such as intelligence.

MacLean's concept of the triune brain (1951, 1973, this volume) also stresses a differential evolutionary advance of neural tissue, but an evolution along three general lines. The thalamic data presented here suggest further refinements of this concept as well. The thalamus is developed from the diencephalon, yet it cannot be relegated to any one of the three brains. The human nuclei range in phylogenetic status from conservative to divergent, and the variation in number of neurons will obviously affect both cortical and other subcortical, including brain stem, sources and targets. The thalamic changes are not merely added to lower structures, but interconnections, whether mono- or polysynaptic, integrate the new with the old. The form the integration takes depends the particular selection pressures in the organism's history.

A better concept for understanding either the evolutionary history of the brain of a particular organism, such as humans, or a particular taxonomic group, such as primates, is mosaic evolution. Mosaicism is a well respected concept of evolutionary studies. Many researchers have noted a differential expansion of parts of the primate brain during evolution (e.g., Le Gros Clark, 1959; Holloway, 1968; Radinsky, 1979), but the explicit application of mosaicism to the brain has been limited and frequently used only when the brain is taken as one unit (e.g., McHenry, 1975). If brain size is the only dimension studied, a less complete and less accurate picture of a given phylogenetic history emerges. But if the concept of mosaicism is understood to apply to the various parts of the brain, a better appreciation of selective forces will be forthcoming.

Acknowledgments: I thank Drs. H. Stephan, N. Moskowitz, C. Noback, and W. Welker for the generous use of their primate brain collections. My thanks also go to R. Elston for statistical advice and W. Welker, L. Radinsky, and D. Falk for valuable comments on an earlier version of this paper. The work in this paper was supported in part by a grant from the Southern Regional Education Board and NIH BRSG 449-89-5148.

REFERENCES

Amaral, D.G. and Cowan W.M., 1980. Subcortical afferents to the hippocampal formation in the monkey. J. Comp. Neurol., 189:573-591.

Andrew, J. and Watkins, E.S., 1969. A Stereotaxic Atlas of the Human Thalamus and Adjacent Structures: a Variability Study, Williams and Wilkins, Baltimore.

Andy, O.J. and Stephan, H., 1976. Septum development in primates. In, The Septal Nuclei, J.F. De France, ed., Adv. in Behav. Bio., Vol. 20, Plenum, New York, pp. 3-36.

Armstrong, E., 1976. A quantitative comparison of the hominoid thalamus. Unpublished Ph. D. dissertation: Columbia University.

Armstrong, E., 1979. A quantitative comparison of the hominoid thalamus: I. Specific sensory relay nuclei. Am. J. Phys. Anthrop. 52:405-419.

Armstrong, E., 1980a. A quantitative comparison of the hominoid thalamus: II. Limbic nuclei anterior principalis and lateralis dorsalis. Am. J. Phys. Anthrop. 52:43-54.

Armstrong, E., 1980b. A quantitative comparison of the hominoid thalamus: III. A motor substrate -- the ventrolateral complex. Am. J. Phys. Anthrop. 52:405-419.

Armstrong, E., 1981. A quantitative comparison of the hominoid thalamus: IV. Posterior association nuclei -- the pulvinar and lateral posterior nucleus. Am. J. Phys. Anthrop., 55:369-383.

Bauchot, R., 1963. L'architectonique comparée, qualitative et quantitative, du diencéphale des insectivores. Mammalia, 27: Suppl. 1, pp. 1-400.

Bauchot, R., 1979. Indices encéphaliques et distances interspecifiques chez les insectivores et les primates. II. Diencéphale et thalamus. Mammalia, 43:407-426.

Blinkov, S.M., and Glezer, I.I., 1968. The Human Brain in Figures and Tables, Plenum Press, New York, p. 227.

Blinkov, S., and Zvorykin, V.P., 1950. Dimensions of the auditory cortex and the medial geniculate body in man and monkeys. Referred to in, The Human Brain in Figures and Tables, S.M. Blinkov and I.I. Glezer, eds., Basil Haigh, trans., Plenum Press, New York, 1968, pp. 225-226 and 410-412.

Bok, S.T., 1959. Histonomy of the Cerebral Cortex, Van Nostrand-Rinehold, Princeton, N.J.

Brady, J.V., 1960. Emotional behavior. In, Handbook of Physiology, Section I, J. Field, ed., Neurophysiology, Vol. III, American Physiological Society, Washington, D.C., pp. 1529-1552.

Chalupa, L.M., 1977. A review of cat and monkey studies implicating the pulvinar in visual function. Behav. Biol., 20:149-167.

Chow, K.L., 1951. Numerical estimates of the auditory central nervous system of the rhesus monkey. J. Comp. Neurol., 95:159-175.

Chow, K.L., Blum, J.S., and Blum, K.A., 1950. Cell ratios in the thalamo-cortical visual system of *Macaca mulatta*. J. Comp. Neurol., 92:227-239.

Curtiss, S., 1977. Genie: A Psycholinguistic Study of a Modern Day "Wild Child," Academic Press, New York.

Elder, J.H., 1934. Auditory acuity of the chimpanzee. J. Comp. Psychol., 17:157-183.

Farrer, D.N., and Prim, M.M., 1965. A preliminary report on auditory frequency threshold comparison of humans and pre-adolescent chimpanzees. Aeromedical Research Laboratory Technical Report, Holloman AFB, New Mexico, pp. 65-66.

Farrer, D.N., and Young, F.A., 1970. Chimpanzee color vision, acuity, and ocular components. In, The Chimpanzee, Vol. 2, G. Bourne, ed., Karger, Basel, pp. 16-25.

Feremutsch, K., 1963. Thalamus. Primatologia, 2:1-226.

Freud, S., 1930. Civilization and Its Discontents, Hogarth Press, London.

Gould, J.C., and Gould, C.G., 1981. The instinct to learn. Science 81, 2:44-50.

Grether, W.F., 1941. Spectral saturation curves for chimpanzees and man. J. Exp. Psychol., 28:419-427.

Grether, W.F., 1942. The magnitude of simultaneous color contrast and simultaneous brightness contrast for chimpanzee and man. J. Exp. Psychol., 30:69-83.

Harting, J.K., Hall, W.C., and Diamond, I.T., 1972. Evolution of the pulvinar. Brain Behav. Evol., 6:424-452.

Hassler, R., 1959. Anatomy of the thalamus. In, Introduction to Stereotaxis with an Atlas to the Human Brain. Grune and Stratton, New York, pp. 230-290.

Haug, H., 1972. Stereological methods in the analysis of neuronal parameters in the central nervous system. J. Micros., 95:165-180.

Heiner, J.R., 1960. A reconstruction of the diencephalic nuclei of the chimpanzee. J. Comp. Neurol., 114:217-238.

Hirsh, R., Davis, R.E., and Holt, L., 1979. Fornix-thalamus fibers motivational states, and contextual retrieval. Exp. Neurol., 65:373-390.

Holloway, R.L., Jr., 1968. The evolution of the primate brain: some aspects of quantitative relations. Brain Res., 7:121-172.

Holloway, R.L., Jr., 1970. Neural parameters, hunting, and the evolution of the human brain. In, The Primate Brain, C.R. Noback and W. Montagna, eds., Appleton-Century-Crofts, New York, pp. 299-310.

Holloway, R.L., Jr., 1975. The Role of Human Social Behavior in the Evolution of the Brain. 43rd James Arthur Lecture, 1973, American Museum of Natural History, New York.

Holloway, R.L., Jr., 1979. Brain size allometry and reorganization: toward a synthesis. In, Development and Evolution of Brain Size, M.E. Hahn, C. Jensen and B.C. Dudek, eds., Academic Press, New York, pp. 61-88.

Hopf, A., 1965. Volumetrische Untersuchungen zur vergleichenden Anatomie des Thalamus. J. f. Hirnforschung, 8:25-38.

Hsu, F.L.K., 1979. The cultural problem of the cultural anthropologist. Am. Anthrop., 81:517-532.

Jerison, H.J., 1973. Evolution of the Brain and Intelligence, Academic Press, New York.

Kanagasuntheram, J. and Wong, W.C., 1968. Nuclei of the diencephalon of Hylobatidae. J. Comp. Neurol., 134:265-286.

Kievit, J., and Kuypers, H.G.J.M., 1977. Organization of the thalamo-cortical connexions to the frontal lobe in the rhesus monkey. Exp. Brain Res., 29:299-322.

Krayniak, P.F., Siegel, A., Meibach, R.C., Fruchtman, D., and Scrimenti, M., 1979. Origin of the fornix system in the squirrel monkey. Brain Res., 160:401-411.

Kuhlenbeck, H., 1954. The Human Diencephalon, Karger, New York.

Kurepina, M.M., 1938. Structure and phylogenetic development of the thalamus in primates. Arkh. Biol. Nasuk., 49:116. In, The Human Brain in Figures and Tables, S.M. Blinkov and I.I. Glezer, eds., B. Haigh, trans., 1968, Plenum Press, New York.

Le Gros Clark, W.E., 1941. The laminar organization and cell content of the lateral geniculate body in the monkey. J. Anat., 75:419-433.

Le Gros Clark, W.E., 1959. The Antecedents of Man, Edinburgh University Press, Edinburgh.

Lovejoy, C.O., 1981. The origin of man. Science, 21:341-350.

MacLean, P.D., 1952. Some psychiatric implications of physiological studies on the frontotemporal portion of the limbic system (visceral brain). Electroencep. Clin. Neurol., 4:407-418.

MacLean, P.D., 1973. A triune concept of the brain and behavior. In, The Hincks Memorial Lectures, T. Boag and D. Campbell, eds., University of Toronto Press, Toronto, pp. 6-66.

Mayer, O., 1912. Mikrometrische Untersuchungen über Zelldichtigkeit der Grosshirnrinde bei den Affen. Jahrb. u. Psychol. Neurol., 17. Reprinted in, The Human Brain in Figures and Tables, S.T. Blinkov and I.I. Glezer, eds., B. Haigh, trans., 1968, Plenum Press, New York, p. 402.

McDonnell, M.F., and Flynn, J.P., 1968. Attack elicited by stimulation of the thalamus and adjacent structures of cats. Behav., 31:185-202.

McGuinness, C.M., and Krauthamer, G.M., 1980. The afferent connections to the centrum medianum of the cat as demonstrated by retrograde transport of horseradish peroxidase. Brain Res., 184:255-269.

McHenry, H., 1975. Fossils and the mosaic nature of human evolution. Science, 190:425-431.

Mesulam, M.-M., and Geschwind, N., 1978. On the possible role of neocortex and its limbic connections in the process of attention and schizophrenia: Clinical cases of inattention in man and experimental anatomy in monkey. J. Psychiat. Res., 14:249-259.

Mesulam, M.-M., Van Hoesen, G.W., Pandya, D.N., and Geschwind, N., 1977. Limbic and sensory connections of the inferior parietal lobule (Area PG) in the rhesus monkey: A study with a new method for horseradish peroxidase histochemistry. Brain Res., 136:393-414.

Mikol, J., Brian, S., Derome, P., De Pommery, J., and Gallissot, M.C., 1977. Connections of latero-dorsal nucleus of the thalamus. II. Experimental study in *Papio papio*. Brain Res., 138:1-16.

Mirsky, A., Rosvold, H.E., and Pribram, K.H., 1957. Effects of cingulectomy on social behavior in monkeys. J. Neurophys., 20:588-601.

Morison, R.S., and Dempsy, E.W., 1942. A study of thalamo-cortical relations. Am. J. Physiol., 135:281-292.

Namba, M., 1958. Uber die feineren Strukturen des Medio-dorsalen Supranucleus und der Lamella Medialis des Thalamus beim Menschen. J. f. Hirnforsch., 4:1-42.

Niimi, K., and Kuwahara, E., 1973. The dorsal thalamus of the cat and comparison with monkey and man. J. f. Hirnforsch., 14:303-325.

Niimi, K., Katayama, K., Karaseki, T., and Morimoto, K., 1960. Studies on the derivation of the centremedian nucleus of Luys. Tokushima J. Exp. Med., 6:261-268.

Ojemann, G.A., 1974. Speech and short-term verbal memory alterations evoked from stimulation in pulvinar. In, The Pulvinar-LP Complex, I.S. Cooper, M. Riklan, and P. Rakic, eds., Charles C Thomas, Springfield, pp. 173-184.

Olszewski, J., 1952. The Thalamus of the *Macaca Mulatta*, Karger, New York.

Papez, J.W., 1937. A proposed mechanism of emotion. Arch. Neurol. Psychiat., 38:725-744.

Passingham, R.E., 1973. Changes in the size and organization of the brain in man and his ancestors. Brain Behav. Evol., 11:73-90.

Passingham, R.E., 1979. Specialization in the language areas. In, Neurobiology of Social Communication in Primates, H.O. Steklis and M.J. Raleigh, eds., Academic Press, New York, pp. 221-256.

Phillips, C.G., and Porter, R., 1977. Corticospinal neurons: Their role in movement. In, Monographs of the Physiol. Soc., Vol. 34, Academic Press, New York.

Radinsky, L.B., 1979. The Fossil Record of Primate Brain Evolution. 49th James Arthur Lecture on the Evolution of the Human Brain, American Museum of Natural History, New York.

Rakic, P., 1974. Embryonic development of the pulvinar-LP complex in man. In, The Pulvinar-LP Complex, I.S. Cooper, M. Riklan, and P. Rakic, eds., Charles C Thomas, Springfield, pp. 3-30.

Rakic, P., and Sidman, R.L., 1969. Telencephalic origin of pulvinar neurons in the fetal human brain. Z. Anat. Entwickl. -Gesch., 129:53-82.

Reynolds, V., 1976. The Biology of Human Action, Freeman, San Franscisco.

Riesen, A.H., 1970. Chimpanzee visual perception. In, The Chimpanzee, Vol. 2, G. Bourne, ed., Karger, Basel, pp. 1-15.

Riss, W., Halpern, M., and Scalia, F., 1969. Anatomical aspects of the evolution of the limbic and olfactory systems and their potential significance for behavior. In, Experimental Approaches to the Study of Emotional Behavior, E. Toback, ed., Ann. N.Y. Acad. Sci., 159:1096-1111.

Rockel, A.J., Nivens, R.W., and Powell, T.P.S., 1980. The basic uniformity in structure of the neocortex. Brain, 103:221-244.

Shariff, G.A., 1953. Cell counts in the primate cerebral cortex. J. Comp. Neurol., 98:381-400.

Singer, M., 1980. Signs of the self: an exploration in semiotic anthropology. Am. Anthrop., 82:485-507.

Spiegel, E.A., Wycis, H.I., Freed, H., Orchinik, C., 1951. The central mechanisms of the emotions. Am. J. Psychiat., 108:426-431.

Stebbins, W.C., 1971. Hearing. In, Behavior of Nonhuman Primates, Vol. 3, A.M. Schrier and Stollnitz, eds., Academic Press, New York, pp. 159-192.

Stephan, H., 1969. Quantitative investigations on visual structure in primate brains. Proc. 2nd Int. Congr. Primat., 3:34-42.

Stephan, H., and Andy, O.J., 1964. Quantitative comparisons of brain structures from insectivores to primates. Am. Zool., 4:59-74.

Stephan, H., and Andy, O.J., 1974. Comparative primate neuroanatomy of structures relating to aggressive behavior. In, Primate Aggression Territoriality and Xenophobia: A Comparative Perspective, R. Holloway, ed., Academic Press, New York, pp. 305-330.

Stephan, H., Bauchot, R., and Andy, O.J., 1970. Data on size of the brain and of various brain parts in insectivores and primates. The Primate Brain, C.R. Noback and W. Montagna, eds., Appleton-Century-Crofts, New York, pp. 289-298.

Sullivan, P.R., Kuten, J., Atkinson, M.S., Angevine, J.B., and Yakovlev, P.I., 1958. Cell count in the lateral geniculate nucleus of man. Neurol., 8:566-567.

Thompson, R., Gates, C.E., and Gross, S.A., 1979. Thalamic regions critical for retention of skilled movements in the rat. Physiol. Psychol., 7:7-21.

Tower, D.B., 1954. Structural and functional organization of the mammalian cerebral cortex. The correlation of neuron density with brain size. Cortical density in the finwhale with a note on the cortical neurone density in the Indian elephant. J. Comp. Neurol., 101:19-53.

Tower, D.B., and Young, O.M., 1973. The activities of butyrylcholinesterase and carbonic anhydrase, the rate of anaerobic glycolysis, and the question of a constant density of glial cells in cerebral cortices of mammalian species from mouse to whale. J. Neurochem., 20:269-278.

Tsubokawa, T., and Moriyasu, N., 1978. Motivational slow negative potential shift (CNV) related to thalamotomy. Appl. Neurophysiol., 41:202-208.

Van Buren, J., and Borke, R., 1972. Variations and Connections of the Human Thalamus, Vol. I, II, Springer-Verlag, New York.

Victor, M., Adams, R.D., and Collins, H.G., 1971. The Wernicke-Korsakoff Syndrome, E. A. Davis Co., Philadelphia.

Vinagradova, O.S., 1975. Functional organization of the limbic system in the process of registration of information: Facts and hypotheses. In, The Hippocampus, II, R.L. Isaacson and K.H. Pribram, eds., Plenum Press, New York, pp. 3-69.

Vogt, B.H., Rosene, D.L., and Pandya, D.N., 1979. Thalamic and cortical afferents differentiate anterior from posterior cingulate cortex in the monkey. Science, 204:205-207.

Vyshinskaya, G.A., 1961. Personal communication. In, The Human Brain in Figures and Tables, S.M. Blinkov and I.I. Glezer, eds., B. Haigh, trans., Plenum Press, New York, p. 44.

Walker, A.E., 1938. The Primate Thalamus, University of Chicago Press, Chicago.

Wilson, W.P., 1975. Sociobiology: The New Synthesis, Belknap Press, Cambridge, Massachusetts.

BRAIN ORGANIZATION AND TAXONOMIC RELATIONSHIPS IN INSECTIVORA AND PRIMATES

Roland Bauchot*

The brain's anatomical criteria are seldom considered when systematicians determine the relationships of different taxa within a classification. This fact is as much because of the difficulties in studying these criteria as because of a strong anatomical stability shown by the brain. The conservatism is so strong that qualitative differences are not found either between species of the same genus or between genera of the same family. The only way to avoid this difficulty is to use quantitative methods that can show size differences that are not visible either macro- or microscopically. The need for a strict protocol and the amount of work necessary before results are obtained limit the numbers of this kind of study. Such problems, however, are not insurmountable.

METHODS

Because accurate averages of body and brain weights were needed for each species studied, the animals were taken as often as possible in the field, and each brain was fixed without delay. The brain was perfused with Bouin's solution, and it was then serially sectioned and stained with a Nissl stain. The volume of the different brain structures were measured by weighing the relevant parts of photographs taken from 50 equidistant sections from the tip of the olfactory bulb to the end of the fourth ventricle (Bauchot, 1963; Stephan et al., 1970). Sixty-three species of Insectivora and Primates were studied.

The volumes of the different brain structures were difficult to compare because the species were of vastly different sizes. They ranged

*From the Laboratory of Comparative Neuroanatomy, University of Paris 7, Paris Cedex 05 France

TABLE 1. Encephalic Indices

Genus and Species	m.	MO	Ce	Me	Et	Sb	Ht	Re	Pt	An	Ml	Mn	Do	Ve	GL	GM	BO	Pa	Se	St	Sc	Hi	Ne
01 *Sorex minutus*	46	83	20	44	65	26	67	41	15	24	15	52	4	37	18	44	133	99	72	28	53	67	5
02 *Sorex araneus*	53	83	21	50	84	32	76	46	18	33	21	60	7	40	17	48	139	121	88	31	69	85	7
03 *Crocidura russula*	49	71	24	53	79	29	66	47	23	23	18	45	6	39	25	46	154	107	80	24	56	63	6
04 *C. occidentalis*	57	98	30	68	76	27	61	53	21	32	20	53	8	41	23	49	192	131	83	26	66	82	7
05 *Suncus murinus*	42	71	23	51	54	20	56	40	17	18	14	33	5	34	18	40	151	93	57	20	53	57	6
06 *Echinops telfairi*	40	78	16	53	54	16	45	38	17	17	10	32	3	37	22	24	157	107	62	18	22	50	3
07 *Hemicentetes s.*	43	71	28	45	56	17	49	33	14	19	11	33	3	36	19	24	193	115	62	19	35	55	4
08 *Setifer setosus*	51	71	20	44	74	20	62	35	15	19	15	41	4	41	21	27	253	154	74	21	49	49	4
09 *Tenrec ecaudatus*	35	54	21	36	48	17	44	28	13	10	10	26	3	33	17	22	159	92	49	17	37	33	4
10 *Erinaceus europaeus*	43	49	24	36	62	22	64	31	18	16	17	36	6	35	23	23	185	101	58	22	62	49	7
11 *Aethechinus algirus*	53	77	29	46	74	35	81	47	23	30	28	57	9	48	25	32	180	115	60	31	66	68	8
12 *Solenodon paradoxus*	54	81	49	51	73	34	49	40	15	24	18	72	7	41	16	27	245	132	35	37	70	65	7
13 *Nesogale talazaci*	68	94	37	79	88	28	62	52	23	53	20	63	8	39	25	46	257	181	91	33	69	143	8
14 *Limnogale mergulus*	61	152	48	76	68	37	77	72	21	126	24	72	10	36	17	107	100	76	107	42	82	76	11
15 *Potamogale velox*	54	151	44	76	48	33	67	78	21	28	29	48	15	39	16	115	55	45	76	26	87	78	16
16 *Neomys fodiens*	61	103	35	64	92	47	73	61	21	34	22	73	9	34	17	101	121	103	115	44	66	105	10
17 *Talpa europaea*	64	81	54	63	92	56	83	58	18	35	28	91	11	38	12	71	157	111	93	49	101	96	11
18 *Galemys pyrenaicus*	92	149	68	95	125	87	117	96	30	59	55	151	28	43	22	127	124	127	150	73	137	144	26
19 *Desmana moschata*	72	106	50	87	86	80	99	82	31	33	39	103	19	38	22	114	117	100	83	57	130	93	20
20 *Chlorotalpa stuhl.*	75	88	35	82	121	53	74	63	21	38	28	106	13	40	13	74	245	199	108	54	83	100	13
21 *Elephantulus fus.*	120	122	68	180	138	73	148	135	273	142	25	44	8	43	218	118	203	145	126	44	144	228	18
22 *Rhynchocyon st.*	135	125	72	195	175	85	171	140	287	125	31	57	12	47	225	129	327	204	148	53	164	188	20
23 *Tupaia glis*	131	124	80	212	167	79	126	187	172	129	133	107	46	93	117	254	214	162	128	69	137	97	41
24 *Urogale everetti*	118	128	68	167	152	79	106	163	115	111	118	110	42	95	72	261	209	154	120	69	135	82	36
25 *Microcebus murinus*	109	87	87	112	125	82	115	103	123	154	96	125	59	78	140	120	133	117	121	96	126	138	61
26 *Cheirogaleus med.*	96	99	71	105	112	68	84	106	95	142	82	102	50	90	90	139	149	103	112	68	87	111	45
27 *C. major*	98	96	89	88	105	80	100	89	92	126	87	110	55	79	117	121	126	97	123	84	113	113	58
28 *Lepilemur rufic.*	71	71	66	70	68	60	80	77	63	93	67	81	39	81	53	125	67	55	69	61	98	81	40
29 *Hapalemur simus*	67	63	67	61	55	65	63	75	68	105	63	76	42	84	56	129	32	34	59	55	84	89	42
30 *Lemur fulvus*	125	99	153	100	117	145	95	114	114	154	146	215	133	96	109	148	88	93	110	137	161	114	115
31 *L. variegatus*	121	106	117	115	114	138	120	126	115	186	143	182	100	98	97	150	87	86	113	114	137	137	86
32 *Avahi laniger*	95	94	88	89	87	104	98	111	103	149	107	119	67	99	88	154	43	51	98	73	106	106	56
33 *Propithecus ver.*	87	81	97	76	70	100	74	91	91	117	93	137	74	91	88	140	35	45	80	93	81	92	68
34 *Indri indri*	85	63	91	67	76	108	82	85	95	114	96	112	73	89	86	137	20	45	75	78	112	92	69
35 *Daubentonia mad.*	172	118	184	106	177	233	153	137	112	266	215	357	163	105	86	162	168	141	146	208	246	181	130
36 *Loris gracilis*	112	76	86	108	131	117	123	119	145	110	144	174	100	95	124	147	89	98	98	110	98	81	86

37 *Nycticebus coucang*	124	94	85	93	143	140	129	117	128	[illegible]	146	211	98	98	126	151	91	107	121	131	114	132	83
38 *Perodicticus potto*	105	96	87	83	122	91	92	95	75	[illegible]	109	148	79	94	65	144	136	109	118	96	139	111	70
39 *Galago demidovii*	158	137	122	165	179	108	149	165	183	[illegible]	175	221	117	100	191	155	212	172	153	132	146	161	95
40 *G. senegalensis*	132	119	114	144	125	98	137	154	161	[illegible]	152	161	94	103	145	158	119	89	123	94	151	160	75
41 *G. crassicaudatus*	91	82	85	90	95	79	89	95	84	[illegible]	96	120	67	88	77	135	96	82	89	83	92	92	63
42 *Tarsius syrichta*	151	142	118	198	123	102	152	203	264	[illegible]	203	178	144	108	269	166	43	86	118	99	149	139	109
43 *Callithrix jacchus*	130	125	103	167	145	120	155	158	185	[illegible]	156	70	134	185	191	93	33	99	137	120	96	117	123
44 *Saguinus oedipus*	122	97	98	143	138	120	166	140	186	[illegible]	157	63	133	177	188	89	21	93	141	113	84	95	127
45 *S. tamarin*	130	136	110	151	96	134	149	157	192	[illegible]	167	75	158	189	228	95	16	75	101	126	107	114	141
46 *Aotus trivirgatus*	112	110	104	110	105	116	121	119	113	[illegible]	134	74	130	183	109	92	32	79	117	129	132	107	120
47 *Callicebus moloch*	133	140	96	145	112	132	142	159	165	[illegible]	174	91	166	205	167	103	11	102	98	142	131	129	138
48 *Pithecia monacha*	142	121	170	139	116	177	138	147	163	[illegible]	186	110	210	191	163	96	14	97	107	186	101	131	188
49 *Alouatta seniculus*	79	83	105	71	54	97	84	81	72	[illegible]	89	56	95	161	71	81	6	50	51	112	58	78	101
50 *Cebus sp.*	170	151	256	143	132	248	122	159	178	[illegible]	223	159	332	197	209	99	13	114	128	260	108	96	290
51 *C. albifrons*	170	141	228	167	95	269	133	175	212	[illegible]	235	159	339	196	227	98	9	78	88	262	90	100	297
52 *Saimiri sciureus*	142	131	150	144	116	175	152	147	185	[illegible]	197	100	226	181	230	91	16	96	121	178	92	82	212
53 *Ateles geoffroyi*	110	72	177	87	94	165	91	88	104	[illegible]	126	87	183	152	109	76	11	75	93	186	86	70	208
54 *Lagothrix lagotr.*	139	90	227	111	107	209	132	116	141	[illegible]	184	119	247	166	145	83	13	92	92	233	93	94	257
55 *Macaca mulatta*	115	94	154	98	83	144	95	99	117	[illegible]	138	88	204	156	143	78	12	68	94	184	91	83	225
56 *Cercocebus albig.*	114	107	154	105	83	135	103	108	117	[illegible]	146	81	197	168	137	84	15	76	85	157	75	76	203
57 *Cercopithecus tal.*	197	176	173	178	152	283	177	197	263	[illegible]	280	171	352	201	346	101	11	133	137	273	110	138	298
58 *C. ascanius*	129	105	136	112	110	157	111	116	148	[illegible]	157	87	213	168	185	84	20	88	118	167	134	100	218
59 *C. mitis*	103	91	112	92	82	118	92	101	117	[illegible]	128	62	167	170	132	85	16	67	82	119	84	81	170
60 *Colobus badius*	97	86	134	86	69	110	77	87	91	[illegible]	105	62	146	155	103	78	7	50	90	130	105	93	163
61 *Pan troglodytes*	101	74	203	72	58	134	77	73	74	[illegible]	121	63	219	137	83	69	10	41	80	151	78	63	280
62 *Gorilla gorilla*	61	44	149	39	42	76	46	42	39	[illegible]	61	31	105	122	40	61	6	34	51	83	49	37	152
63 *Homo sapiens*	185	94	491	119	135	229	74	101	87	[illegible]	165	97	500	119	74	60	3	105	188	270	181	132	740
	m.	100	100	100	100	100	100	100	100	[illegible]	100	100	100	100	100	100	100	100	100	100	100	100	100
	σ	28	76	45	35	65	36	46	74	[illegible]	70	59	104	57	78	50	84	38	30	69	39	37	119

Abbreviations: An, Anterior nuclei; BO, Olfactory bulb; Ce, Cerebellum; Di, Diencephalon; Do, Lateral nuclei; Et, Epithalamus; GL, Lateral geniculatum; GM, Medial geniculatum; Hi, Hippocampus; Ht, Hypothalamus; Me, Mesencephalon; Ml, Dorsomedial nucleus; Mn, Midline nuclei; MO, Medulla oblongata; Ne, Neocortex; Pa, Palaeocortex + amygdala; Pt, Pretectum; Re, Nucleus reticularis; Sb, Subthalamus; Sc, Schizocortex; Se, Septum; St, Striatum; Te, Telencephalon; Th, Thalamus; Ve, Ventral nuclei.

from the pygmy shrew, *Sorex minutus*, with a body weight of 5.3 grams, to the gorilla, *Gorilla gorilla*, with body weight of 125,000 grams. The volumes of the brain and of its parts, however, are linked to the body weight by an allometric relationship, the coefficient of which is about 2/3. Therefore, for each structure and each species an isoponderal index, I, which is the encephalization index (or the index for the mesencephalon, for the anterior nucleus, etc.), was determined by the two formulae:

$$I = 100\, k/k_0 \quad \text{and} \quad k = E/S^{\alpha}$$

in which S is the body (somatic) weight; E the brain (encephalic) weight or the volume of any brain structure; a the coefficient of allometry, taken here as the value 2/3 (0.667); k the coefficient of encephalization and k_0 the same for the centroid or the point with coordinates $\bar{x}$ and $\bar{y}$ of the 63 species. The isoponderal index was so calculated that the 63 indices for each structure were automatically weighted. Table 1 gives the indices for the 22 structures studied. The telencephalic, diencephalic, encephalic, and thalamic indices were not used in Table 1 because they are represented by the other 22 indices. The coefficient of correlation (r) and the coefficient of allometry (α) link the volumes of all 25 structures to the body and the brain weights (Table 2).

The indices can be used directly to calculate the interspecific distances. I determined the Euclidean distance (d) between two species i and j by the formula:

$$d = (^1/N\Sigma(I_i-I_j)^2)^{1/2}$$

The dendrograms were based on the 22 indices included in Table 1. To avoid redundancy, I did not use the telencephalic, diencephalic and thalamic indices.

The calculation of the distance between each of the 63 species yielded a distance matrix that could not be reproduced here because of its size. It was used, however, to construct a dendrogram, in which the shortest distance between any two species is determined, then, with those two considered as a single figure, the next shortest distance between taxa is calculated, and so on. Many methods for constructing these dendrograms are known; the best one for use in this study was that of Ward (1963). The results of the dendrogram are given in Figures 1, 2, and 3.

In 1970, Stephan, Bauchot, and Andy gave the volumes of the five encephalic structures (medulla oblongata MO, cerebellum Ce, mesencephalon Me, diencephalon Di, and telencephalon Te), and those of seven telencephalic subdivisions (olfactory bulb BO, paleocortex + amygdala Pa, septum Se, striatum St, schizocortex Sc, hippocampus Hi, and neocortex Ne). I recently added (Bauchot, 1979-1980) the volumes of six diencephalic structures (epithalamus Et, thalamus Th, subthalamus Sb, hypothalamus Ht, nucleus reticularis thalami Re, and pretectal region Pt) as well as those of seven thalamic subdivisions (anterior nuclei An, dorsomedial nucleus Ml,

TABLE 2. Multiple Correlations Between Encephalic Structures[a,b]

(1-r) .10³

	rS	αS	rE	αE	GM	GL	Ve	Do	Mn	Ml	Mn	Pt	Re	Ht	Sb	Th	Et	Se	Te	Ne	Hi	Sc	St	Pa	BO	Di	Me	Ce	MO
GM	.926	.79	.942	.83		71	59	61	42	43	38	53	29	42	45	41	47	52	64	59	44	33	55	104	526	40	38	46	53
GL	.897	.95	.961	1.00	71		39	37	81	31	27	5	28	37	39	26	53	65	37	32	58	69	42	114	676	28	30	46	68
Ve	.966	.88	.988	.93	59	39		25	52	22	44	43	19	17	19	12	29	29	13	22	37	43	15	69	648	12	23	18	25
Do	.908	1.24	.983	1.30	61	37	25		43	4	28	34	23	37	16	7	47	50	14	2	52	53	12	117	734	12	34	23	55
Mn	.938	.75	.962	.79	42	81	52	43		30	39	70	32	35	24	31	28	34	42	46	32	26	26	73	517	28	43	29	43
Ml	.920	.99	.981	1.04	43	31	22	4	30		18	26	14	26	13	5	34	42	18	8	43	40	13	101	668	9	25	22	46
An	.902	.86	.961	.91	38	27	44	28	39	18		17	20	36	28	19	40	51	39	29	32	40	33	150	612	22	30	36	62
Pt	.895	.94	.961	.99	.53	5	43	34	70	26	17		19	33	33	22	46	60	39	30	50	57	40	115	658	24	23	42	66
Re	.950	.76	.986	.80	29	28	19	23	32	14	20	19		9	0	7	17	21	18	19	19	20	15	68	600	5	3	13	18
Ht	.969	.70	.985	.73	42	37	17	37	35	26	36	33	9		1	15	10	14	18	30	16	17	16	46	562	9	10	14	15
Sb	.950	.92	.993	.97	45	39	19	16	24	13	28	33	10	11		6	16	20	9	12	20	22	3	67	632	2	15	8	25
Th	.943	.95	.994	1.00	41	26	12	7	31	5	19	22	7	15	6		22	25	6	5	27	29	5	79	656	1	13	9	28
Et	.968	.67	.982	.70	47	53	29	47	28	34	40	46	17	10	6	22		21	22	39	15	13	20	25	503	15	14	16	21
Se	.977	.66	.984	.70	52	65	29	50	34	42	51	60	21	14	20	25	21		21	41	13	13	21	31	520	18	18	12	13
Te	.951	1.00	.999	1.06	64	37	13	14	42	18	39	39	18	18	9	6	22	21		8	29	35	4	63	668	6	18	7	27
Ne	.920	1.23	.990	1.29	59	32	22	2	46	8	29	30	19	30	12	5	39	41	8		43	46	9	106	725	9	27	16	47
Hi	.965	.66	.976	.70	44	58	37	52	32	43	32	50	19	16	20	27	15	13	29	43		10	26	49	527	20	19	17	20
Sc	.963	.72	.971	.76	33	69	43	53	26	40	40	57	20	17	22	29	13	13	35	46	10		30	50	500	22	23	17	23
St	.950	.93	.996	.98	55	42	15	12	26	13	33	40	15	16	3	5	20	21	4	9	26	30		66	648	4	20	7	27
Pa	.962	.57	.943	.60	108	114	69	117	73	101	150	115	68	46	57	79	25	31	63	106	49	50	66		424	66	54	56	43
BO	.509	.46	.352	.49	526	676	648	734	517	668	612	658	600	562	632	656	503	520	668	725	527	500	648	424		632	585	607	556
Di	.952	.89	.995	.93	40	28	12	12	28	9	22	24	5	9	2	1	15	18	6	9	20	22	4	66	632		10	6	22
Me	.953	.71	.986	.75	38	30	23	34	43	25	30	23	3	10	15	13	14	18	18	27	19	23	20	54	585	10		14	16
Ce	.965	.94	.996	.99	46	46	18	23	29	22	36	42	13	14	8	9	16	12	7	16	17	17	7	56	607	6	14		18
MO	.979	.65	.980	.69	53	68	25	55	43	46	62	66	18	15	25	28	21	13	27	47	20	23	27	43	556	22	16	18	
m	.927	.83	.954	.88	70	72	56	63	60	55	62	67	44	45	46	46	46	50	51	59	51	51	48	89	599	43	46	46	56
σ	.091	.19	.126	.20	98	131	127	145	98	132	120	128	119	111	126	131	98	101	132	144	102	97	129	78	78	126	115	120	108

[a]The matrix is composed of values representing $(1 - r) \times 10^3$, so that higher the correlation, the lower number. The correlations are derived from 63 couples of values.

[b]rS, coefficient of correlation in reference to body weight; αS, correlation of allometry in reference to body weight; rE, coefficient of correlation in reference to brain weight; αE, correlation of allometry in reference to brain weight. Other abbreviations are given in Table 1.

midline and intralaminar nuclei Mn, dorsal nuclei Do, ventral nuclei Ve, lateral geniculatum GL, and medial geniculatum GM). These data and encephalic indices for each of the 63 species of Insectivora and Primates are listed in Table 1.

FUNCTIONAL CORRELATIONS

The various encephalic volumes of the 63 species had a total of 300 calculable correlations. Table 2 lists the correlations between the volumes, as well as the correlations of these 25 structures with the brain and body weights. For the latter two correlations I also give both the coefficient of allometry (α) and the coefficient of correlation (r). In the correlation matrix of the 25 structures, I give the value $(1-r) \times 10^3$ rather than the value of the coefficient of correlation p. Thus the matrix is more easily read; the lower the number, the better the correlation. With the exception of the olfactory bulb, all the structures correlated well with brain and body weights.

What can we infer from an excellent correlation between two structures? Do not forget that these correlations are calculated from 63 couples of values, belonging to animals vastly different in morphology and behavior. A good correlation can hardly be due to chance, yet it cannot automatically be assumed to mean functional connections. A good correlation therefore should be regarded as a hypothesis that must be verified experimentally. Because the coefficients of correlations are often good, most of the discussion will be limited to the best ones.

The medial geniculate body is without good correlations. The lateral geniculate body has a good correlation (r = 0.995) with the pretectal area because both are linked to visual inputs. The ventral thalamic nuclei are linked to the telencephalon (r = 0.987), especially to the striatum (r = 0.985). The dorsal thalamic nuclei are closely linked (r = 0.998) to the neocortex, a correlation that was already indicated in insectivores (Bauchot, 1966). This excellent correlation may indicate that the dorsal nuclei project to the whole neocortex.

The midline and intralaminar nuclei are linked to the subthalamus (r = 0.976) and to the striatum (r = 0.974). Connections are well known among these structures. The dorsomedial complex has stronger correlations to the dorsal nuclei (r = 0.996) and neocortex (r = 0.992), than to olfactory centers. The complex seems to be more "advanced" than was first supposed from studies of the insectivores (Bauchot, 1963). Interestingly the anterior thalamic nuclei do not have particularly good correlations with the structures with which they are related (e.g., hippocampus, hypothalamus). The thalamic reticular nucleus has an excellent correlation (r = 0.997) with the mesencephalon, which is to be expected because of its connections, but is also surprising that the correlation is so strong.

The hypothalamus, subthalamus, epithalamus, and septum have no special relations. The hippocampus is linked to the schizocortex (r = 0.990). The paleocortex has relatively low coefficients with every other structure, except the epithalamus (r = 0.975), septum (r = 0.969), and medulla oblongata (r = 0.957). Of these, the first two structures belong to the general olfactory sphere. The olfactory bulb is poorly correlated with other structures because of the strong differences between the macrosmatic insectivores and the microsmatic apes and man. The only correlations are with centers that are more or less olfactory: paleocortex (r = 0.576), epithalamus (r = 0.497), and septum (r = 0.480).

Because the diencephalon is a good relay center, it is linked to almost every other structure. The best correlations, however, are with the striatum (r = 0.996) and the cerebellum (r = 0.994). The cerebellum, in addition to its good correlation to the diencephalon, is related to the striatum (r = 0.993) and to the subthalamus (r = 0.992), that is, to the more important extrapyramidal centers.

Most of the results shown in Table 2 confirm connections already known. The value of this method will probably be shown in its extension to measurements and analyses of interrelationships among structures of less well known taxa than mammalian ones, such as those of the lower vertebrates--fish, frogs, and snakes.

DENDROGRAMS

Whereas the matrix of correlations indicate functional relations between brain structures, the matrix of interspecific distances figured by the dendrograms may clarify the relations between species, genera, families, or orders inside the big Insectivora-Primate phylum. The systematic positions of some of the groups in this work are problematic. Tree shrews, for example, are considered insectivores by some authors, and prosimians by others. Similar problems concern the positions of Macroscelididae (elephant-shrews) and Tarsiiformes. Our encephalic criteria will generate more arguments about these classifications.

We constructed dendrograms from the distance matrix. Because the entire dendrogram gives the relationships of all 63 species, it is difficult to present directly. Consequently, I give first, in Figure 1, the dendrogram of the major groupings, designated A to H. Each of these major groupings is then given with every species in Figures 2 (A to D) and 3 (E to H). In these figures, the different species are numbered 01 to 63 as specified in Table 1.

As seen in Figure 1, all Insectivora (1-20) are closely associated (groups A and B). The next grouping (C-D-E) is that of Prosimii and Menotyphla, and the last grouping (F-G-H) is that of Anthropoidea. Thus the major groupings of these species, based on their encephalic indices, coincide with their taxonomic affinities.

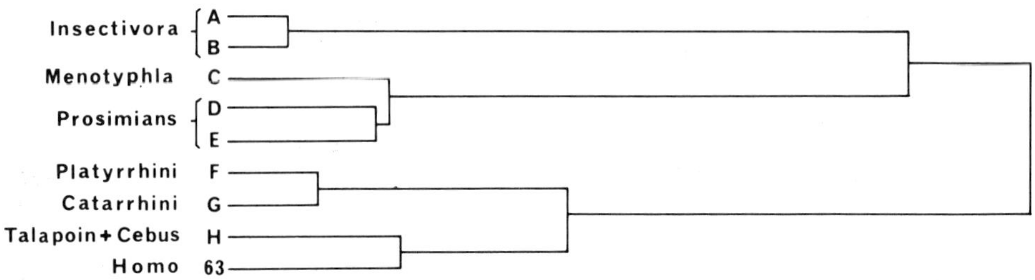

Fig. 1. Detail of the different systematic groups of Insectivora and Primates.

There are some interesting, if ambiguous, differences from the typical taxonomy. The C grouping, which we call "Menotyphla," consists of both the Macroscelididae and Tupaiiformes. Interestingly, although the families are distantly related, and have a different mode of life so that convergence is not possible, the encephalic criteria of these four species are closer to each other than to any other species. Also, the Macroscelididae, which are taken for insectivores by most authors, are more closely related to prosimians and Tupaiiformes than to other insectivores. The nearness of the Macroscelididae and Tupaiiformes favors the bringing together of these two taxonomic groups, and this group in turn appears as close to prosimians as to insectivores. For this reason, I returned to the old distinction of Haeckel (1866), who divided the Insectivora into Lipotyphla (Insectivora *sensu stricto*) and Menotyphla or Macroscelididae. The results suggest that this group of Menotyphla could incorporate, at least according to encephalic criteria, the Tupaiiformes. In addition, the Tupaiiformes, whose taxonomic position is debated, have encephalic indices that are closer to prosimians than to insectivores, thereby supporting the theory of those authors who consider Tupaiiformes to be prosimians, despite their shrew-like appearance.

Also in Figure 1, the E grouping of prosimians contains the Tarsiiforme species. In this analysis *Tarsius* is not intermediate between prosimians and anthropoids.

The last gouping, that of Anthropoidea, includes New World monkeys (Platyrrhini, grouping F) and Old World monkeys (Catarrhini, grouping G). They are more similar to each other than to prosimians. The two apes studied (*Gorilla*, 61, and *Pan*, 62) are nearer monkeys than humans. The human species (*Homo*, 63) although greatly separated from any other species has its closest associations with three curious species: two New World monkeys (genus *Cebus*), and the Old World talapoin monkey, *Cercopithecus talapoin*. Because these connections are slight, however, I give no great importance to them.

The distinction between the species grouped in A and B (Fig. 2) is more a matter of adaptation than systematics. The B grouping consists of aquatically adapted species plus the mole (17). The two aquatic Tenrecidae are similar (*Limnogale mergulus*, 14, and *Potamogale velox*, 15), whereas the pyrenean desman (*Galemys pyrenaicus*, 18) is far different from the other B species (see Bauchot and Stephan, 1959).

The subgroups of the A grouping are interesting in that here one can see the similarity of data on the four best-encephalized nonaquatically adapted insectivores (*Setifer setosus*, 8, *Solenodon paradoxus*, 12, *Nesogale talzaci*, 13, and *Chlorotalpa stuhlmanni*, 20). This observation suggests that the progression of encephalization influences the different parts of the brain similarly.

In the C grouping (Fig. 2), the two Macroscelididae on one hand and the two Tupaiiformes on the other are clearly related by encephalic criteria, but the two families are far from being closely associated. The data suggest that the two families should be kept separate, but that as a group they

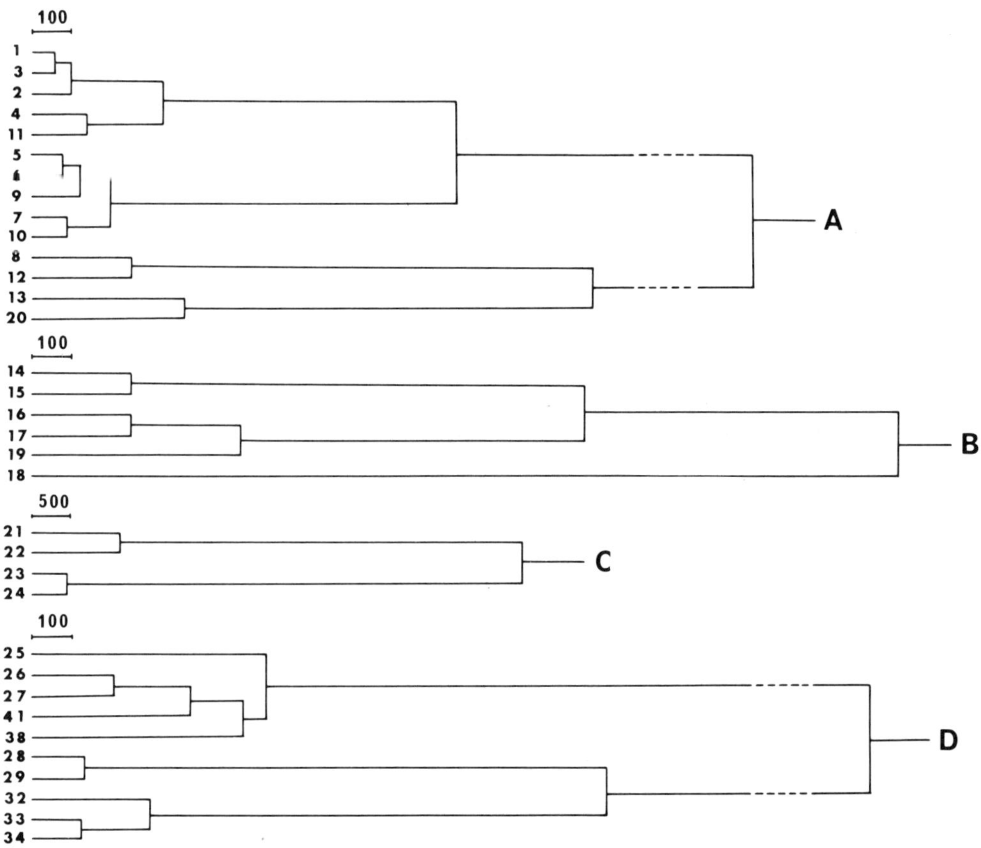

Fig. 2. Detailed dendrogram of Insectivora (A and B), Menotyphla (C) and part of Prosimii (D). Each number refers to a species (see Table 1).

should either both be brought into prosimians or both kept in a grouping distinct from either Insectivora or Prosimii.

The first prosimian grouping D (Fig. 2) has several good associations. Two species (*Lepilemur ruficaudatus*, 28, and *Hapalemur simus*, 29) that we had isolated long ago into a subfamily Lepilemurini solely on the consideration of their encephalization (Bauchot and Stephan, 1969) continue to show close associations. Another set of close associations is among Indridae (32 to 34). The third grouping is that of the two *Cheirogaleus* species plus *Microcebus* and two of the six Lorisidae (*Perodicticus potto*, 38, and *Galago crassicaudatus*, 41). Other criteria also bring cheirogalids and lorises together (Tattersall and Schwartz, 1975).

In the second prosimian grouping (Fig. 3), species in the *Lemur* genus (30-31) are connected with two other lorisids (*Nycticebus coucang*, 37, and *Loris gracilis*, 36), whereas two *Galago* species (39, 40) are linked, but loosely, to the Tarsiiforms. The position of the aye-aye (*Daubentonia madagascariensis*, 35) is far from any other prosimians, its encephalization being much higher.

In summary, apart from the good groupings of Lepilemurini with Indridae and the remoteness of the aye-aye, the relationships among prosimians as determined by encephalic indices are more likely to reflect their overall level of encephalization than their taxonomy. This result is similar to our grouping of the aquatic-adapted Insectivora suggesting that brain organization reflects different ecological adaptations among closely related animals.

Finally, the two groupings of anthropoid species, F and G, (Fig. 3) are not as satisfactory as the preceding ones. Group F is preponderantly platyrrhine, although *Cercopithecus ascanius* (58) is included. Similarly, although group G is preponderantly catarrhine and ape, it also includes *Aotus trivirgatus* (46), *Alouatta seniculus* (49) and the spider monkey (*Ateles geoffroyi*, 53). These associations, plus that connecting the human with two *Cebus* species and the talapoin monkey, reflect the relative narrowness or closeness of relationships among all anthropoids.

The distances given in Figure 1 are a little inaccurate in that distances between species with high indices are likely to be larger than those between species of low brain weights. If we take this fact into account (by dividing the main distances by the average index in the first column of Table 1), we get the following results:

Group Intervals	uncorrected distances	corrected distance
A - B	9963	17903
D - E	23895	21515
F - G	14855	12890
H - 63	27627	15306
C - DE	26408	23210
FG - H63	54850	42962

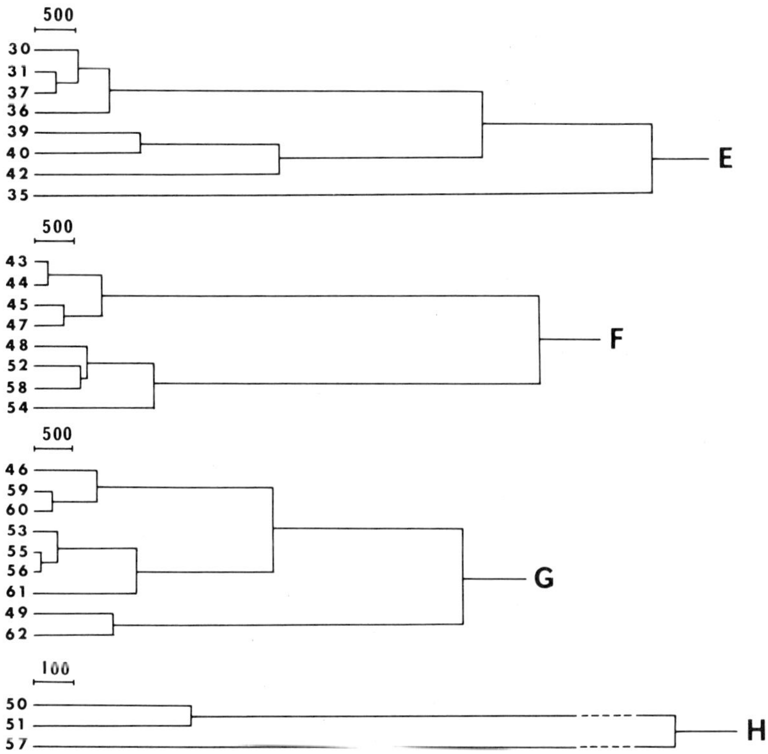

Fig. 3. Detailed dendrogram of part of Prosimii (E), and Platyrrhini (F), Catarrhini (G) and other simian species (H). Each number refers to a species (see Table 1).

The differences are not significant, and the corrected distance between anthropoid groupings (F-G and H-63) is less than the insectivoran (A-B) or prosimian (D-E) ones.

CONCLUSION

The correlations among each of the measured volumes of 25 brain structures in 63 species of Insectivora and Primates are given in the matrix of Table 2. The best correlations agree with our present knowledge of the brain connections, the most interesting ones being those of the dorsal thalamic nuclei and medial (=dorsomedian) nucleus with the neocortex.

From the calculated indices of these 22 structures, we determined the distance between each of the 63 species and converted the corresponding matrices into dendrograms (Figs. 1 through 3). Systematicians should consider the results of quantitative shifts in brain volumes on the same level with criteria taken from dentition, skull or biochemistry. The major associations conform to the established taxonomic divisions. The grouping of aquatic-adapted Insectivora suggests, however, that within these subdivisions, biological adaptations may also influence the brain organization. Indeed inside each of the three main groupings (Insectivora, Prosimii, and Anthropoidea), the relationships do not always agree with the taxonomy, either because of different adaptations or because of the general level of encephalization. The dendrograms support other evidence that positions *Tarsius* inside the prosimian group, and suggest that elephant shrews (Macroscelididae) and tree shrews (Tupaiidae) should be removed from Insectivores and placed in a menotyphlan group. The latter is encephalically closer to prosimian than to insectivoran orders.

REFERENCES

Bauchot, R., 1963. L'architectonique comparée, qualitative et quantitative, du diencéphale des Insectivores. Mammalia, 27 (Suppl. 1): 1-400.

Bauchot, R., 1966. Le développement phylogénétique du thalamus des Insectivores, In, Evolution of the Forebrain, R. Hassler and H. Stephan, eds., pp. 346-355.

Bauchot, R., 1979a. Indices encéphaliques et distances interspécifiques chez les Insectivores et les Primates. I. Encéphale et télencéphale. Mammalia, 43: 173-189.

Bauchot, R., 1979b. Indices encéphaliques et distances interspécifiques chez les Insectivores et les Primates. II. Diencéphale et thalamus. Mammalia, 43:407-426.

Bauchot, R., and Stephan, H., 1959. Le cerveau de *Galemys pyrenaicus* Geoffroy 1811 (Insectivora Talpidae), et ses modifications dans l'adaptation à la vie aquatique. Mammalia, 23:1-18.

Bauchot, R., and Stephan, H., 1966. Données nouvelles sur l'encéphalisation des Insectivores et des Prosimiens. Mammalia, 30:160-196.

Bauchot, R., and Stephan, H., 1968. Etude des modification encéphaliques observées chez les Insectivores adaptés à la recherche de nourriture en milieu aquatique. Mammalia, 32:228-275.

Bauchot, R., and Stephan, H., 1969. Encéphalisation et niveau évolutif chez les Simiens. Mammalia, 33:225-275.

Bauchot, R., and Stephan, H., 1970. Morphologie comparée de l'encéphale des Insectivores Tenrecidae. Mammalia, 34:514-541.

Haeckel, E., 1866. Generelle Morphologie der Organismen, Berlin.

Stephan, H., and Bauchot, R., 1968. Vergleichende Volumenuntersuchungen an Gehirnen europäischer Maulwürfe (Talpidae). J. f. Hirnforsch., 10:247-258.

Stephan, H., Bauchot, R., and Andy, O.J., 1970. Data on size of the brain and of various parts in Insectivores and Primates. In, The Primate Brain. Advances in Primatology, 1, pp. 289-297.

Tattersall, I., and Schwartz, J.H., 1975. Relationships among the Malagasy Lemurs: the craniodental evidence. In, Phylogeny of the Primates.

A Multidisciplinary Approach, W.P. Luckett and F.S. Szalay, eds., pp. 299-312.

Ward, J. H., 1963. Hierarchial grouping to optimize an objective function. S. Am. Stat. Assoc., 58: 236.

QUANTITATIVE CYTOARCHITECTONICS OF THE CEREBRAL CORTICES OF SEVERAL PROSIMIAN SPECIES

Karl Zilles*
Heinz Stephan
Axel Schleicher

CYTOARCHITECTONICS OF THE ISOCORTEX

The classical architectonic studies represented by the works of Meynert (1872), Betz (1881), Campbell (1905), Elliot Smith (1904, 1907), Vogt (1904, 1906, 1910), Vogt and Vogt (1907), Brodmann (1908, 1909), Economo and Koskinas (1925) and many others reached a "golden age" in the first decades of our century. These studies were based mainly on qualitative descriptions of structural forms, nerve cell morphology, and myelinization in different regions of the cortex. The investigations were mostly done on histological sections stained with Nissl-methods (*cytoarchitectonic studies*) and with methods demonstrating myelin (*myeloarchitectonic studies*). Important results were obtained, when (1) brains of many different mammalian species were compared (e.g., Brodmann, 1909) or (2) data from morphological and physiological studies were compared (e.g., Vogt and Vogt, 1907). Unfortunately these approaches were not followed by most later researchers, with the result that subjective and nonreproducible descriptions of minute morphological differences in the laminae or in cell structure increasingly prevailed. The discussions (1) about "haarscharfe Grenzen" (= boundaries fine as a hair) between brain regions, and (2) about the exact number of laminae in a given region of the cortex finally left the realm of science and were not even semantic problems. The consequence of this "development" was severe criticism of cytoarchitectonic studies (Lashley

---

*From the Anatomical Institute (KZ and AS), University Kiel, Kiel, FRG and the Max-Planck-Institute for Brain Research (HSt), Frankfurt/M., FRG.

Dedicated to Professor Herbert Haug, Lubeck, on the occasion of his 60th birthday.

and Clark, 1946; Bailey and v. Bonin, 1951). Since that time workers using cytoarchitectonic methods have had to overcome a wall of more or less justified suspicion. A general suspicion about this type of brain research, however, is not justified, as the results of Hassler on the substantia nigra (1937) and the thalamic nuclei (1959) and the work of Stephan on the allocortex (1975) demonstrate. In these and other cases, the important traditions of classical investigations are apparent--the significance of the morphological structures are ascertained by comparisons among species and with the results of independent physiological studies. The mass of modern axonal transport and electrophysiological studies clearly show by the coincidence of architectonic and functional entities that cytoarchitectonic studies can be a useful tool in working out the function of a brain structure, as for instance, vertical structures such as ocular dominance columns (Hubel and Wiesel, 1963, 1977), barrel fields (Woolsey and Van der Loos, 1970), and dendritic bundles (Fleischhauer, 1974; Fleischhauer et al., 1972).

Where strongly contrasting laminar structures are found, as in the *allocortical* regions, maps produced convincing results (e.g., Stephan and Andy, 1970; Stephan, 1975). Mapping of the *isocortex* (neocortex) by purely qualitative (non-measuring) methods on Nissl- or myelin-stained material is much more problematic and only in a few cases have these results been confirmed by independent methods (Sanides and Krishnamurti, 1967). Pigmentarchitectonic methods (Braak, 1972a,b; 1976a,b; 1977), axonal transport studies, and microelectrode mapping cannot replace cytoarchitectonic analyses in all cases, because in many instances either in vivo and/or consuming methods cannot be used (human; seldom and difficult to breed primate species) or else histochemical studies of pigment pattern is not possible (e.g., in lower primates and in nonsenile mammals).

As a means of preserving the advantages of cyto- and myelo-architectonic studies and to overcome the disadvantages of a purely qualitative approach, *quantitative* methods have been introduced (e.g. Haug, 1956, 1958; Ryzen and Campbell, 1955; Hopf, 1966). This work was done with nonautomatic methods and most of these workers have not continued these studies, probably because this work is enormously time-consuming and often boring.

Therefore *semiautomatic* methods have been developed for the study of certain human and nonhuman cortical areas (Haug 1972, 1976; Haug et al. 1971). We introduced an *automatic*, quantitative method for cytoarchitectonic studies using a television-image analyser, a computer-controlled scanning procedure, and a graphical presentation of the measurements by a relatively simple computing procedure (Schleicher et al., 1978; Zilles et al., 1978 a,b,c; 1979a,b; 1980; Zilles and Schleicher, 1980). These measurements were compared with the histological (mostly Nissl-stained) sections and the results of these studies are therefore based on an objective, i.e., quantitative, work-up of the histological material.

Cytoarchitectonic studies of the isocortex provide an important and useful scientific approach, if the following demands are fulfilled:

1. Comparative neuroanatomical aspects must be studied to establish criteria for identifying common and differing isocortical patterns among species.

2. Results of microelectrode mapping and axonal transport studies must be carried out either on the same species or, if this is not possible, in related species.

3. Quantitative methods must be an integral part of cytoarchitectonic studies to give an objective basis for the delineation of areas and laminar structures. It then becomes possible to take account of interindividual variability.

METHODS

Stereological methods have been developed to determine quantitatively such variables as volume density, size, shape, and the spatial arrangement of the cellular elements (Weibel, 1979). These measuring procedures are time-consuming and at present cannot carried out by fully automated analysers from standard histological preparations mainly because of artifacts caused by the thickness of the histological sections. On the other hand, automatic measuring devices like television image analyzers in combination with calculators offer new possibilities for data accumulation and analysis. Therefore, our studies are based on the grey-level index (GLI), a parameter that is not an unbiased estimate of one of the variables mentioned above, but that can be measured fully automatically by image analyzers and that has been shown to be a powerful tool in cytoarchitectonic analyses.

The GLI is measured from routine histological preparations (Nissl-stain, 20 μm). The microscopic image (magnification x 80, Universal, Carl Zeiss, FRG) is projected into a television camera and the electrical signals are passed into the image analyzer (Microvideomat II, Carl Zeiss, FRG). Figure 1a shows the cellular elements as seen on the television screen. The image comprises grey levels ranging from black to white. To extract image elements corresponding to cellular structures, the grey-level threshold appropriate to the boundaries between these dark elements and the bright background is established by the following procedure (Fig. 2). The grey-level threshold is adjusted from black (0 scale divisions on the discriminator of the image analyzer) to white (1000) by 100 steps (increment 10) and the change of the areal fraction occupied by image elements darker than this grey-level threshold is plotted. The appropriate threshold is located at the local minimum of the curve (Prewitt and Mendelssohn, 1966) at 610 scale divisions. The cellular boundaries defined by this threshold are shown in Figure 1b. For histological sections of constant thickness and staining

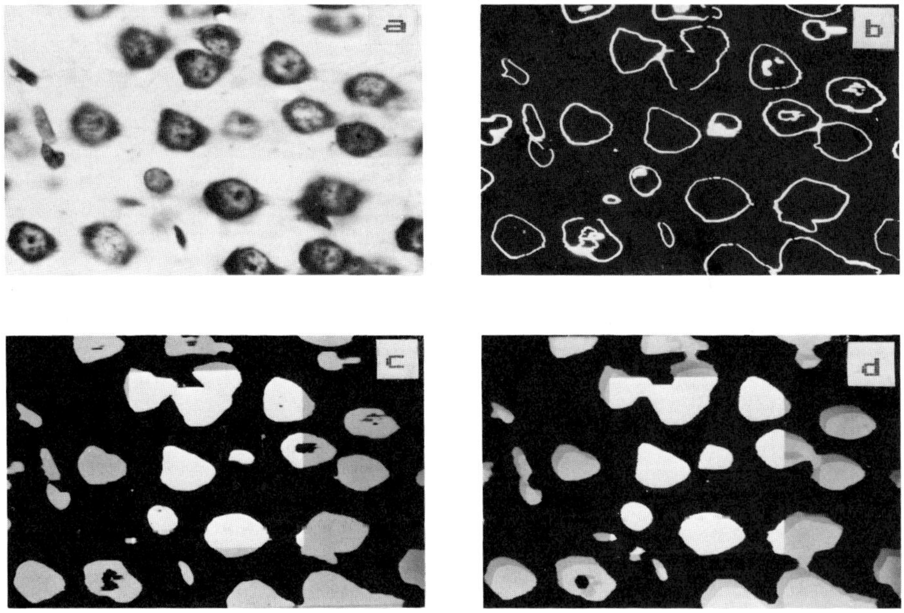

Fig. 1. Processing of the microscopic image by the image analyser.

a: Processing of the microscopic image on the television screen.

b: Boundaries of the cellular elements.

c: Binary image, thresholded by the grey-level value demonstrated in b.

d: Image processed by the operation "closing" (4 picture points).

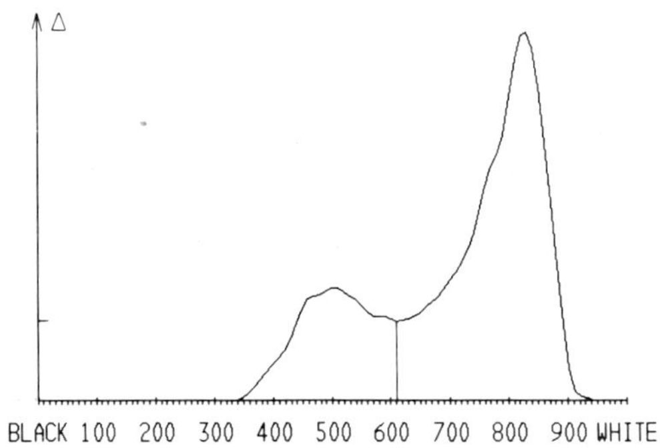

Fig. 2. Dependence of the change of the areal fraction (ordinate) from the grey-level threshold (abscissa, arbitrary scale units) to indicate the grey-level value of the particle boundaries.

intensity, this adjustment of the image analyzer is constant during measurements on one brain.

The grey-level index (GLI) is defined as the areal fraction of image elements on the area of a measuring field of fixed size (50 x 50 µm in Fig. 1c). Depending on the staining intensity, some cellular elements may not be represented in this binary image, e.g., pale cell nuclei. Processing the image with the operation "closing" (Serra, 1972) can fill these holes (Fig. 1d).

The parameter GLI should not be confused with the "grey-cell coefficient," which represents an unbiased estimate of the volumetric density of the cellular structures (Haug, 1956). The difference between these two parameters is illustrated in Figure 3. Images generated by sectioning the tissue with test planes are a prerequisite for measuring the grey cell coefficient. The areal fraction of the cellular profiles (dashed areas) on the area of the test plane (square field) is an unbiased estimate of the volume density of the cellular structures. The GLI parameter, however, uses sections of thickness T in which all elements within the depth of focus (t) are projected onto a plane (shaded areas; the area of the profiles of Figure 3a are superimposed for comparison). These projections are greater than or at least equal to the profiles; and hence, for a given structure, the grey-level index is expected to be greater than the value of the grey-cell coefficient.

The histological section or selected parts of it are scanned by adjoining quadratic measuring fields by means of a computer (Wang 2200 T)-controlled scanning procedure (Fig. 4). For each field, the GLI (0< GLI < 100%) is measured, transferred to the desk calculator, and stored on floppy discs for further processing. During the scanning process, automatic

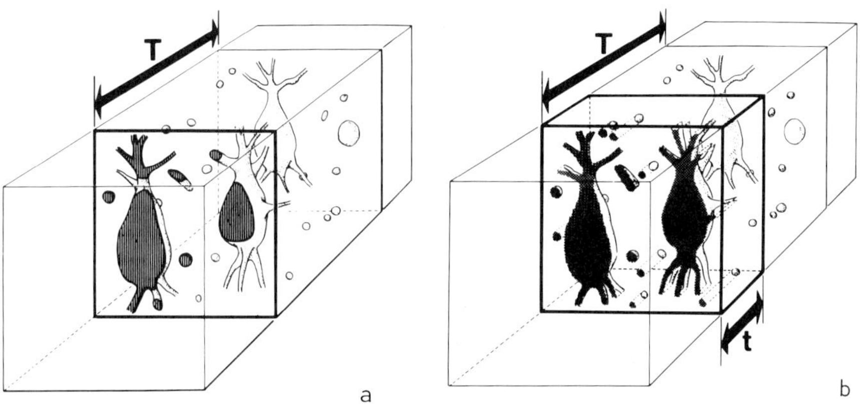

Fig. 3. Illustration of the difference between the grey-cell coefficient (a) and the grey-level index (b).

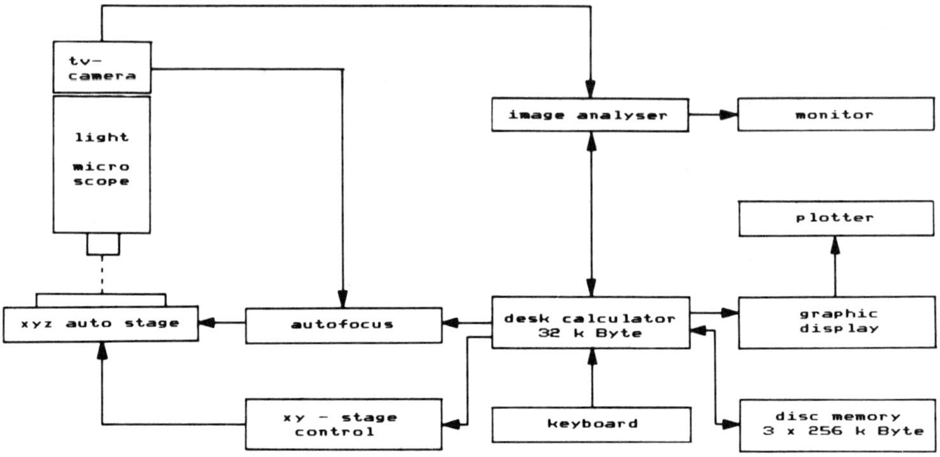

Fig. 4. Image-analysing and data-processing system.

focusing is triggered after measuring a previously defined number of fields. Approximately six fields are measured within one second.

The spatial resolution of the scanning procedure is determined by the size of the measuring fields. For the analysis of the areal pattern, measuring fields of 50 x 50 μm have shown to give good resolution and to keep the amount of data within workable limits. With measuring fields of this size, only laminae of equal or greater width than 50 μm in width can be recognized (Prewitt, 1965). For some applications, like a detailed analysis of the laminar pattern, higher resolutions (up to 10 x 10 μm) are necessary.

Data processing for the analysis of the areal pattern of a cerebral cortex should indicate the boundaries between adjacent areas. These boundaries are marked by a change in the laminar pattern, which is determined by a change in the volume density of the cellular structures and is reflected in the GLI values. Histological sections of known position are scanned completely. The data are smoothed by local averaging (Rosenfeld and Kak, 1976), i.e., each GLI value is replaced by the average of the original value and the values of its eight neighbors. The graphical presentation of the data marks the positions of the measuring fields with GLI values within a selected interval by points (Fig. 5). Thus, laminae or regions with low, medium, or high GLI values are distinguished. Different intervals must be chosen to find all quantitatively definable boundaries. The exact topographic position of these boundaries does not change when the intervals change because the borderlines have the character of a none-or-all feature. Several intervals may be combined in one plot (see p. 186).

The laminar pattern is analyzed in detail by a high-resolution scanning procedure (measuring field 10 x 10 μm). The grey-level indices are

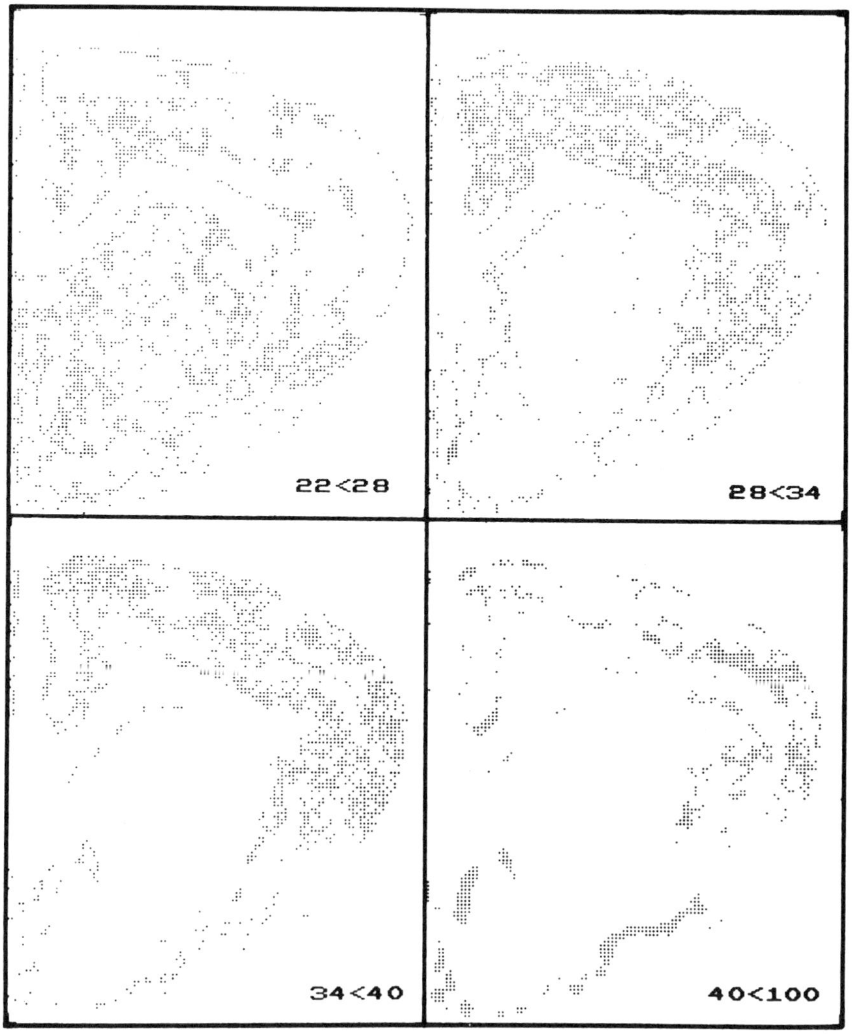

Fig. 5. Computer plots of the spatial distribution of grey-level index values in a coronal section of a *Microcebus* brain. Four adjoining intervals to indicate areal boundaries.

measured in adjacent measuring fields forming stripes perpendicular to the cortical surface from the pial surface to the white matter. Because of the high resolution, the GLI values depend on the random position of the stripe. Local inhomogeneities within the same laminae are excluded by measuring ten adjacent stripes and averaging the values. The number of stripes to be averaged is limited by the curvature of the cortical surface and the resulting shifting of laminar boundaries (Hudspeth et al., 1976). The values of these averaged stripes are plotted as a profile curve (see page 193). The boundaries between the laminae occur at points where the absolute difference between two adjacent measurements is maximal. These positions are determined by the maxima of the differential quotient curve, i.e., the curve of the absolute differences between two adjacent values of the profile curve. Only those maxima that reappear in profiles from different parts of the same cortical area and that can be distinguished as boundaries in the histological preparation are considered to be "real" boundaries. To compare profile curves from cortical areas of unequal thickness and staining intensity, the width of the whole cortex is standardized to 100% and the area beneath the curve to 100%, too.

We have standardized the original measuring values in two respects:

1. The width of the whole cortex is standardized to 100%. Therefore, the cortices of all species have the same length on the abscissa. We are speaking of the *relative width*.

2. The area under the profile curve is also standardized to 100%. The *standardized grey level* index is plotted along the ordinate axis.

In the present investigation no attempt was made to estimate tissue shrinkage due to histological processing because we are working exclusively with relative values, the GLI and the relative width. Absolute values like the volume of brain regions were not determined.

For a broadly founded comparative analysis of the isocortex (and other regions) it is necessary to have a sensitive and efficient method that uses routine histological preparations. This is particularly important for primates because many primate species are not suitable for experimental studies. Serially sectioned brains of nearly all species are available in our collection.

COMPARATIVE ASPECTS OF ISOCORTICAL AREAS IN PROSIMIANS

Lateral views of the hemispheres of four different prosimian genera are shown in Figure 6. Five cortical regions were chosen for a comparison. We have already published detailed descriptions of the cortical areal pattern of *Microcebus* and *Galago* (Zilles et al., 1979a,b). The topology of these five regions is comparable in all representatives of the four subfamilies that we have studied (Cheirogaleinae, Galaginae, Lemurinae, Lorisinae).

This comparable overall pattern means that the primary motor cortex (= M) is positioned in the dorsal half of the frontal third of the hemisphere and is followed by the primary somatosensory cortex (= S). S is always separated from the motor cortex by a small area that can be identified as an intermediate type of cortical laminar pattern (Fig. 7). In our former publications we delineated this small stripe as Area Poc 2 (Zilles et al.,

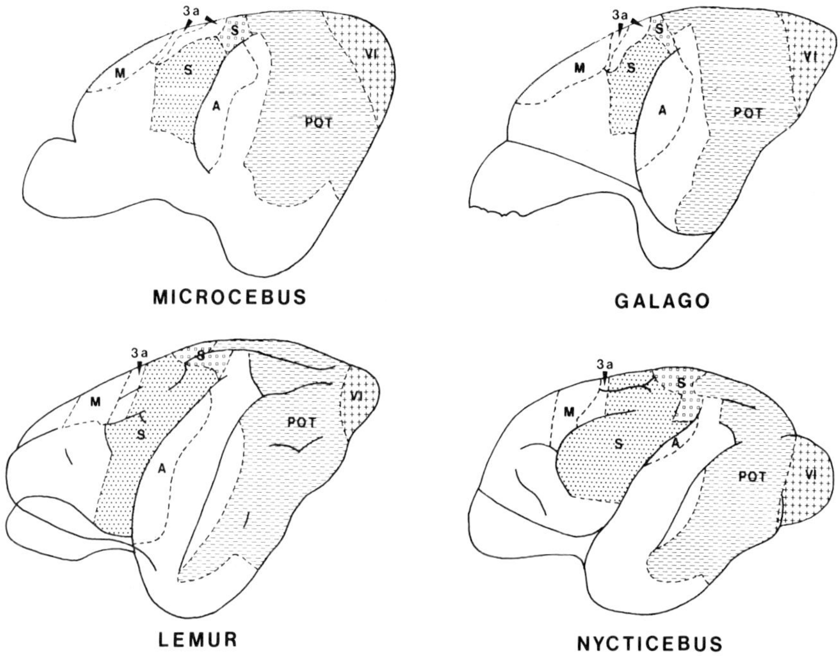

Fig. 6. Lateral views of four prosimian brains. The brain sizes are standardized. Abbreviations: A = primary auditory cortex; M = primary motor cortex; POT = parieto-occipito-temporal cortex (=third visual tier +Oc 2.1); S = primary somatosensory cortex; VI = primary visual cortex (striate area); 3a = Area 3a (= intermediate motorsensory cortex, = Poc2)

1979a,b). The Area Poc 2 can be identified as Area 3a of Vogt and Vogt (1919), which they described as a region with a higher content of granule cells than Area 4 and with a higher content of larger pyramidal cells than Area 3b. The supragranular layers (laminae above layer IV) of this region show somewhat higher grey-level indices than do the layers in the more anterior motor cortex and lower grey-level indices than in the caudally bordering somatosensory cortex (Fig. 7). In some cases Area 3a can be subdivided into two stripes (Figs. 6, 7). The lateral subarea 3a has a structure more closely related to S, and the medial subarea 3a shows more similarity to the laminar pattern of M. The interpretation of Poc 2 as Area 3a is also in agreement with the microelectrode mapping of the somatic sensory cortex by Carlson and Welt (1980) and Sur et al. (1980) in *Galago senegalensis* and *Galago crassicaudatus*.

Sanides and Krishnamurti (1967) identified an ISM-field in *Nycticebus*, which is identical with our Area 3a in our map of the slow loris cortex. Later Krishnamurti et al. (1976) corroborated the existence of this field as a

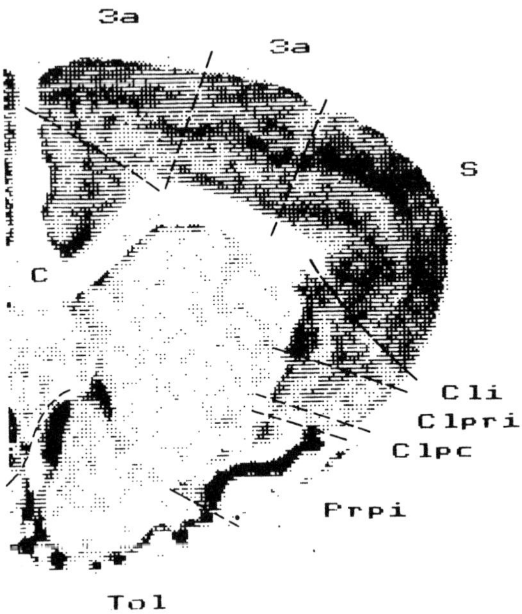

Fig. 7. Coronal section of a *Microcebus* brain after GLI measurement. Four increasing different GLI intervals are indicated by four different plotting densities (see Fig. 5). Abbreviations: C = (below line) = cingulate cortex; Cli = isocortical claustrocortex; CLpc = periallocortical claustrocortex; Clpri = priosocortical claustrocortex; Prpi = Regio praepiriformis; S = primary somatosensory cortex; Tol =tuberculum olfactorium; 3a = Area 3a (= Poc2)

"high threshold tap" area in a microelectrode study. Area 3a, identified as an intermediate type of cortex, has also been found in squirrel and cebus monkeys (Sanides, 1968; Jones, 1975; Lucier et al., 1975), in baboon and rhesus monkeys (Phillips et al., 1971; Schwarz et al., 1973; Wiesendanger, 1973; Heath et al., 1976; Hore et al., 1976), in the tree shrew (Zilles et al., 1978a), in the rat (Zilles et al., 1980), and in the guinea pig (Wree et al., 1981). A broad discussion of the Area 3a problem can be found in a recent review (Jones and Porter, 1980). The topological relations of the motor cortex, Area 3a, and the somatic sensory cortex appear to be similar in all mammals.

In addition to this common feature our quantitative cytoarchitectonic results show a clear difference between the motor cortices in *Microcebus* and *Galago* on one side and *Lemur* and *Nycticebus* on the other side, in that M is separable into two areas in the latter two species, but not in the former pair. In *Lemur* and *Nycticebus* only the caudal part of M contains the typically huge pyramidal Betz cells, whereas in *Microcebus* and *Galago*, M is a single more or less homogeneous area without typical Betz cells. In Figure 6 only the caudal Betz cell containing part of M is shown for *Lemur* and *Microcebus*. Our delineation of the Betz cell area fits well the description of Mott and Halliburton (1908) and Mott and Kelly (1908) of the same area in *Lemur mongoz*. The present material does not indicate whether this feature is a morphological sign of (1) a closer systematic relationship between *Microcebus* and *Galago* (Martin, 1972; Cartmill, 1975; Zilles and Schleicher, 1980) rather than the accepted one that places them in different suborders (Lemuriformes and Lorisiformes after Gregory, 1915); or (2) a functional adaptation common to both species (Zilles and Schleicher, 1980); or (3) only a reduction in cell size within smaller brains.

The somatosensory cortex (= S) of all four species shown in Figure 6 can be divided into two principal regions. The bigger one is positioned more rostrolaterally, the smaller one is inserted between the bigger part of S and the parieto-occipito-temporal cortex (= POT). As our measurements show, this cortical area, compared with the more rostral part of S, has (1) a clearly reduced granular layer, (2) an increased layer V, and (3) a lower overall packing density of neurons in the supragranular layers. Sanides and Krishnamurti (1967) identified a completely comparable region in the slow loris as parakoniocortex (= park) and Krishnamurti et al. (1976) in a microelectrode study described a region like Area 3a as a high-threshold tap region. We were able to corroborate the delineation of such a region in slow loris (unpublished observation), *Lemur* (Fig. 8), *Galago* (Zilles et al., 1979a), *Microcebus* (Zilles et al., 1979b), and *Tupaia* (Zilles et al., 1978a). We designated this area as Pa 1. Therefore we can conclude that the somatosensory cortex, in all prosimian species so far analyzed, shows a common areal pattern with a large, rostrolaterally positioned and highly granular koniocortex (probably Areas 3, 1, 2) and a smaller caudally attached area described as parakoniocortex.

To finish the description of common cortical areal patterns in our four prosimian examples we have only to refer to Figure 6, which shows the similarity in the position and in the extension of the primary auditory cortex (= A), the primary visual cortex (= V I), and the parieto-occipito-temporal cortex (= POT). The primary auditory cortex in *Lemur* (Te 1 in Fig. 8) can be divided into two subareas, which are comparable to the subdivision of R and AI in the owl monkey (Imig et al., 1977). POT includes the visual area II (= Area 18) and the third visual tier areas. The different areas of POT, demonstrated by our cytoarchitectonic method, will be discussed below using the brain of *Lemur*. Region V I corresponds to the striate area (= Area 17).

In conclusion, we have discussed only the topology and extension of five comparable cortical areas in four prosimians. The internal structure (laminar pattern, cell types) of these areas shows some degree of variation between the different species and, following a more detailed description of the larger visual region in one prosimian, *Lemur fulvus*, we will give a more detailed analysis of the variability on the primary visual center in the last section of this paper.

THE ISOCORTEX OF *LEMUR FULVUS* AND THE THIRD VISUAL TIER

A cortical map of the *Lemur* cortex analyzed with our quantitative methods and differing in many features from Brodmann's description (1908) is given in Figure 8. Region Prc 1.1 corresponds to the Betz-cell area (Area

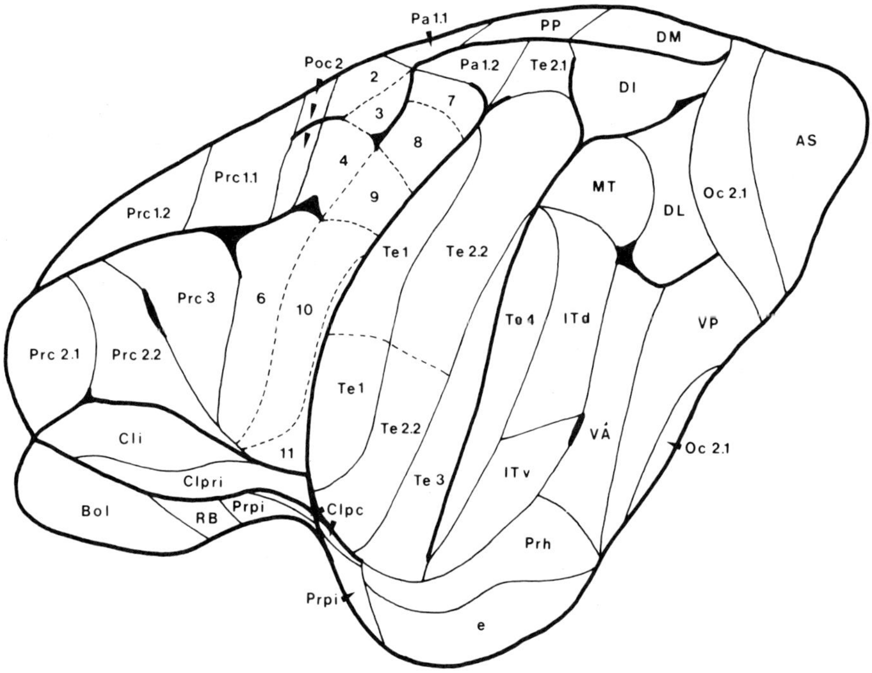

Fig. 8. Cortical map of the lateral view of *Lemur's* brain. Abbreviations: AS = primary visual cortex (= striate area); Bol = bulbus olfactorius; Cli = isocortical claustrocortex; Clpc, periallocortical claustrocortex, Clpri = proisocortical claustrocortex; DI = dorsointermediate area of the third visual tier; DL = dorsolateral area of the third visual tier; DM = dorsomedial area of the third visual tier; e = area entorhinalis; ITd, ITv = dorsal and ventral parts of the infratemporal area of the third visual tier; MT = middle temporal area of the third visual tier; Oc 2.1 = secondary visual cortex (= VII = Area 18); Pa 1.1; Pa 1.2 = parietal areas (parakoniocortical part of the somatosensory cortex); Poc 2 = intermediate motorsensory cortex (=Area 3A); PP = postparietal area of the third visual tier; Prc 1.1, 1.2, 2.1, 2.1, 3 = frontal isocortical areas; Prh = area perirhinalis; Prpi = regio praepiriformis, RB = regio retrobulbaris; Te 1, 2.2, 3, 4 = temporal isocortical area; VA = ventroanterior area of the third visual tier; VP = ventroposterior area of the third visual tier; 2, 3, 4, 6, 7, 8, 9, 10, 11 = subfields of primary somatosensory cortex.

4); the numbers 2-4 and 6-11 indicate subareas (out of a total of 13 subareas) in the koniocortical part; Pa 1.1 and Pa 1.2 are subareas of the parakoniocortical part of the somatosensory cortex; Te 1 is the primary auditory cortex; Te 2.1 and Te 2.2 belong to the auditory belt regions; Oc 2.1, DM, DI, DL, MT, VP, VA, ITd, ITv, and PP are areas of the third visual tier. The topology of these areas fits well with the position of the respective areas in *Galago* (Allman et al., 1973, 1979; Raczkowski and Diamond, 1978; Symonds and Kaas, 1978) and in the owl monkey (Allman and Kaas, 1971a,b, 1974a,b, 1975, 1976; Graham et al., 1979). This is a further corroboration of the assumption of Allman et al. (1979, this volume) that several of the third visual tier areas are a feature common to all primates. Figure 9 shows the computer plot of a frontal section in the range of the third visual tier.

Moreover, we think that this is also an example of the efficiency of our quantitative cytoarchitectonic method, since our analyses agree with the delineations of these areas made by microelectrode mapping and histological studies of myelinated fibres. Previously cytoarchitectonic

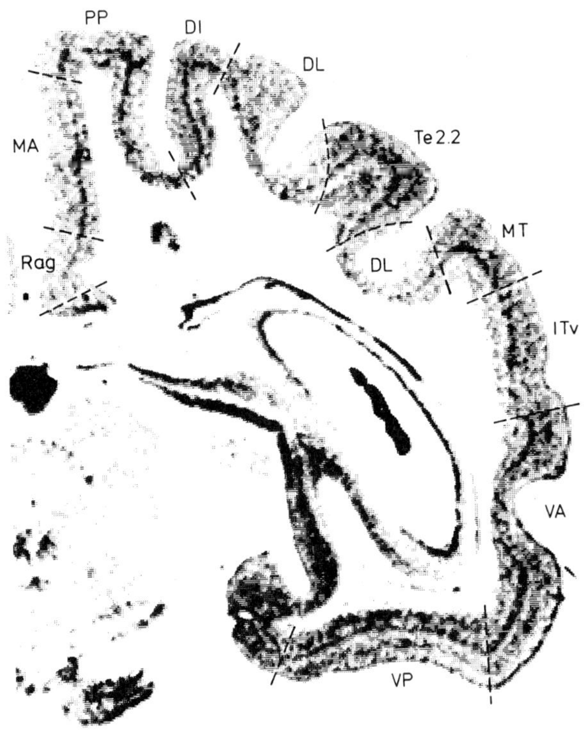

Fig. 9. Coronal section of a *Lemur* brain after GLI measurement at a level of the third visual tier. Four increasing GLI intervals are indicated by four different plotting densities. Abbreviations: MA = medial area of the third visual tier; Rag = regio retrosplenialis agranularis. See Fig. 8 for the other abbreviations.

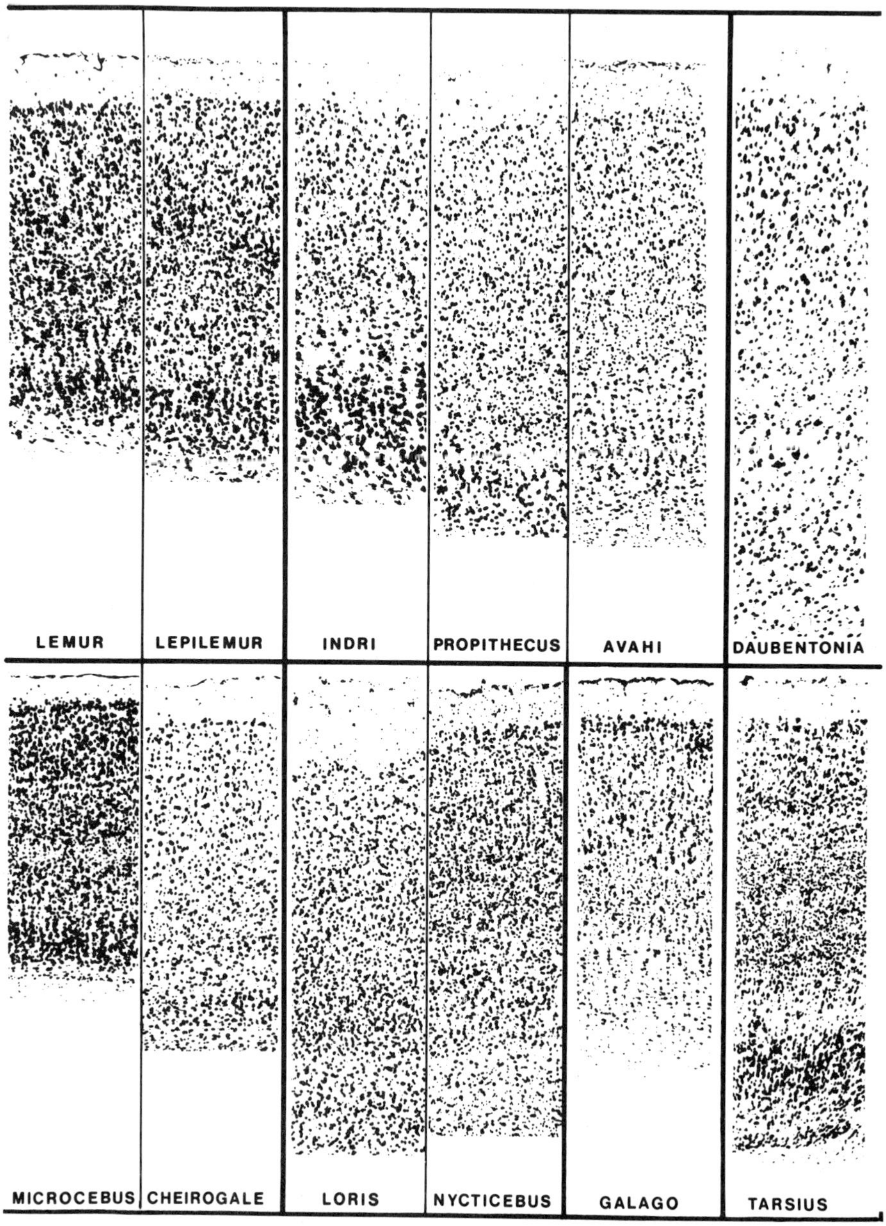

Fig. 10. Micrographs of the striate area in 12 different prosimian species at the same magnification.

patterns have been described but not used for delineation. Since experiments cannot be performed on many primates, comparisons of the third visual tier (and other regions) will benefit from quantitative methods that can use routine histological preparations.

THE LAMINAR PATTERN OF THE STRIATE AREA IN SOME PROSIMIAN SPECIES

This last section will show that even the striate area (= V I), which is without any doubt homologous in all primates, has a considerable degree of variability in laminar pattern. One aim of our methods is to establish quantitative data that can describe the extent of functional adaptation or phylogenetic change. A purely qualitative description of the striate areas in different species could establish morphological *differences*, but could not produce insights into the *trends* and developmental dynamics.

Micrographs of Nissl-stained material show the laminar pattern of Area 17 in twelve different prosimian species (Fig. 10). Computer plots of the grey-level index distribution in a comparable part of the striate area in these prosimians, in man, and in a basal insectivore are shown in Figure 11. The pial surface of the cortex is at the top of all the insets and the cortex-white matter boundary forms the lower end of the insets. The upper arrowheads in Figure 11 indicate the boundaries between the supragranular and granular layers, and the lower arrowheads indicate the boundaries between the granular and infragranular layers. The plots show that the different laminar patterns in different species are like fingerprints. It is difficult, however, to extract quantifiable parameters from this sort of graphical presentation. Therefore, we performed a variation of our usual data presentation (for details see Methods section).

The GLI measurement of the primary visual cortex of *Tarsius* is represented as a profile curve in Figure 12. Although we are able to determine all laminar boundaries and some additional laminar structures (broken lines), for the following discussion we will identify only supragranular, granular, and infragranular layers (solid lines). This type of profile curve permits the extraction of some quantifiable and nonsubjective measurements. We have used standardized values in this analysis:

1. In *relative width* the whole cortex is standardized to 100%. Therefore, the cortices of all species have the same length on the abscissa.

2. The area under the profile curve is also standardized to 100%. The *standardized grey level* index is plotted along the ordinate axis.

Figure 13 compares the relative widths of the supragranular and granular striate layers of day- and night-active prosimians and those of man. The Mann-Whitney test produces significant differences between these two groups at the 1%-level. The four day-active primate species, indicated by black dots, have relatively wider supragranular and narrower granular layers than do the eight night-active species, indicated by rectangles. The infragranular layers show no statistically significant difference between the

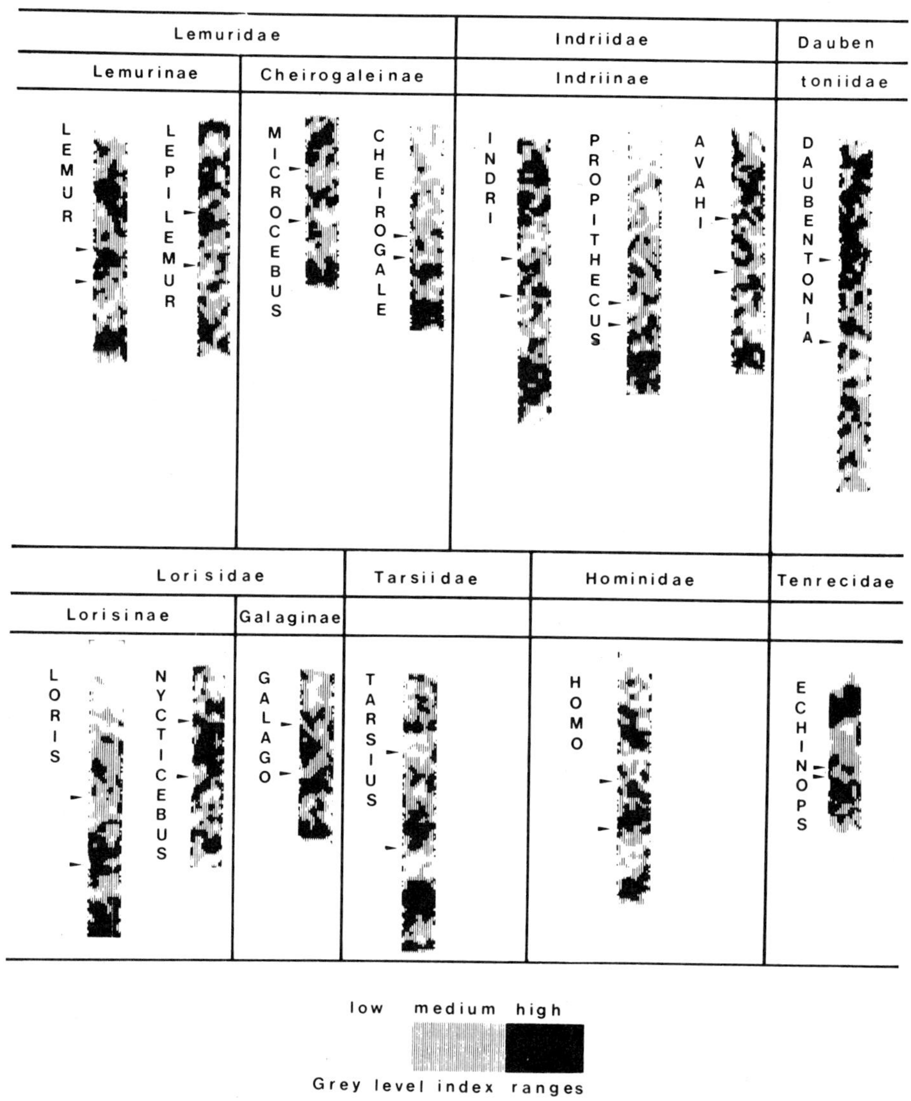

Fig. 11. Plots of GLI - ranges in the striate area of prosimian and nonprosimian species after measurement with 10 x 10 μm field. The pial surface is above, the cortex - white matter boundary below.

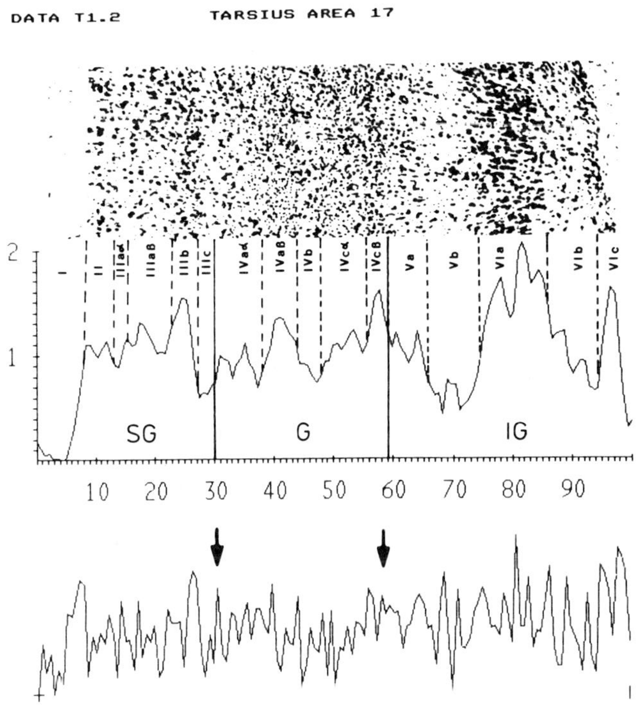

Fig. 12. Micrograph, standardized GLI profile curve and the respective differential quotient curve of the striate area in *Tarsius*. The pial surface is positioned at the left and the cortex-white matter boundary at the right.

groups. The size of the granular layer, the target of the retino-geniculo-cortical system, may be correlated with the larger eyes and extremely broad retinal receptor layers of the night-active primates. The supragranular layers, which are engaged in intracortical information processing, are relatively wider in day-active primates. We conclude that information processing of retinal input is relatively more advanced in striate cortices of day-active primates. The fact that *Tarsius* has the largest relative width of the granular, but together with *Microcebus*, the smallest width of the supragranular layer is of some interest, given the functional interpretation of these two variables. Despite the striate area's enormous volume in *Tarsius* (Stephan, 1969), this cortex is not much advanced in a functional sense compared with those of the other primates because the large cortical volume is primarily engaged in information perception not processing. This aspect of laminar structure reflects *functional adaptation* and not a general phylogenetic progress.

The *phylogenetic aspects* of the laminar patterns are analyzed by correlating variables of the laminar structure with the progression index of

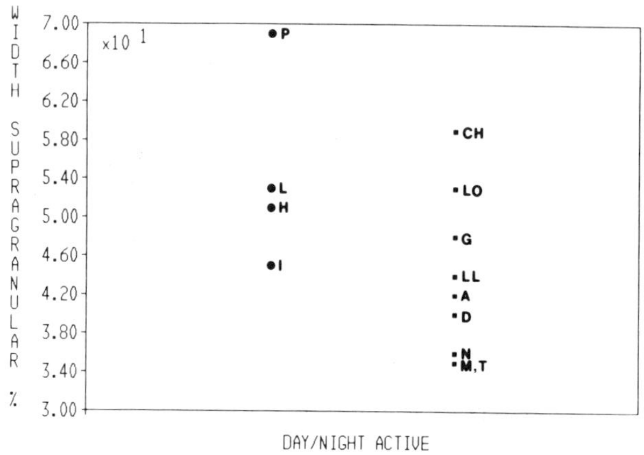

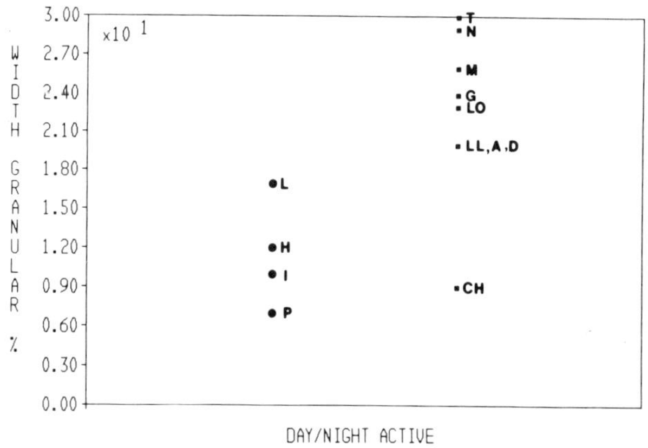

Fig. 13. Relative widths of the supragranular and granular layers of the striate area in day and night active primates (A = *Avahi*; C H = *Cheirogale*; D = *Daubentonia*; G = *Galago*; H = *Homo*; I = *Indri*; L = *Lemur*; LL = *Lepilemur*; LO = *Loris*; M = *Microcebus*; N = *Nycticebus*; P = *Propithecus*; T = *Tarsius*).

the striate area. The progression index of the striate area indicates how many times the volume of a given species' striate area is larger than this region's volume in a hypothetical *Lepilemur* of equal body size (Stephan, 1969). The definition and allometric calculations of this index were published by Stephan (1969). The progression index is indicated on the abscissa of Figures 14 to 15. *Lepilemur* was chosen as the basal reference because this species was determined to have the lowest striate area volume in primates (Stephan, 1969). In the primary visual cortex the mean standardized grey-level index (= GLI) of the granular layer is plotted against the progression index (Fig. 14). The data from *Tarsius* and man are not

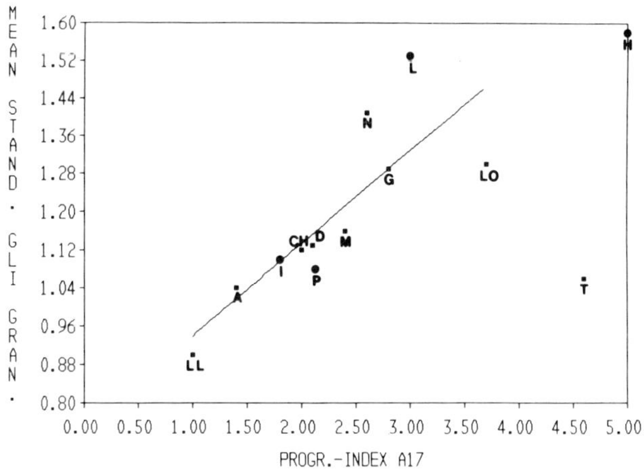

Fig. 14. Mean standardized GLI's of the granular layers in 13 primate species depending on the progression indices of the striate areas.

included in the calculation of these regression lines because this line is based only on prosimians and *Tarsius* seems to be an example of extreme and perhaps unusual specialization in its striate area. Nevertheless, the data for *Tarsius* and man are plotted because their comparative position is of interest. The regression line in Figure 14 demonstrates a significant increase in the mean standardized GLI with increasing progression index ($r = 0.81$; $p < .01$). This means that with an increasing evolutionary level of the striate area, the nerve cells in the granular layer become more densely packed. We can speak of a tendency toward increasing granularization. The supragranular layer, however, shows a tendency of decreasing GLI with increasing progression index. A trend toward a progressive development of the supragranular layers is thereby evident because a decreasing GLI is a sign of an increasing neuropil volume. An increasing neuropil volume means an increasing functional capacity because the network of dendrites, axons, and synapses form a considerable part of neuropil volume. The mean standardized GLI of the infragranular layer shows no clear relationship with the progression index of the striate area. This may indicate both a stabilization of the cortical output system and much individual variation.

The tendency toward the progressive development of intracortical information processing structures (supragranular layers) paralleling increasing progression index of the striate area is summarized in Figure 15. The ratio of the mean standardized GLI's of the supragranular to granular layers, a parameter linked with the supragranular neuropil volume per unit volume of the granular layer, increases. Therefore, a negative regression line ($r = 0.79$; $p < .01$) demonstrates more supragranular neuropil volume per unit volume of granular layer neuropil as progression index increases. In

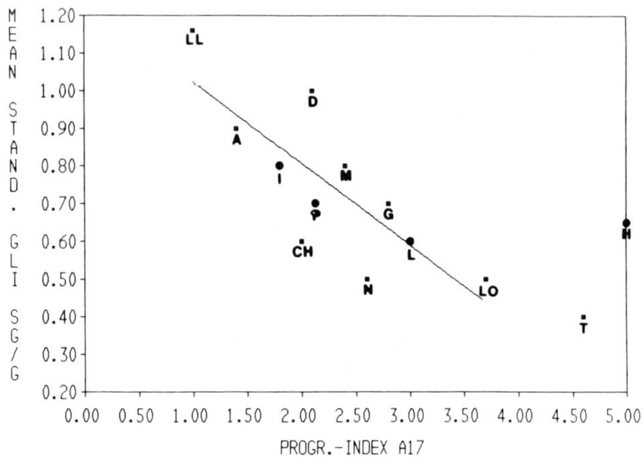

Fig. 15. Ratios of the mean standardized GLI's of the supragranular/granular layers in the primate species depending on the progresion indices of the striate areas.

other words there is a clear morphological tendency toward increasing intracortical information processing in the striate area from *Lepilemur* to *Lemur*.

Although the striate area is certainly a homologous structure in prosimians, its morphological structure shows great variability. The variability can be successfully quantified by our cytoarchitectonic method. The quantitative results are related in part to functional adaptations and in part to evolutionary processes.

Acknowledgments: We are grateful to Ursula Blohm, Helga Grobecker, Monika Martin, Ingrid Riettiens and Clare Roberg for their excellent technical assistance; to Jutta Schlahn for typing the manuscript; and to Ingrid Bertelsen and John Nelson for checking the English text. This study was supported by Deutsche Forschungsgemeinschaft Zi 192/4-3 (KZ).

REFERENCES

Allman, J.M., and Kaas, J.H., 1971a. A representation of the visual field in the caudal third of the middle temporal gyrus of the owl monkey (*Aotus trivirgatus*). Brain Res., 31:85-105.

Allman, J.M. and Kaas, J.H., 1971b. Representation of the visual field in striate and adjoining cortex of the owl monkey (*Aotus trivirgatus*). Brain Res., 35:89-106.

Allman, J.M., and Kaas, J.H., 1974a. The organization of the second visual area (V II) in the owl monkey: a second order transformation of the visual hemifield. Brain Res., 76:247-265.

Allman, J.M., and Kaas, J.H., 1974b. A crescent-shaped cortical visual area surrounding the middle temporal area (MT) in the owl monkey (*Aotus trivirgatus*). Brain Res., 81:199.

Allman, J.M., and Kaas, J.H., 1975. The dorsomedial cortical visual area: a third tier area in the occipital lobe of the owl monkey (*Aotus trivirgatus*). Brain Res., 100:473-487.

Allman, J.M., and Kaas, J.H., 1976. Representation of the visual field on the medial wall of occipital-parietal cortex in the owl monkey. Science, 191:572-575.

Allman, J.M., Campbell, C.B.G., and McGuinness, E., 1979. The dorsal third tier area in *Galago senegalensis*. Brain Res., 179:355-361.

Allman, J.M., Kaas, J.H., and Lane, R.H., 1973. The middle temporal visual area (MT) in the bushbaby, *Galago senegalensis*. Brain Res., 57:197-202.

Bailey, P., and v. Bonin, G., 1951. The Isocortex of Man, University of Illinois Press, Urbana.

Betz, W., 1881. Über die feinere Struktur der Grosshirnrinde des Menschen. Centralbl. f. d. mediz. Wiss., 19:193-195; 209-213; 231-234.

Braak, H., 1972a. Zur Pigmentarchitektonik der Grosshirnrinde des Menschen. I. Regio entorhinalis. Z. Zellforsch, 127:407-438.

Braak, H., 1972b. Zur Pigmentarchitektonik der Grosshirnrinde des Menschen. II. Subiculum. Z. Zellforsch, 131:235-254.

Braak, H., 1976a. A primitive gigantopyramidal field buried in the depth of the cingulate sulcus of the human brain. Brain Res., 109:219-233.

Braak, H., 1976b. On the striata area of the human isocortex. A Golgi- and pigmentarchitectonic study. J. Comp. Neurol., 166:341-364.

Braak, H., 1977. The pigmentarchitectonic of the human occipital lobe. Anat. Embryol., 150:229-250.

Brodmann, K., 1908. Beitrage zur histologischen Lokalisation der Grosshirnrinde. VII. Mitteilung: Die cytoarchitektonische Cortexgliederung der Halbaffen (Lemuriden). J. Psychol. Neurol., 10:287-334.

Brodmann, K., 1909. Vergleichende Lokalisationslehre der Grosshirnrinde in ihren Prinzipien dargestellt auf Grund des Zellenbaus. J.A. Barth, Leipzig.

Campbell, A.W., 1905. Histological Studies on the Localisation of Cerebral Function, University Press, Cambridge.

Carlson, M., and Welt, C., 1980. Somatic sensory cortex (SmI) of the prosimian primate *Galago crassicaudatus*: organization of mechanoreceptive input from the hand in relation to cytoarchitecture. J. Comp. Neurol., 189:249-271.

Cartmill, M., 1975. Strepsirhine basicranial structures and affinities of the cheirogaleidae. In, Phylogeny of the Primates, W.P. Luckett and F.S. Szalay, eds., Plenum Press, New York and London, pp. 313-354.

Economo, C. von, and Koskinas, G.N., 1925. Die Cytoarchitektonik der Hirnrinde des erwachsenen Menschen, Springer, Berlin.

Elliot-Smith, G., 1904. The morphology of the occipital region of the cerebral hemisphere in man and apes. Anat. Anz., 24:436-451.

Elliot-Smith, G., 1907. A new topographical survey of the human cerebral cortex, being an account of the distribution of the anatomically distinct cortical areas and their relationship to the cerebral sulci. J. Anat. Physiol., 41:237-254.

Fleischhauer, K., 1974. On different patterns of dendritic bundling in the cerebral cortex of the cat. Z. Anat. Entwickl. Gesch., 143:115-126.

Fleischhauer, K., Petsche, H., and Wittkowski, W., 1972. Vertical bundles of dendrites in the neocortex. Z. Anat. Entwickl. Gesch., 136:213-223.

Graham, J., Lin, C.S., and Kaas, J.H., 1979. Subcortical projections of six visual cortical areas in the owl monkey, *Aotus trivirgatus*. J. Comp. Neurol., 187:557-580.

Gregory, W.K., 1915. On the classification and phylogeny of the Lemuroidea, Bull. Geol. Soc. Am., 26:426-446.

Hassler, R., 1937. Zur Normalanatomie der Substantia nigra. Versuch einer architektonischen Gliederung. J. Psychol. Neurol. Lpz., 48:1-55.

Hassler, R., 1959. Anatomie des Thalamus. In, Einführung in die sterotaktischen Operationen, G. Schaltenbrand and P. Bailey, eds., Thieme Verlag, Stuttgart, pp. 230-290.

Haug, H., 1956. Remarks on the determination and significance of the gray cell coefficient. J. Comp. Neurol., 104:473-492.

Haug, H., 1958. Quantitative Untersuchungen an der Sehrinde, Thieme Verlag, Stuttgart.

Haug, H., 1972. Stereological methods in the analysis of neuronal parameters in the central nervous system. J. Microscopy, 95:165-180.

Haug, H., 1976. Die verschiedenen Verfahren zur Werteerfassung in der biologischen Morphometrie und Sterologie. Microscopica Acta, 78:197-220.

Haug, J., Kebbel, J., and Wiedermeyer, G.L., 1971. Die Messung der mittleren Zelldichte und ihre Verteilung in Geweben mit erheblichen Zelldichteunterschieden (Auswertung am Cortex cerebri als Beispiel). Microscopica Acta, 71:121-128.

Heath, C.J., Hore, J., and Phillips, C.G., 1976. Inputs from low threshold muscle and cutaneous afferents of hand and forearm to areas 3a and 3b of baboon's cerebral cortex. J. Physiol. Lond., 257:199-227.

Hopf, A., 1966. Über eine Methode zur objectiven Registrierung der Myloarchitektonik der Hirnrinde. J. Hirnforsch., 8:301-313.

Hore, J., Preston, J.B., Durkovic, R.G., and Cheney, P.D., 1976. Responses of cortical neurons (areas 3a and 4) to rump stretch of hindlimb muscles in the baboon. J. Neurophysiol., 39:484-500.

Hubel, D.H., and Wiesel, T.N., 1963. Shape and arrangement of columns in cat's striate cortex. J. Physiol., 165:559-568.

Hubel, D.H., and Wiesel, T.N., 1977. Functional architecture of macaque monkey visual cortex. Proc. R. Soc. Lond., B198:1-59.

Hudspeth, A.J., Ruark, J.E., and Kelly, J.P., 1976. Cytoarchitectonic mapping by microdensitometry. Proc. Natl. Acad. Sci. 73:2928-2931.

Imig, T.J., Ruggero, M.A., Kitzel, L.M., Javel, E., and Brugge, J.F., 1977. Organization of auditory cortex in the owl monkey (*Aotus trivirgatus*). J. Comp. Neurol., 171:111-128.

Jones, E.G., 1975. Lamination and differential distribution of thalamic afferents within the sensory motor cortex of the squirrel monkey. J. Comp. Neurol., 160:167-204.

Jones, E.G., and Porter, R., 1980. What is Area 3a? Brain Res. Reviews, 2:1-43.

Krishnamurti, A., Sanides, F., and Welker, W.I., 1976. Microelectrode mapping of modality-specific somatic sensory cerebral neocortex in slow loris. Brain Behav. Evol., 13:267-283.

Lashley, K.S., and Clark, G., 1946. The cytoarchitecture of the cerebral cortex of *Ateles*: a critical examination of architectonic studies. J. Comp. Neurol., 85:223-305.

Lucier, G.E., Ruegg, D.C., and Wiesendanger, M., 1975. Responses of neurons in motor cortex and in area 3a to controlled stretches of forelimb muscles in cebus monkey. J. Physiol. Lond., 251:833-853.

Martin, R.D., 1972. Adaptive radiation and behavior of the Malagasy lemurs, Philos. Trans. R. Soc. London, Ser B 264:295-352.

Meynert, R., 1872. Der Bau der Grosshirnrinde und seine örtlichen Verschiedenheiten, nebst einem pathologisch-anatomischen Corollarium, JH Heuser, Neuwied.

Mott, E.W., and Halliburton, W.D., 1908. Localization of function in the lemur's brain. Proc. Royal Soc. London, Ser B 80:136-147.

Mott, E.W., and Kelly, A.M., 1908. Complete survey of the cell lamination of the cerebral cortex of the lemur. Proc. Royal Soc. London, Ser B 80:488-506.

Phillips, C.G., Powell, T.P.S., and Wiesendanger, M., 1971. Projection from low threshold muscle afferents of hand and forearm to area 3a baboon's cortex. J. Physiol. Lond., 217:419-446.

Prewitt, J.M.S., 1965. The selection of sampling rate for digital scanning. IEEE transact. Bio-Med. Engineering, 1:14-21.

Prewitt, J.M.S., and Mendelssohn, M.L., 1966. The analysis of cell images. Ann. New York Acad. Sci., 128:1035.

Raczkowski, D., and Diamond, I.T., 1978. Connections of the striate cortex in *Galago senegalensis*. Brain Res., 144:383-388.

Rosenfeld, A., and Kak, A., 1976. Digital Picture Processing, Academic Press, New York.

Ryzen, M., and Campbell, B., 1955. Organization of the Cerebral Cortex II. Cortex of *Sorex pacificus*. J. Comp. Neurol., 102:365-424.

Sanides, F., 1968. The architecture of the cortical taste nerve areas in squirrel monkey (*Saimiri sciureus*) and their relationships to insular, sensorimotor and prefrontal regions. Brain Res., 8:97-124.

Sanides, F., and Krishnamurti, A., 1967. Cytoarchitectonic subdivisions of sensorimotor and prefrontal regions and of bordering insular and limbic fields in slow loris (*Nycticebus coucang coucang*). J. Hirnforsch., 9:225-252.

Schleicher, A., Zilles, K., and Kretschmann, H.J., 1978. Automatische Registrierung und Auswertung eines Grauwertindex in histologischen Schnitten. Anat. Anz., 144 (Erg-Heft): 413-415.

Schwarz, D.W.F., Deecke, L., and Frederickson, J.M., 1973. Cortical projection of group I muscle afferents to areas 2, 3a, and the vestibular field in the rhesus monkey. Exp. Brain Res., 17:516-526.

Serra, J., 1972. Stereology and structuring elements. J. Microscopy. 95:93.

Stephan, H., 1969. Quantitative investigation on visual structures in primate brains. In, Proc. 2nd Int. Congr. Primatol. Soc., Atlanta GA 1968, Karger, Basel New York, Vol 3, pp. 34-42.

Stephan, H., 1975. Allocortex. In, Handbuch der mikroskopischen Anatomie des Menschen, W. Bargmann, ed., 4. Bd Nervensystem, Teil 9, Springer, Berlin.

Stephan, H., and Andy, O.J., 1970. The Allocortex in Primates. In, The Primate Brain, C.R. Noback, and W. Montagna, eds., Appleton-Century-Crofts, New York, pp. 109-135.

Sur, M., Nelson, R.J., and Kaas, J.H., 1980. Representation of the body surface in somatic koniocortex in the prosimian *Galago*. J. Comp. Neurol., 189:381-402.

Symonds, L.L., and Kaas, J.H., 1978. Connections of striate cortex in the prosimian, *Galago senegalensis*. J. Comp. Neurol., 181:477-512.

Vogt, C., and Vogt, O., 1907. Zur Kenntnis der elektrisch erregbaren Hirnrindengebiete bei den Säugetieren. J. Psychol. Neurol., 8:277-456.

Vogt, C., and Vogt, O., 1919. Allgemeinere Ergebnisse unserer Hirnforschung, J. Psychol. Neurol., (Lpz.), 25:279-462.

Vogt, O., 1906. Über strukturelle Hirnzentra, mit besonderer Berücksichtigung der strukturellen Felder des Cortex pallii. Anat. Anz., 29:(Erg Bd) 74-114.

Vogt, O., 1910. Die myeloarchitektonische Felderung des menschlichen Stirnhirns. J. Psychol. Neurol., 15:221-232.

Weibel, E.R., 1979. Stereological Methods, Volume 1: Practical methods for biological morphometry, Academic Press, London.

Wiesendanger, M., 1973. Input from muscle and cutaneous nerves of the hand and forearm to neurones of the precentral gyrus of baboons and monkeys. J. Physiol. (Lond)., 228:203-219.

Woolsey, T.A., and Van der Loos, H., 1970. The structural organization of layer IV in the somatosensory region (SI) of mouse cerebral cortex. Brain Res., 17:205-242.

Wree A., Zilles K. and Schleicher, A., 1981. A quantitative approach to cytoarchitectonics. VII. The areal pattern of the cortex of the guinea pig. Anat. Embryol. 162:81-103.

Zilles, K., Schleicher, A., and Kretschmann, H.J., 1978a. A quantitative approach to cytoarchitectonics. I. The areal pattern of the cortex of *Tupaia belangeri*. Anat. Embryol., 153:195-212.

Zilles, K., Schleicher, A., and Kretschmann, H.J., 1978b. A quantitative approach to cytoarchitectonics. II. The allocortex of *Tupaia belangeri*. Anat. Embryol., 154:335-352.

Zilles, K., Schleicher, A., and Kretschmann, H.J., 1978c. Quantitative Darstellung cytoarchitektonischer Areale im Cortex von *Tupaia belangeri* und SPF-Katze. Anat. Anz., 144(Erg-Heft):409-411.

Zilles, K., Rehkamper, G., Stephan, H., and Schleicher, A., 1979a. A quantitative approach to cytoarchitectonics. IV. The areal pattern of the cortex of *Galago demidovii* (E Geoffroy, 1976), (Lorisidae, Primates). Anat. Embryol., 157:81-103.

Zilles, K., Rehkamper, G., and Schleicher, A., 1979b. A quantitative approach to cytoarchitectonics. V. The areal pattern of the cortex of *Microcebus murinus* (E Geoffroy, 1828), (Lemuridae, Primates). Anat. Embryol., 157:269-289.

Zilles, K., and Schleicher, A., 1980. Similarities and differences in the cortical areal patterns of *Galago demidovii* (E Geoffroy, 1796), (Lorisidae, Primates) and *Microcebus murinus* (E Geoffroy, 1828), (Lemuridae, Primates). Folia Primatol., 33:161-171.

Zilles, K., Zilles, B., and Schleicher, A., 1980. A quantitative approach to cytoarchitectonics. VI. The areal pattern of the cortex of the albino rat. Anat. Embryol., 159:335-360.

Zilles, K., and Schleicher, A., 1980. Quantitative Analyse der laminären Struktur menschlicher Cortexareale, Anat. Anz., (Erg-Heft),146:725-726.

ROLE OF ARCHITECTONICS AND CONNECTIONS IN THE STUDY OF PRIMATE BRAIN EVOLUTION

Albert M. Galaburda*
Deepak N. Pandya

In the study of primate brain evolution a variety of gross anatomical features is customarily examined to trace changes in the brain among the different species. Thus, brain size and shape are measured, and aspects of cortical folding into lobes and gyri are carefully noted. Analyses of this type are particularly useful, if not altogether necessary, in the study of extinct brains by means of cranial endocasts. Such studies are remarkably limited, however, in their ability to provide information about interspecies homology. In smoother (lissencephalic) brains, for instance, this limitation is obvious. But even endocasts showing surface features of cortical folding can be an insurmountable challenge to statements of homology. On the one hand, the shape, depth and height of the folds varies considerably within a given species; on the other hand, different species often have similar surface topography. Furthermore, information is still insufficient concerning the functional meaning of shifts in surface markings, and even of gyri and sulci present in one species but not in others. Finally, the possibility exists that gyri having the same shape and occupying the same location in two species do, in fact, subserve different functions by nature of their different architecture and connectivity.

Although in the case of extinct species no other means of study are available, in extant primates the relationship between gross anatomical landmarks on the brain's surface and their microscopic structure may be studied. The emergence of governing principles of cortical topography from such studies can then be adapted to the study of endocasts. In particular,

*The Neurological Units of the Beth Israel and Boston City Hospitals, The Department of Neurology of Harvard Medical School, Boston, MA, and The Charles A. Dana Research Institute (AMG); Edith Nourse Rogers Memorial Veterans Hospital, Bedford, Massachusetts, and Boston University School of Medicine (DNP).

correlating gross anatomy with architectonics and connections of the cortex becomes relevant because these microscopic features more closely reflect the functional organization of the brain.

Architectonics refers to the arrangement of cells in a nervous structure. In the cortex it stresses the cellular arrangement in layers and columns. On the basis of the specific layered appearance, cell sizes, packing densities, and special cellular features, the cortex can be divided into areas of different architectonic structure. If the cellular perikarya are stained, the analysis is termed *cytoarchitectonics*; if differences in patterns of myelination are sought it is termed *myeloarchitectonics*. In effect the cortex can be stained to show the architectonic distribution of a particular constituent, such as blood vessels, glia, chemoreceptors, and various pigments. Areas obtained by cytoarchitectonic parcellation tend to correlate well with those obtained by neurophysiologic mapping and by studies of cortical fiber connectivity, thus emphasizing their probable functional significance.

Evolutionary Trends

A considerable amount of work has been done on comparative architectonics in primates. Sanides (1970) described in great detail some of the changes that occur in architectonic differentiation and shifts in architectonic topography in primitive insectivores, prosimians, and other primates including man. According to Sanides (1970), primates have the highest development of the isocortex (neocortex, six-layered cortex), as manifested by its greater differentiation and size. The isocortex is also highly variable, and the most apt to show interspecies differences. The isocortex derives its phylogenetic and ontogenetic origin from primitive cellular moieties making up the cortical plate (Fig. 1). These primitive moieties comprise archicortical (hippocampal) and paleocortical (olfactory and insular) layered structures that are already present in amphibians. Both archi- and paleocortex give rise to nearly six-layered cortex (proisocortex), that is present in reptiles. Proisocortex in turn gives rise to isocortex (true six-layer cortex) first seen indisputably in mammals.

Within the mammalian line isocortex is largest and most differentiated in primates and its growth promotes the formation of the temporal lobes and the remarkable fissuration of the brains of anthropoid primates. The large brains of whales and elephants are less differentiated (Kesarev et al., 1977). According to Sanides (1970) isocortical differentiation proceeds phylogenetically along the following lines: adjacent to the perihippocampal and insular proisocortex appear belts of isocortex that retain some of the primitive features of their respective neighboring proisocortex. Within these belts are regions of greater differentiation, pointing the way to newer stages of isocortex. These stages are found in the peri-insular and perihippocampal location. The second motor (MII), the second sensory (SII), and the second auditory regions (AII) are the peri-insular structures; the supplementary motor, supplementary sensory, and the second visual areas (VII) are the perihippocampal ones. The remainder of the isocortex is disposed between these primitive isocortical belts and is further subdivided. It contains highly specialized cores of primary cortices (motor, sensory, auditory, and visual) surrounded by belts of less differentiated cortices (paramotor or parakonio). The primary and para-primary units in turn lie in beds of undifferentiated, yet highly evolved cortex characterized by a well ordered, layered appearance. The latter are the eulaminate, homotypical cortices of the frontal, parietal, temporal and occipital lobes, which make up the massive expansions of the primate brain and are particularly vast in the human brain. Thus, during primate evolution, a tendency for specialization was manifested by the greater definition and growth of primary and paraprimary cortical fields, and growth of the less differentiated eulaminate areas

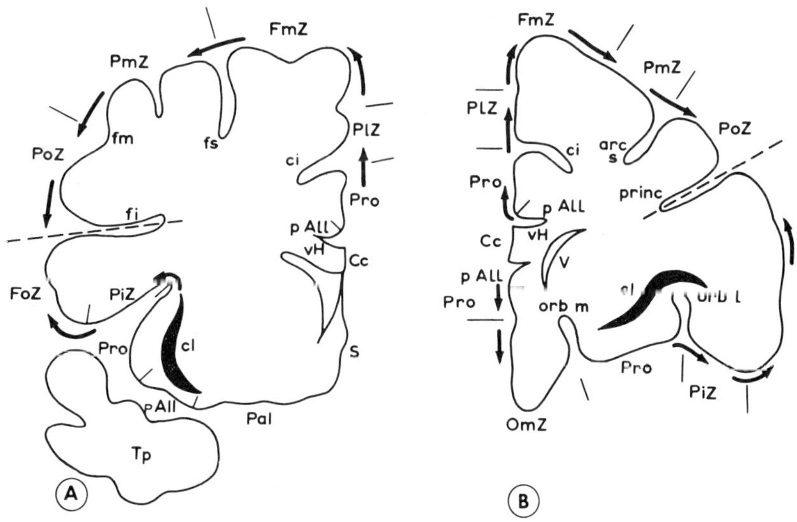

Fig. 1. Evolutionary cortical trends in frontal lobe transverse sections in man (A) and rhesus monkey (B) according to Sanides (1970). The isocortex is strung between its two allocortical roots: the olfactory-insular Pal and the hippocampal vestige (vH). Pal is not visible in this section in monkey. Progressive stepwise specialization and eulamination away from allocortex is achieved beginning with periallocortex (pAll), and continuing through proisocortex (Pro), the parainsular (PiZ) and paralimbic (PlZ) zones, the fronto-opercular (FoZ) and frontomotor (FmZ) zones, and finally the paraopercular (PoZ) and paramotor (PmZ) zones. Note that growth of the frontal convexity in the human brain has resulted in a vertical claustrum (cl). Other abbreviations: arc s, arcuate sulcus superior limb; Cc, corpus callosum; ci, sulcus cinguli; fi, sulcus frontalis inferior; fm, sulcus frontalis medius; fs, sulcus frontalis superior; orb l, sulcus orbitalis lateris; orb m, sulcus orbitalis medius; OmZ, orbitomedial zone; princ, sulcus principalis; Tp, temporal pole. (From Sanides, F., 1970. In, The Primate Brain. Advances in Primatology, vol. 1. Courtesy of Appleton-Century-Croft.)

appeared. Much of the shift of cortical topography that occurs among the primate species results from this growth (Fig. 1).

Cortico-cortical Connections

It is generally accepted that areas differing in architecture also have a separate set of connections. One might say in effect that the difference in architectonic appearance is the result, at least in part, of a different synaptic and connectional organization. It might follow, then, that a change in a given pattern of connections will result in a change in architectonics. For instance it has been shown (Leonard, 1969) that the mediodorsal nucleus of the thalamus in the rat projects to the dorsal bank of the rhinal sulcus and to the medial frontal cortex. In primates the same thalamic nucleus projects to the frontal granular cortex. The rat frontal fields are not equivalent in granularity to the primate prefrontal cortex, so direct comparisons are not possible; however, it is conceivable that a shift in the mediodorsal-frontal connections away from the rhinal sulcus is correlated with a change in the cellular architecture of the prefrontal cortex in primates *vis à vis* the rat. Therefore with a change in the topography of the connections, a change in the architectonic appearance of the receiving area occurred. On the other hand a shift in topography of connections may simply result in a shift in the topography of the architectonic areas, without significantly changing the architectonic appearance. An example of this phenomenon is found within the primate species. Visual projections to a visual temporal field (MT) on the middle temporal gyrus of the owl monkey have been shown (Allman and Kaas, 1971), whereas equivalent projections have been described in the rhesus monkey to an area similar in architectonics to MT (area OAa), but which is located in the depth of the superior temporal sulcus (Seltzer and Pandya, 1978). Thus, in this situation a shift in the direction of the projections does not result in a change in the architectonic appearance of the receiving area, but simply in a change in the location of this area. A change in location may not be found just across members of the same order, but perhaps across orders as well, since an area similar to MT and OAa is found in the parietal lobe of the cat (Clare and Bishop, 1956). Changes in location and architectonics after shifts in connectional topography illustrate the principle that functional areas may be much less bound to gross anatomical landmarks than to specific connectional (architectonic) units. (See also Campbell, this volume.)

Language Areas

One of the greatest barriers to the progress of research on language-structure relationships has been the absence of suitable animal models. Recent demonstrations of language capability and cerebral lateralization in nonhuman primates (Premack, 1976; Petersen et al., 1978; LeMay and Geschwind, 1970), however, present a fresh opportunity to study structure-function relationships and cerebral lateralization in animals, for it is likely that similar anatomical substrates underlie these functions in man and nonhuman primates. (See Noback, this volume.) In man the auditory region in the temporal lobe, and within it architectonic area Tpt, play a noteworthy role in language function (Galaburda et al., 1978; Galaburda and Sanides, 1980). First, area Tpt is located in a part of the superior temporal gyrus where lesions are apt to result in a particular language disturbance known as Wernicke's aphasia. Second, Tpt appears to be larger on the left side, a finding in favor of left-sided specialization for language function (Galaburda et al., 1978). The frontal opercular formations are also involved in language functions (Broca, 1861) with area 44 (of Brodmann, 1909) making up a crucial portion of this anterior language region (Galaburda, 1980). The temporal and frontal language regions are connected by a fiber bundle whose integrity appears necessary for normal language (Geschwind, 1970).

In the rhesus monkey the auditory region in the temporal lobe is organized in a manner similar to that in man (Pandya and Sanides, 1973; Galaburda and Sanides, 1980) and an area Tpt can also be identified in this animal. Frontal lobe architectonic analyses have disclosed granular motor fields similar to those found in the opercular formations in man albeit in a different location within the frontal lobe (see below). The frontal operculum in monkey contains a different type of cortex, more akin to the dysgranular proisocortical and primitive isocortical fields found lying adjacent to the anterior insula in man. The relevance of the granular motor fields of the monkey to primate communication is not known at the present time, but it is tempting to presume that these areas are precursors to those found in man and subserve scaled-down communicative functions in these animals.

Superior Temporal Gyrus. The auditory regions of both man and rhesus monkey make up a major portion of the superior temporal gyrus and are similarly organized. The primitive root of the superior temporal gyrus is anterior and medial in the temporopolar proisocortex. In man the temporopolar proisocortex is all medial, whereas in the macaque it extends laterally to cover the most rostral and inferior tip of the temporal pole (Fig. 2). In both species this proisocortex is characterized by a barely discernible lamina IV and by large-celled, almost fused laminae V and VI (Fig. 2). The latter are primitive features, and they become progressively less evident as one proceeds caudally on the superior temporal gyrus (Fig. 2). As mentioned above, increased lamination and specialization are both features of an evolved isocortex, and both of these characteristics are increasingly manifest caudally on the superior temporal gyrus.

The greatest degree of specialization in the temporal lobe is achieved by the auditory koniocortex (primary cortex) and the differentiation is characterized by an overwhelming burst in the number of granule cells. A parakoniocortex located immediately lateral to the koniocortex (PaAlt in monkey, PaAi in man) is specialized in its striking development of the pyramidal cells in lamina IIIc (Fig. 2). Conversely, a tendency toward the development of a truly eulaminate cortex such as found in the inferior parietal lobule is realized in the most caudal region, area Tpt (Fig. 2). Thus

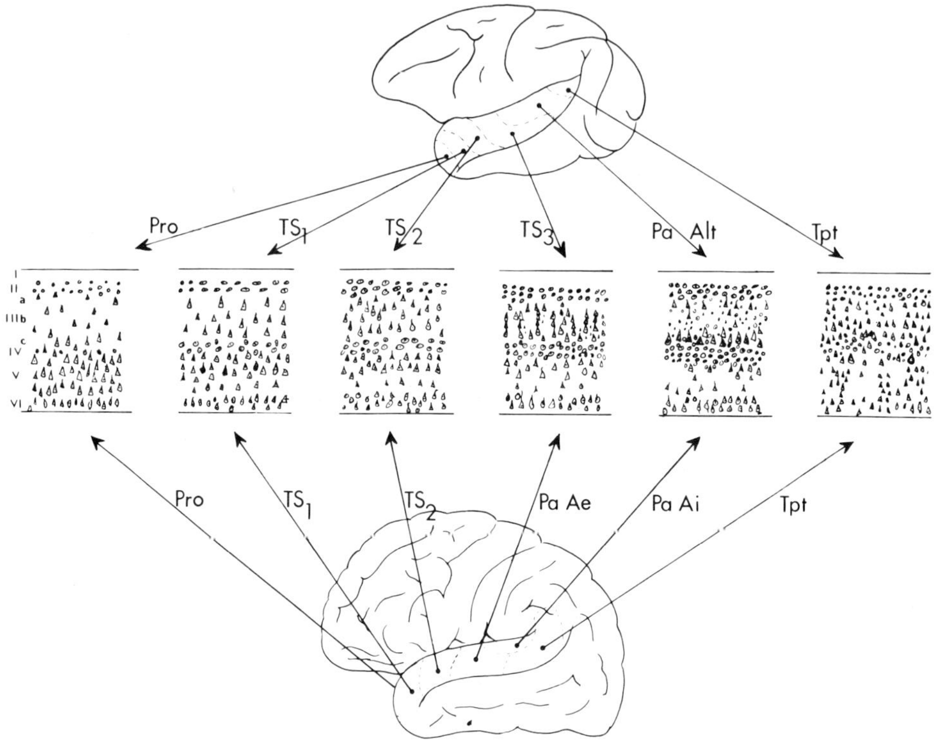

Fig. 2. Diagram of the superior temporal gyrus in monkey (above) and man (below) to show rostro-caudal trends of architectonic differentiation. The temporopolar proisocortex (Pro) is rostral in the monkey and rostro-medial in the human. With caudad movement there is stepwise differentiation of layer III and increased breadth in layer IV, which reach a peak in the monkey's lateral parakoniocortex (PaAlt) and in human internal parakoniocortex (PaAi). Temporoparietal Tpt shows a tendency toward eulamination, common to the neighboring homotypical cortex of the inferior parietal lobule. Note the Pro is barely a six-layered cortex with near fusion of layer V and VI.

interposed between the proisocortical temporal pole and the nearly eulaminate caudal end one can find areas of progressive specialization as well as progressive eulamination as illustrated by the trend from Pro to Tpt shown in Figure 2. Features of specialization are illustrated by a development of the granular layer, or the enlargement of the pyramidal neurons of layers III and V. These features may accompany increased eulamination seen in modern cortices, but may obscure the visual impression of eulamination, thus giving highly specialized cortices a strikingly uneven (heterotypical) appearance.

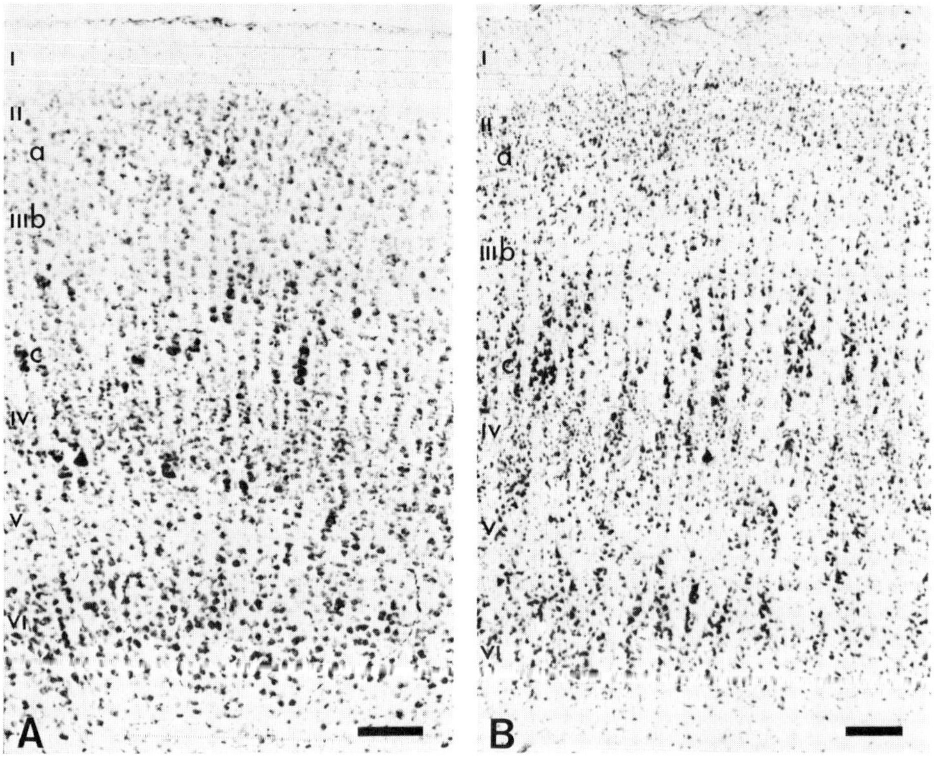

Fig. 3. Area Tpt in the monkey (A) and in man (B). (Also see text.) Note that similar features are present in this area in both species: layer III pyramids in rows; a relatively broad granular layer IV with indistinct borders with III and V; occasional large pyramids in Va; a relatively clear, cell-poor Vb; and a dense, broad layer VI. Also note difference in magnification needed to equalize the cortical thickness. Bars: A) 150 μm; B) 200 μm.

Area Tpt is most peculiar because it exhibits both temporal and parietal features. On the one hand, it shares a vertical columnar appearance with the adjacent temporal field TS_3 (PaAe in man), a similarity that led von Economo (1925) to classify these two cortices together (TA_1 and TA_2). On the other hand, it has a eulaminate structure similar to that of the inferior parietal areas. In both man and rhesus monkey Tpt also has a poorly defined layer IV because of pyramidal cells encroaching from laminae IIIc and Va and because of occasional large pyramids in lamina Va. Also the separation between laminae V and VI is indistinct in spots because vertical columns bridge the two laminae (Figs. 3A and 3B).

The Frontal Lobe. As in the case with the superior temporal gyrus, the frontal lobe exhibits equivalent trends of cortical differentiation and eulamination. In this lobe the proisocortical roots are found on the olfactory-insular and cingular regions, and the frontal isocortex is strung between these two regions (Fig. 1). Language-relevant areas are found in relation to both roots: the supplementary speech area in the medial part of the supplementary motor area is adjacent to the cingular regions, and the opercular speech areas are related to the olfactory-insular cortex. Differentiation and eulamination proceed in stages away from the proisocortical roots and culminate with area 4 as the most specialized and areas 8 and 9 and their subdivisions as the most eulaminate. The opercular regions exhibit intermediate stages of eulamination and of specialization. Thus they can have a clear separation of laminae V and VI (eulamination); and a narrow yet distinct lamina IV and relatively well developed pyramids in lamina IIIc (specialization). Area 44 in the human pars opercularis is more specialized

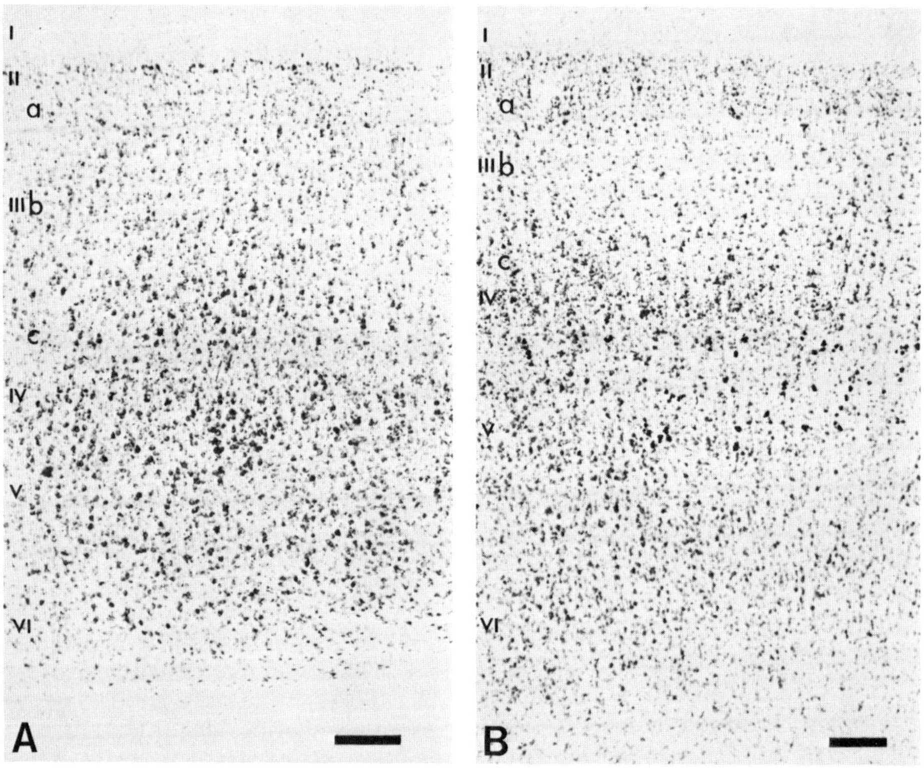

Fig. 4. Area 44 (see text) in monkey (A) and in man (B). Note that similar features are present in this area in both species: layer III contains large pyramids on the border with IV, which is indistinct; layer IV is narrow but clearly present; layer Va contains numerous large pyramids, often in rows, encroaching on the deeper border of layer IV; Vb is relatively sparse for this part of the frontal lobe; and layer VI is not easily separable from Vb. Also note the difference in magnification needed to equalize the cortical thickness. Bars: A) 150 μm; B) 200 μm.

than adjacent area 45 because it contains larger IIIc pyramids and has a lesser degree of eulamination (Fig. 4a). On the other hand, the premotor cortex, area 6, adjacent but posterior and superior to area 44, is even more specialized and less eulaminate. In the monkey an area exhibiting similar architectonic characteristics as the human 44 is found in a different location in the frontal lobe (see below) (Fig. 4b).

Although architectonic characteristics of a given cortical area are similar in homologous areas in two species, there are important differences. These differences are, however, of a histological rather than an architectonic nature, that is cellular size, shape and packing density vary, but the internal relationships of layers and columns are nearly the same (Figs. 3 and 4). It appears, furthermore, that the total number of cells through the depth of the cortical plate does not vary among primates, at least in the visual cortex, even though cortical thickness itself differs (Rockel et al., 1980).

TOPOGRAPHICAL DIFFERENCES

Architectonics

As already discussed, the general distribution and arrangement of architectonic areas on the superior temporal gyrus in the human and monkey brains are similar (Pandya and Sanides, 1973; Galaburda and Sanides, 1980). What is important to note and of greatest relevance to this paper is the fact that shifts occur without their being necessarily reflected in changes in the gross surface topography of the brain. In both species more primitive fields are located rostrally, and either more specialized or eulaminate fields are found caudally. The differences between species consist of the relative sizes of individual areas and some shifts in areal topography.

Anteriorly in the temporal pole the macaque proisocortex occupies the most rostral, and slightly ventral portion of the superior temporal gyrus. In man this cortex is found to occupy the dorsomedial corner of the superior temporal plane, just lateral to the rostralmost portion of the insular proisocortex. It is as though the expanded growth of TS_1 had pushed the proisocortex medially.

In the back of the superior temporal gyrus, the topography of area Tpt shifts considerably. In the monkey, Tpt is restricted to the caudalmost portion of the Sylvian fissure, extending between it and the climbing superior temporal sulcus (Fig. 5B). Tpt extends medially on the superior temporal plane for a variable distance. Thus Tpt in the rhesus monkey is well defined by the Sylvian fissure and the superior temporal sulcus, making recognition during surgery for connectional and physiological studies an easy task. In the human brain Tpt is relatively larger, making up a bigger proportion of the posterior auditory region (Galaburda and Sanides, 1980; Galaburda, unpublished observations). In addition, Tpt is not restricted to the temporal lobe. In most specimens, and particularly on the left side, this

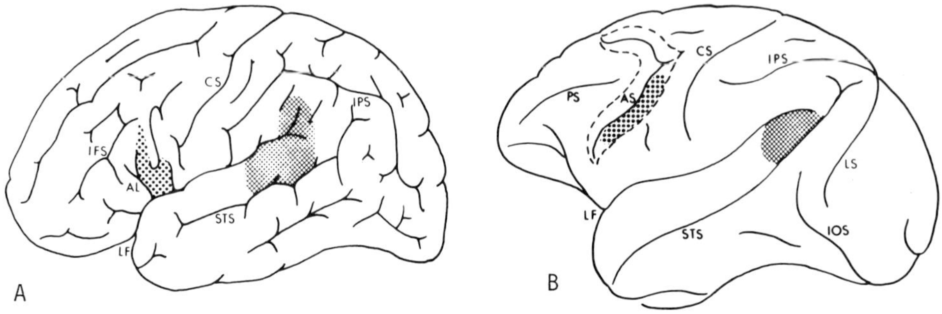

Fig. 5. Diagram of the cerebral convexity in man (A) and monkey (B) to show the topographic distribution of areas Tpt (crosshatched) and 44 (closed circles). Note the extension of Tpt to the parietal lobe in man. Also note the shift of area 44 toward the frontal operculum from monkey to man. The similar relationship of 44 to an anterior sulcus in both species suggests homology for the ascending limb of the Sylvian fissure in man (AL) and the inferior limb of the arcuate sulcus in the monkey (AS). Other abbreviations: CS, central sulcus; IFS, inferior frontal sulcus; IOS, inferior occipital sulcus; IPS, intraparietal sulcus; LF, lateral (Sylvian) fissure; LS, lunate sulcus; PS, principal sulcus; STS, superior temporal sulcus.

architectonically defined area extends onto the inferior parietal lobule (Galaburda et al., 1978) (Fig. 5A). The undersurface of the inferior parietal lobule (the parietal operculum) also contains a variable amount of Tpt that extends from the superior temporal gyrus along the caudal end of the Sylvian fissure. Thus, in the human brain, area Tpt becomes opercularized in the parietal lobe as well as in the temporal lobe.

In the frontal lobe we find a difference between the two species in the topography of area 44. In the rhesus monkey area 44 is found on the caudal bank of the inferior limb of the arcuate sulcus, entirely in the depth of the sulcus (Fig. 5B). The area assumes an oblong configuration and occupies a major portion of the length of the sulcus. In man area 44 is found on the pars opercularis of the frontal operculum. It extends from the ascending limb of the Sylvian fissure anteriorly to the precentral sulcus posteriorly, and from the inferior frontal sulcus superiorly downward around the opercular lip and into the Sylvian fossa towards the peri-insular region (Fig. 5A). Area 44 appears to have expanded in the human brain, extending outside of the monkey arcuate sulcus, and becoming opercularized, thus pushing more primitive areas medially around the insula. Also, given the comparable relationship of area 44 to an anterior sulcus in both species, the inferior limb of the arcuate sulcus in the monkey may have become the ascending limb of the Sylvian fissure in the human brain.

Connections

Once homologous areas are identified on architectonic grounds, further differences and similarities can be ascertained by studying fiber connections. Furthermore, the discovery that areas thought to be homologous purely on architectonic grounds also have similar connectional patterns strengthens the notion that the areas are indeed homologous. In the human brain connections may be studied in postmortem material through the use of gross dissection techniques and special stains for nerve fibers. In addition, in brains that have sustained lesions during life, fiber connections may be traced by staining for degenerated fibers (Mesulam, 1979). By and large these methods are not entirely satisfactory for the uncovering of detailed connectional patterns, although sufficient clues to general connections can be obtained. The fiber organization can then be tested more accurately in homologous areas of experimental animals.

Modern experimental techniques for the study of fiber connections in animals use tracer molecules that are transported along axons in living neural tissue. The administration of radioactively labeled amino acids into an area containing neuronal cell bodies results in their uptake by the cell, the incorporation of the amino acids into cellular proteins, and the distribution of the protein throughout the extent of the neuron and its processes. Thus, the site of termination of the axons can be identified and the anterograde connection outlined (Cowan et al., 1972). Conversely, horseradish peroxidase may be injected into an area containing neuronal terminals that incorporate the enzyme and transport it in a retrograde fashion towards the cell body. Afferent connections to a given area can thus be discerned (Mesulam, 1976).

In man the posterior superior temporal gyrus and the frontal operculum are thought to be connected by the arcuate fasciculus of Burdach (Fig. 6A). Except for its opercular portion this fascicle is not compact. It arises from the superior and middle temporal gyri as far anteriorly as the temporal pole, throughout the length of these gyri and as far back as the inferior parietal lobule and preoccipital regions. The fascicle then condenses into an arcuate structure convex upwards, which travels in the parietal operculum near the upper margin of the insula. According to Meynert (1865) the fiber bundle terminates in the Rolandic operculum, (Brodmann's area 43), and in the foot of the third frontal convolution, or pars opercularis (Area 44) (Fig. 6A).

In the rhesus monkey injection of radioactive tracer into area Tpt shows that this area indeed sends fibers to area 44 in the arcuate sulcus. The fibers thus displayed do not travel in the parietal operculum as they do in man. Instead the major portion of the fibers course deeply under the posterior lip of the insula to run along with the fiber systems underlying the insula in the extreme and internal capsules (Fig. 6B). A similar projection appears to exist in the squirrel monkey (Forbes and Moskowitz, 1977). Thus, the major difference in topography of this fiber system between human and

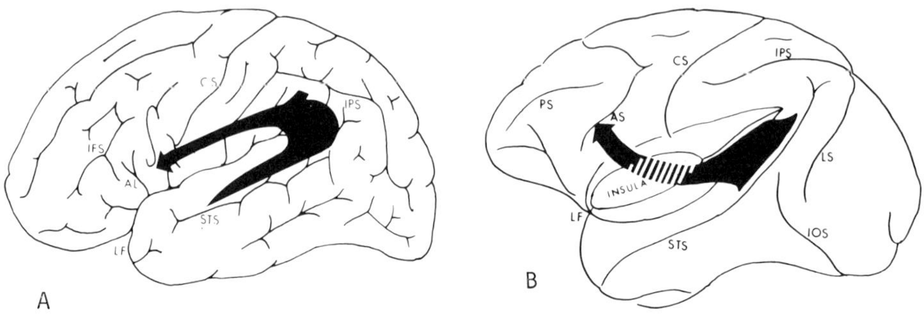

Fig. 6. Diagram of the cerebral convexity in man (A) and monkey (B) to show the origin, course, and termination of the fiber pathway connecting areas Tpt and 44. In man this pathway forms a portion of the arcuate sulcus, which is only partially shown in the diagram to show primarily the superior temporal gyrus origins. The pathway becomes a compact bundle only in the operculum (see text) and terminates in the subcentral and opercularis portions of the frontal operculum (arrow). In the rhesus monkey radioactively labeled tracer in area Tpt results in labeled fibers and terminations along the course of the arrow shown, passing beneath the insula (interrupted) and terminating in the caudal bank of the inferior limb of the arcuate sulcus corresponding to area 44. For abbreviations see Figure 5 legend.

monkey brains is that in man the fiber tract as well as the architectonic areas it connects has become opercularized.

CONCLUSIONS

With the massive expansion of eulaminate and highly specialized isocortical formations in the human brain, the surface topography of certain areas shifted. In the human frontal lobe, area 44, part of Broca's speech area, appears to have grown out of the vertical arcuate sulcus and down to form an ample opercular formation in the frontal lobe, pushing more primitive cortices medially toward the insula. In conjunction with this shift the arcuate sulcus has been displaced downward and become an ascending branch of the Sylvian fissure. It is likely that a downward shift has occurred because of a parallel burst of growth in the prefrontal granular cortex above area 44. An enlargement of the speech-relevant area, Tpt, in man's superior temporal gyrus resulted in the encroachment of this area into the adjacent parietal operculum. With the opercularization of frontal area 44 and temporal Tpt, there has also been an adjunctive displacement of fiber pathways into the opercular formations. The identification of homologies via architectonic analysis and fiber connections provides for a better understanding of changes in the brain's configuration among different primate species and helps in the recognition of homologous gross landmarks on the brain's surface, knowledge that can then be applied to the study of brain endocasts from extinct primates.

Acknowledgments: This study was supported by NIH grants NS14018, NS09211, V.A. Research Project #6901, and a biomedical support grant to the Beth Israel Hospital

REFERENCES

Allman, J.M., and Kaas, J.H., 1971. A representation of the visual field in the caudal third of the middle temporal gyrus of the owl monkey (*Aotus trivirgatus*). Brain Research, 31:85-105.

Broca, P., 1861. Remarques sur le siège de faculté de langage articule suivies d'une observation d'aphémie (perte de la parole). Bull. Soc. Anat. Paris, pp. 330-357.

Brodmann, K., 1909. Vergleichende Lokalisationslehere der Grosshirnrinde, Barth, Leipzig.

Clare, M.H., and Bishop, G.H., 1956. Potential wave mechanisms in the cat cortex. Electroencephalog. and Clin. Neurophysiol., 8:583-602.

Cowan, W.M., Gottlieb, D.I., Hendrikson, A.E., Price, J.C., and Woolsey, T.A., 1972. The autoradiographic demonstration of axonal connections in the central nervous system. Brain Research, 37:21-51.

Economo, C. von, 1925. The Cytoarchitectonics of the Human Cerebral Cortex, Oxford University Press, London.

Forbes, B.F., and Moskowitz, N., 1977. Cortico-cortical connections of the superior temporal gyrus in the squirrel monkey. Brain Research, 136:547-552.

Galaburda, A.M., 1980. La région de Broca: Observations anatomiques faites un siècle après la mort de son découvreur. Rev. Neurol., 136:609-616.

Galaburda A., and Sanides, F., 1980. Cytoarchitectonic organization of the human auditory cortex. J. Comp. Neurol., 190:597-610.

Galaburda, A.M., Sanides, F., and Geschwind, N., 1978. Cytoarchitectonic left-right asymmetries in the temporal speech region. Arch. Neurol., 35:812-817.

Geschwind, N., 1970. The organization of language and the brain. Science, 170:940-944.

Kesarev, V.S., Malofeyeva, L.I., and Trikova, O., 1977. Structural organization of the cerebral cortex in cetaceans. Arkhiv. Anatomii, Gistologii i Embriologii 73:23-30.

LeMay, M., and Geschwind, N., 1975. Hemispheric differences in the brains of great apes. Brain Behav. Evol., 11:48-52.

Leonard, C.M., 1969. The prefrontal cortex of the rat. I. Cortical projections of the mediodorsal nucleus. II. Efferent connections. Brain Research, 12:321-343.

Mesulam, M-M., 1976. Tetramethyl benzidine for horseradish peroxidase neurohistochemistry: incubation parameters and visibility. J. Histochem. Cytochem., 24:1273-1280.

Meynert, T., 1865. Anatomie der Hirnrinde und ihre Verbindungsbahnen mit den empfindenden Oberflächen und den bewegenden Massen. M. Leidesdorf's Lehrbuch der psychiat. Krankheiten, Erlangen.

Pandya, D.N., and Sanides, F., 1973. Architectonic parcellation of the temporal operculum in rhesus monkey and its projection pattern. Z. Anat. Etwickl. -Gesch., 139:127-161.

Petersen, M.R., Beecher, M.D., Zoloth, S.R., Moody, D.B., and Stebbins, W.C., 1978. Neural lateralization of species-specific vocalizations by Japanese macaques (*Macaca fuscata*). Science, 202:324-326.

Premack, D., 1976. Language and intelligence in ape and man. Am. Scientist, 64:674.

Rockel A.J., Hiorns, R.W., and Powell, T.P.S., 1980. The basic uniformity in structure of the neocortex. Brain, 103:221.

Sanides, F., 1970. Functional architecture of motor and sensory cortices in primates in the light of a new concept of neocortex evolution. In, The Primate Brain, Advances in Primatology, vol. 1, C.R. Noback and W. Montagna, eds., Appleton-Century-Crofts, New York, pp. 137-208

Seltzer, B., and Pandya, D.N., 1978. Afferent cortical connections and architectonics of the superior temporal sulcus and surrounding cortex in the rhesus monkey. Brain Research, 149:1-24.

MAPPING FOSSIL ENDOCASTS

Dean Falk*

One of the major themes of this volume is to underscore current efforts to synthesize the fields of comparative neurology and paleoneurology. In recent years, primate paleoneurologists have used cortical maps determined by neurophysiologists to interpret fossil endocasts (Falk, 1981; Gurche, this volume; Radinsky, 1972, 1975, 1979). As Galaburda and Pandya (this volume) point out, new information on architectonics and connections derived from comparative neurology should be incorporated into functional interpretations of fossil endocasts. The purpose of this paper is to review and, where necessary reinterpret, sulcal patterns of fossil primate endocasts in light of neurophysiological (Welker, 1976) and architectonic (Galaburda and Pandya, this volume) data.

CORTICAL MAPS

Using the method of evoked potentials, Welker and Campos (1963) mapped primary somatic sensory cortices of representatives of five genera of procyonids. They determined that sulci consistently separate cortical projection areas of major body subdivisions, such as hindlimb, forelimb, and head. In the raccoon, for example, various segments of the triradiate sulcus separate cortical representations of individual digit and palm pad areas. This enlarged tactile representation of forepaw in the raccoon cortex reflects the fact that this animal "uses its paw to an unusual degree in the exploration of its environment" (Welker and Campos, 1963).

In a study on the slow loris (*Nycticebus coucang coucang*), Sanides and Krishnamurti (1967) give evidence that cortical sulci, grooves, dimples, and

*From the Department of Anatomy and Caribbean Primate Research Center, University of Puerto Rico School of Medicine, San Juan, Puerto Rico 00936

fissurets delimit cytoarchitectonic fields. This conclusion has been supported by microelectrode mapping of cortex in slow loris (Krishnamurti et al., 1976). Although Galaburda and Pandya (this volume) agree that architectonics and connections reflect the functional organization of the brain, their cytoarchitectonic work on the superior temporal gyrus in human and rhesus monkey brains suggests that shifts in architectonic topography "occur without their being necessarily reflected in changes in the gross topography of the brain" (this volume). Zilles et al. (this volume) also reports that laminar patterns in clearly homologous cortical areas (e.g., striate area) show a considerable degree of variability in different primate species. These findings should act as a caution to workers who "read" sulcal patterns from fossil endocasts. Nevertheless, an extensive literature on cortical maps in ceboid (Fulton and DeBarenne, 1933; Hirsch and Coxe, 1958; Coxe et al., 1957; Pubols and Pubols, 1971) and cercopithecoid (Fulton and DeBarenne, 1933; Vogt and Vogt, 1919; Werner and Whitsel, 1968; Woolsey, 1958) monkeys suggests that face, forelimb, hindlimb, and tail (in the case of prehensile-tailed New World monkeys, Radinsky, 1972; Falk, 1980b) representations of primary somatic sensory and motor cortices *are* delimited by sulci (Fig. 1). (See Falk, 1978b, for a review of primate cortical maps.)

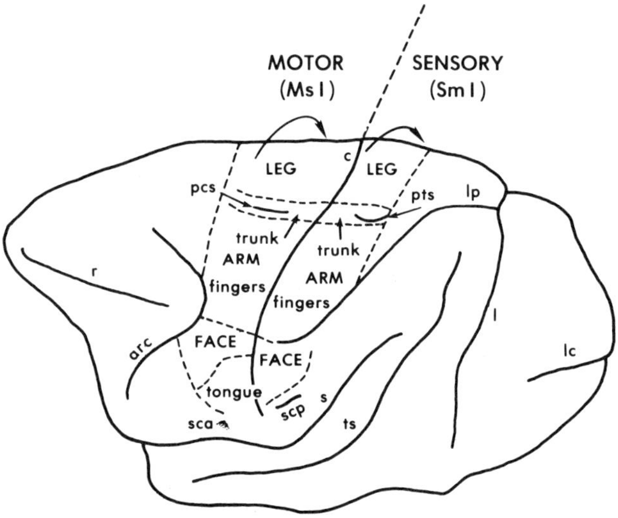

Fig. 1 Hindlimb, forelimb, and face representations relative to sulci of cercopithecoid monkeys. Left, lateral view of cerebral cortex of *Macaca*, modified after Woolsey, 1964. Abbreviations: arc, arcuate sulcus; c, central s.; ip, intraparietal s.; l, lunate s.; lc, lateral calcarine s.; pcs, precentral superior s.; pts, postcentral superior s.; r, rectus s.; s, Sylvian fissure; sca, subcentral anterior s.; scp, subcentral posterior s.; ts, superior temporal s. Note that l approximates the rostral boundary of the striate cortex (area 17).

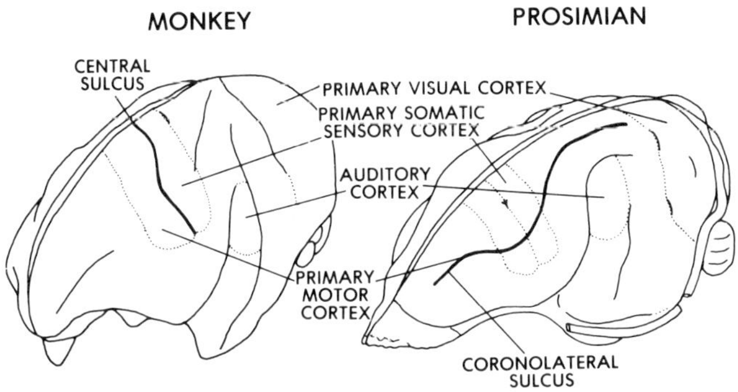

Fig. 2 Differences in sulcal patterns in prosimians and monkeys. Note the central sulcus of monkey that separates primary somatic sensory from motor cortices and the coronolateral (= coronal) sulcus of prosimian that separates the representation of head from that of forelimb and see text for discussion. After Radinsky, 1975.

Finally, the lunate sulcus approximates the rostral boundary of the primary visual cortex (area 17) in some prosimians, monkeys, apes, and (when present) man. See Figure 1 and Falk (1978b; 1980a) for details.

FUNCTIONAL INTERPRETATIONS OF SULCAL PATTERNS

Three sulci will be discussed as examples of how primate cortical maps have been used to interpret sulcal patterns of fossil primate endocasts: 1) the coronolateral (=coronal) sulcus found in fossil prosimans, 2) the central sulcus of fossil *Theropithecus* baboons, and 3) the lunate sulcus of early australopithecine hominids.

Coronolateral Sulcus

Radinsky (1975) illustrated an important difference in the sulcal patterns of extant prosimians and anthropoids (Fig. 2). The primary somatic sensory and motor cortices of most prosimian brains is crossed rostrocaudally between face and forelimb representations by the long coronolateral sulcus whereas there is little or no indication of a central sulcus. On the other hand, a marked central sulcus separates primary somatic sensory and motor cortices in most anthropoids, which lack a coronolateral sulcus (Figs. 1 and 2). Thus, Radinksy (1975) suggested that one evolutionary trend in primate brain evolution was development of a central sulcus in anthropoids rather than the coronolateral sulcus seen in prosimians. Since a central sulcus is present in the approximately 27,000,000-year-old anthropoid fossil *Aegyptopithecus* this trend (the functional significance of which can only be guessed at) had occurred at least once by the late Oligocene (Radinsky, 1973).

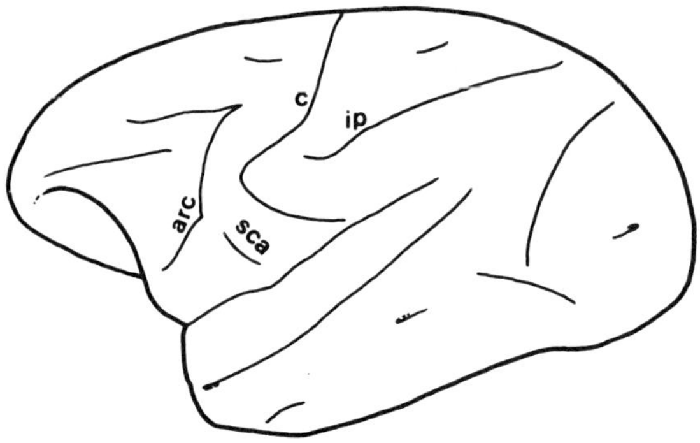

Fig. 3. The sulcal pattern of a fossil baboon, *Theropithecus oswaldi*. Note the configuration of the central sulcus and see text for discussion. After Falk, 1981.

Gurche (this volume) notes that three Eocene lemuriforms exhibit a coronolateral sulcus similar to that found in modern prosimians. He notes, however, that the configuration of the coronolateral sulcus swings more laterally across sensory and motor cortices in extant than in fossil prosimians (see Fig. 2), thus suggesting that modern prosimians have greater areas of forelimb and hindlimb representation than did fossil lemuriforms. Gurche's findings may have implications for interpretations of locomotor behavior during prosimian evolution.

Central Sulcus

Neurophysiologists have determined that the lateral end of the central sulcus separates primary somatic sensory from motor face representation in monkey (Fig. 1). This fact has been useful in interpreting the unique sulcal pattern of an extinct fossil baboon, *Theropithecus oswaldi*, that lived in Africa at least two million years ago (Fig. 3; Falk, 1981). The hook-shaped lateral end of the central sulcus seen in fossil *T. oswaldi*, but not seen in its living relative *T. gelada*, is unlike the much less curved central sulcus that appears in all genera of living New and Old World monkeys (Falk, 1978b, 1980b, 1981). Cortical maps determined from living monkeys suggest that fossil *Theropithecus* had a relatively enlarged cortical sensory and motor face representation that may have been related to masticatory and/or gestural functions. This hypothesis fits well with the literature on the behavior of modern *Theropithecus gelada*, which is known both for "seed-eating" and "lipflip" gestures. Fossil *T. oswaldi*, however, is more derived in its sulcal pattern than extant *T. gelada* and it therefore seems unlikely that *T. oswaldi* was a direct ancestor of modern geladas. Rather, the fossil record suggests that the two groups were separate by 1.8 million years ago. (See Falk, 1981, for details.)

Lunate Sulcus

Due to historical accident, the lunate sulcus is perhaps the best known sulcus in hominid paleoneurology. This sulcus, which approximates the rostral boundary of primary visual cortex (area 17), is present in most monkey and ape brains (Fig. 1). A continuous lunate sulcus was present in about 58% of 120 human hemispheres examined by Connolly (1950). When present, the lunate sulcus of humans is positioned more caudally on the lateral surface of the brain than are the lunate sulci of monkeys or apes. Thus it has been suggested that during the course of human brain evolution, primary visual cortex migrated caudally and medially, possibly because of an expanded parietal association cortex (Falk, 1980a).

The Taung-type specimen for *Australopithecus africanus* was associated with a natural endocast (Dart, 1925). According to Dart, the lunate sulcus occupied a relatively caudal, human-like position in the Taung endocast. The supposed caudal position of the lunate sulcus in australopithecines has been widely accepted and interpreted as indicating reorganization of the australopithecine brain in a human-like direction despite its small size (Holloway, 1972).

Having examined the original specimen (Falk, 1980a), I believe that Dart mistook the lambdoid suture for the lunate sulcus (see also Clark et al., 1936). Contrary to current consensus, my study of all seven of the currently available South African australopithecine natural endocasts indicates that the sulcal pattern of australopithecine brains appeared to be ape-like rather than human-like. It is highly unlikely, however, that australopithecine brains were identical to those of pongids. The fact that australopithecines and pongids differ in posture, locomotion, dentition, other aspects of skeletal morphology, and inferred ecological niche suggests that the nervous systems of these hominoids also differ. These differences are not manifest at the level of cortical sulcal patterns, however, but may have existed at finer neurological levels, perhaps at cellular or molecular levels. Thus, I believe there is little or no evidence with respect to the lunate or other sulci (i.e., superficial evidence) that elucidates the nature of brain evolution in early hominids.

SULCI, CYTOARCHITECTONICS AND LANGUAGE AREAS

There is currently much interest in interpreting gross morphology of extant and fossil primate brains in terms of language origins and functions (Falk, 1978a, 1980c; Galaburda and Pandya, this volume; Galaburda et al., 1978; LeMay, this volume; Noback, this volume). Because the relevant brain areas include association cortices rather than primary cortices (which are fairly well mapped for primates, see above), paleoneurologists interested in language origins are especially likely to benefit from the research of comparative neurologists. As an example of the potential for synthesis between comparative neurology and paleoneurology, the implications of

Galaburda and Pandya's work (this volume) for paleoneurologists studying the evolution of the frontal lobe and/or language origins are discussed below.

As LeMay points out (this volume), human brains and endocasts often show, along the orbital margin of the frontal lobe, a prominence (or so-called "tuber" of Kochetkova, 1978) that is sometimes referred to as Broca's speech area. LeMay suggests that, because Broca's area in humans has ill-defined boundaries, it is misleading to refer to the "tuber" as Broca's area and that it would be especially misleading to apply this label to apparently equivalent (i.e., topographically similar) prominences in nonhuman primates. Nevertheless, Jones and Powell (1970) suggested that the orbitofrontal cortex at the lower end of the arcuate sulcus in monkeys is homologous to areas 44, 45, and 47, which includes Broca's area (i.e., 44 and part of 45, Crosby et al., 1962) on the dominant side, and Holloway (1975) has referred to the third inferior frontal convolution of the Taung endocast as the "so-called Broca's area, which is involved in motor control of speech."

Galaburda and Pandya's architectonic work (this volume) shows that the orbitofrontal cortex in rhesus monkey is *not* homologous with opercular formations in man. Rather, monkey frontal operculum contains cortex akin to the dysgranular prosiocortical and primitive isocortical fields found near the anterior insula in man (Galaburda and Pandya, this volume). Therefore, orbitofrontal cortex in monkey is not homologous with Broca's area in man, as some workers (Jones and Powell, 1970) have suggested.

Based on architectonic criteria, Galaburda and Pandya have determined that area 44 (part of Broca's area in humans) occupies different locations in the frontal lobes of humans and rhesus monkeys (see Fig. 5 of Galaburda and Pandya, this volume). In humans, area 44 occupies the pars opercularis of the frontal operculum and is delimited anteriorly by the ascending limb of the Sylvian fissure. In rhesus monkeys, area 44 is located within the caudal bank of the inferior limb of the arcuate sulcus (Fig. 4) Galaburda and Pandya therefore suggest that the ascending limb of the Sylvian fissure and the inferior limb of the arcuate sulcus are homologous in humans and rhesus monkeys. Further support for this homology is provided by the fact that both sulci define the caudal boundary of area 45 (Walker, 1940). (See Campbell, this volume, for further discussion of methods for determining homologies of nervous system structures.) Thus the work of Galaburda and Pandya requires revision of the generally recognized homology between the transverse portion of the arcuate sulcus in cercopithecoids and the precentral inferior sulcus of man (Falk, 1978b; Ariens-Kappers et al., 1960).

My earlier interpretation of asymmetries in sulcal patterns in frontal lobes of monkeys (Falk, 1978a) must also be revised in light of Galaburda and Pandya's work. A photogrammetric computer analysis showed that the caudal end of the rectus (= principalis) sulcus more closely approximates the arcuate sulcus in left than in right hemispheres of baboons. I suggested that

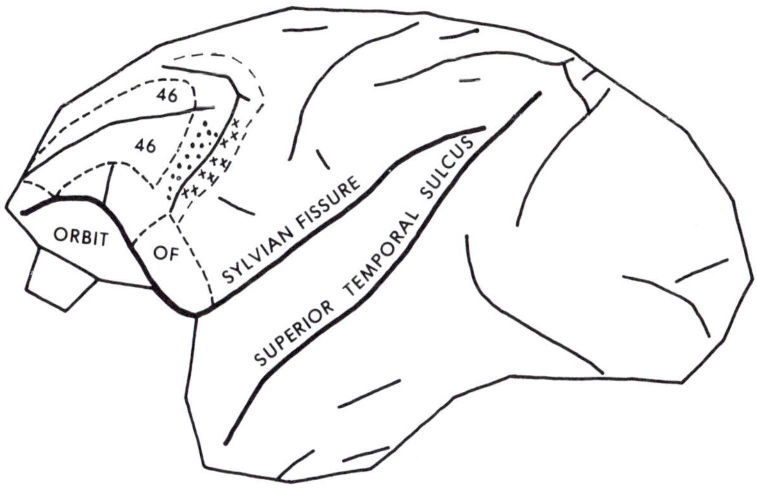

Fig. 4. Cytoarchitectonic areas in frontal lobe of rhesus monkey. Dots = probable homologue of area 45 in humans (Walker, 1940); x's = homologue of area 44 in humans (Galaburda and Pandya, this volume); OF = orbitofrontal cortex. Note that area 45 is found on the "anterior lip of the arcuate sulcus" and also appears on the lateral surface of the brain (Walker, 1940) whereas "area 44 is found on the caudal bank of the inferior limb of the arcuate sulcus, entirely in the depth of the sulcus" (Galaburda and Pandya, this volume). Modified after Falk, 1978b.

this asymmetry was due to expansion in the left hemisphere of the prefrontal granular integrative cortex (Brodmann's area 46, see Fig. 4), which contains the rectus sulcus and which projects to the orbitofrontal cortex "which some workers believe to be homologous with Broca's area in man" (Falk, 1978b). I then suggested that this and other asymmetries seen in sulcal patterns of Old World monkeys were related to vocal communication systems (Falk, 1978a, 1980c). Although the hypothesis that cortical asymmetries in monkeys are related to vocalization systems has since found psychophysical support (Petersen et al., 1978), Galaburda and Pandya's findings suggest the possibility that frontal lobe asymmetries seen in sulcal patterns of monkeys may be due to expansion of areas 44 and 45 in left hemispheres, rather than to expansion of area 46. That is, the asymmetry in question may be due to changes in the configuration of the arcuate sulcus rather than to a caudal elongation of the rectus sulcus. Since the original baboon sample included only eight endocasts (Falk, 1978a), frontal lobe asymmetry needs to be reconfirmed and studied in a much larger sample of endocasts. This is just one example of how comparative neurology can generate new hypotheses that could be tested on large numbers of endocasts (or brains, if available).

Returning to the fossil record, Galaburda and Pandya's findings on the frontal lobe suggest that we should be cautious when making statements about an enlarged Broca's area in australopithecines (Holloway, 1975). I find (Falk, 1980a) that the gross morphology of the orbital area of South African australopithecine natural endocasts that Holloway (1975) calls the "third inferior frontal convolution" resembles the comparable area in African apes rather than the opercular region of human brains. Until we know where area 44 is in pongids, it will be difficult to analyze the so-called "tuber" in australopithecines.

CONCLUSION

Although endocasts from fossil primates have been known for over 100 years (Kochetkova, 1978), cortical maps with which to interpret primate sulcal patterns have only recently been refined and published by neurophysiologists (Welker, 1976). Paleoneurologists have been receptive to this relatively new source of information, as the examples of the coronolateral, central, and lunate sulci discussed above show. Although published neurophysiological maps have helped paleoneurologists interpret the functional significance of expanded primary somatic sensory and motor cortices in fossil primates, their usefulness is limited when it comes to association cortices. Recent comparative neurological research such as the quantitative cytoarchitectonic work of Zilles et al. (this volume) and the architectonic and connectivity work of Galaburda and Pandya (this volume) has the potential for providing new information on association cortices (including language areas) that will be useful to paleontologists. Thus, paleoneurologists can look forward to new information that will help them to continue to refine their understanding of the fossil record of primate, including human, brain evolution.

Acknowledgment: I thank Dr. Leonard Radinsky for criticism of the manuscript.

REFERENCES

Ariens-Kappers, C.U., Huber, G.C., and Crosby, E.C., 1960. The Comparative Anatomy of the Nervous System of Vertebrates, Including Man, Hafner, New York.

Connolly, C.J., 1950. External Morphology of the Primate Brain, Charles C Thomas, Springfield, Illinois.

Coxe, W.S., Hirsch, J.E., Benjamin, R.M., Welker, W.L., Thompson, R.F., and Woolsey, C.M., 1957. Precentral and supplementary motor areas of *Ateles*. Physiologist, I:19 (abstract).

Crosby, E.C., Humphrey, T., and Lauer, E., 1962. Correlative Anatomy of the Nervous System, Macmillan, New York.

Dart, R.A., 1925. *Australopithecus africanus*: The man-ape of South Africa. Nature Lond., 115:195-199.

Falk, D., 1978a. Cerebral asymmetry in Old World monkeys. Acta Anat., 101:334-339.

Falk, D., 1978b. External neuroanatomy of Old World monkeys (Cercopithecoidea). Contrib. Primat., 15:1-95.

Falk, D., 1980a. A reanalysis of the South African australopithecine natural endocasts. Am. J. Phys. Anthrop., 53:525-539.

Falk, D., 1980b. Comparative study of endocranial casts of New and Old World monkeys. In, Evolutionary Biology of the New World Monkeys and Continental Drift, R.L. Ciochon and B. Chiarelli, eds., Plenum, New York, pp. 275-292.

Falk, D., 1980c. Language, handedness and primate brains: did the australopithecines sign? Am. Anthropol., 82:71-78.

Falk, D., 1981. Sulcal patterns of fossil *Theropithecus* baboons: phylogenetic and functional implications. Int. J. Primatol., in press.

Fulton, J.F., and DeBarenne, J.G.D., 1933. The representation of the tail in the motor cortex of primates, with special reference to spider monkeys. J. Cell. Comp. Physiol., 2:399-426.

Galaburda, A.M., LeMay, M., Kemper, T.L., and Geschwind, N., 1978. Right-left asymmetries in the brain. Science, 199:852-856.

Hirsch, J.F., and Coxe, W.S., 1958. Representation of cutaneous tactile sensibility in cerebral cortex of *Cebus*. J. Neurophysiol., 21:481-498.

Holloway, R.L., 1972. Australopithecine endocasts, brain evolution in the Hominoidea, and a model of hominid evolution. In, The Functional and Evolutionary Biology of Primates, R. Tuttle, ed., Aldine, Chicago, pp. 185-203.

Holloway, R.L., 1975. The Role of Human Social Behavior in the Evolution of the Brain. Forty-third James Arthur Lecture on the Evolution of the Human Brain. The American Museum of Natural History, New York.

Jones, E.G. and Powell, T.P.S., 1970. An anatomical study of converging sensory pathways within the cerebral cortex of the monkey. Brain, 93:793-820.

Kochetkova, V., 1978. Paleoneurology, Wiley and Sons, New York.

Krishnamurti, A., Sanides, F., and Welker, W.I., 1976. Microelectrode mapping of modality-specific somatic sensory cerebral neocortex in slow loris. Brain Behav. Evol., 13:267-283.

Le Gros Clark, W.E., Cooper, D.M., and Zuckerman, S., 1936. The endocranial cast of the chimpanzee. J.R. Anthrop. Inst., 66:249-268.

Petersen, M.R., Beecher, M.D., Zoloth, S.R., Moody, D.B., and Stebbins, W.C., 1978. Neural lateralization of species-specific vocalizations by Japenese macaques (*Macaca fuscata*). Science, 202:324-326.

Pubols, B.H., and Pubols, L.M., 1971. Somatotopic organization of spider monkey somatic sensory cerebral cortex. J. Comp. Neurol., 141:63-76.

Radinsky, L.B., 1972. Endocasts and studies of primate brain evolution. In, The Functional and Evolutionary Biology of Primates, Aldine, Chicago, pp. 175-184.

Radinsky, L.B., 1973. *Aegyptopithecus* endocasts: oldest record of a pongid brain. Am. J. Phys. Anthropol., 39:239-248.

Radinsky, L.B., 1975. Primate brain evolution. Am. Sci., 63:656-663.

Radinsky, L.B., 1979. The Fossil Record of Primate Brain Evolution. Forty-ninth James Arthur Lecture on the Evolution of the Human Brain. The American Museum of Natural History, New York.

Sanides, F., and Krishnamurti, A., 1967. Cytoarchitectonic subdivisions of sensorimotor and prefrontal regions and of bordering insular and limbic fields in slow loris (*Nycticebus coucang coucang*). J. Hirnforsch., 9:225-252.

Vogt, C., and Vogt, O., 1919. Allgemeine Ergebnisse unserer Hirnforschung. J. Psychol. Neurol., 25:279-462.

Walker, A.E., 1940. A cytoarchitectural study of the prefrontal area of the macaque monkey. J. Comp. Neurol., 73:59-86.

Welker, W.I., 1976. Mapping the brain. Historical trends in functional localization. Brain Behav. Evol., 13:327-343.

Welker, W.I., and Campos, G.B., 1963. Physiological significance of sulci in somatic sensory cerebral cortex in mammals of the family Procyonidae. J. Comp. Neurol., 120:19-36.

Werner, G., and Whitsel, B.L., 1968. Topology of the body representation in somatosensory area I of primates. J. Neurophysiol., 31:856-869.

Woolsey, C.N., 1958. Organization of somatic sensory and motor areas of the cerebral cortex. In, Biological and Biochemical Bases of Behavior, H. Harlow, and C.N. Woolsey, eds., University of Wisconsin Press, Madison, pp. 63-81.

Woolsey, C.N., 1964. Cortical localization as defined by evoked potiential and electrical stimulation studies. In, Cerebral Localization and Organization, G. Schaltenbrand and C.N. Woolsey, eds., University of Wisconsin Press, Madison, pp. 17-26.

EARLY PRIMATE BRAIN EVOLUTION

John A. Gurche*

In attempting to understand the evolutionary development of the primate brain, we might ultimately wish to identify evolutionary changes (both quantitative and organizational) in the brain during the course of primate evolution and answer questions about adaptive significance and selection pressures involved. How far can the current evidence go toward providing such an understanding? Two main bodies of evidence currently available come from the study of the brains of living primates (comparative neurobiology) and from the study of fossil endocranial casts (paleoneurology). In this paper I will focus on what the fossil record currently can tell us about brain evolution in early prosimian primates.

The only direct record of the brains of early prosimians consists of a disappointingly small number of natural or artificial endocranial casts from Paleocene, Eocene, and Oligocene skulls. There are currently twelve fairly complete specimens and a handful of fragments representing a total of nine genera: *Plesiadapis*, *Ignacius*, *Smilodectes*, *Notharctus*, *Adapis*, *Tetonius*, *Hemiacodon*, *Necrolemur*, and *Rooneyia*. There are also several specimens of *Microsyops*, currently controversial as to its primate status.

In smaller primates, the meningeal membranes surrounding the brain are thin enough to permit imprinting of the brain into the calvarium and thus to enable endocast representation of surface features (such as sulcal patterns) in some detail. In addition, endocasts can provide information about the size of the brain and some of its components, especially the olfactory bulbs, telencephalon, cerebellum, and several areas of the neocortex.

---

*From the Department of Anthropology, University of Kansas, Lawrence, Kansas 66045

ABSOLUTE BRAIN SIZE

The only quantitative information about early primate brains published to date concerns the absolute and relative size of the entire brain. The significance of brain size and its usefulness as a predictor of other statistics have been subjects of controversy (Jerison, 1973; 1975; 1976; 1977; 1980; this volume; Holloway, 1968; 1969; 1975; 1976; 1980; Holloway and Post, this volume; Radinsky, this volume). Changes in brain size were selected for (directly or indirectly), and I have taken the position that any evolutionary changes in the brain, quantitative (including changes in brain size) or organizational, that can be identified are useful in constructing a picture of the course of primate brain evolution, which can then serve as a base for study of selection pressures involved.

Several methods of estimating absolute brain volume for fossil primates have been used recently. Various workers (Stephan, 1960; Bauchot and Stephan, 1967; and see Jerison, 1973) found the ratio of endocast volume to brain volume to be close to 1.0, and the two are commonly used interchangeably. Jerison's (1973; 1979) estimates are derived with the graphic double integration method, using measurements of Radinsky's (1970) dorsal and lateral view illustrations of restored endocasts of five early primates as data. The volume (V) of an endocast is estimated by calculating the volume of an elliptical cylinder with a length equal to the length (l) of the endocast and major and minor axes equal to the mean width ($\bar{w}$) and mean height ($\bar{h}$) of the endocast, using the formula:

$$V = \pi \bar{w} \bar{h} l$$

Errors of between five (Jerison, 1973) and fifteen (Radinsky, 1977) percent have been reported when estimates yielded by this method are checked against water displacement volumetric measurements of the same endocasts. Radinsky (1977) gave estimates based both on this method and on water displacement measurements of clay models, which were based on the same dorsal and lateral view illustrations of restored endocasts of four fossil genera (the *Adapis* endocast was measured directly by water displacement).

I made water displacement measurements directly on six restored endocasts of early primates. Measurements of similarly sized objects of known volume served as a check (ball bearings were chosen because they have a high degree of sphericity and their volumes can be calculated accurately from their diameters), and the deviations of these measurements from the calculated volumes never exceeded 0.3% of the latter. Most of the restoration of hidden or missing portions of the endocasts involved only the assumption that the endocast was bilaterally symmetrical. In addition, where the skull obscured portions of an endocast, it served as a guide and as an upper limit for restoration. Outer-limit estimates were also produced by making improbably large and small restorations of the endocasts, by restoring hidden or missing parts as asymmetrically larger or smaller respectively than their visible counterparts and by assuming unreasonably thin or thick areas of the skull where it covers the endocast. The resulting outer-limit values are highly improbable and are certain to bracket the actual endocast volumes.

Volumetric estimates made by me, Jerison (1979), and Radinsky (1977) are given in Table 1 (also included is an estimate for *Adapis parisiensis* made by Martin (1973) by packing mustard seed into the brain-case). Differences were generally minor when the estimates were made from the same specimen. My measurement of *Necrolemur* was made on an endocast from a skull (Montauban 9) that is much smaller than that in the composite

TABLE 1. Brain Size Estimates for Early Primates

Primate	Source of Estimate[a]	Brain Size in cc's	Volume of Minimum Restoration	Volume of Maximum Restoration
Adapis parisiensis				
BM.M 20192[b]	(JG)	8.31	no	no
	(LR)	9.00	restoration	restoration
	(HJ)	9.40	necessary	necessary
	(RM)	8.80		
Smilodectes gracilis				
YPM 12152	(JG)	9.95	8.66	12.52
USNM 23276	(JG)	9.12	8.84	9.26
Composite restoration	(LR)	9.50		
	(HJ)	9.60		
Notharctus tenebrosus				
Endocast A	(JG)	10.43	8.91	13.29
Tetonius homunculus				
AMNH 4194	(LR)	1.50		
	(HJ)	1.60		
Necrolemur antiquus				
Montabaun 9	(JG)	2.65	2.00	2.86
Composite restoration	(LR)	4.35		
	(HJ)	4.20		
Rooneyia viejaensis				
UT 40688-7	(JG)	7.38	6.50	7.77
		7.50		
		6.40		

[a]Abbreviations: JG, my water displacement measurements of restored endocasts; LR, Radinsky's (1977) estimates by graphic double integration and water displacement measurements of clay models (*Adapis* was measured directly by water displacement); HJ, Jerison's (1979) estimates by graphic double integeration; RM, Martin's (1973) estimate for *Adapis* by packing mustard seed into the cranial cavity.

[b]Endocast identification.

illustration (Radinsky, 1970), which served as a basis for estimates by Radinsky (1977) and Jerison (1979). I did not measure *Tetonius* because the specimen is distorted, and in this case direct volumetric measurement of the endocast could not improve on previous estimates.

BODY SIZE

Because the absolute size of the brain is strongly influenced by body size, absolute volumes must be assessed relative to body size before comparisons can be made. The estimation of body size is one of the most difficult aspects of this type of study. Postcranial bones are fragmentary or lacking for some of the early primate genera. Body size estimates have been based on comparisons of the sometimes scrappy fossil material with skeletal dimensions of modern prosimian genera for which body weight data are available. Differences in the estimation of relative brain size for early primates (Jerison, 1973; 1979; Radinsky, 1977) have been due more to differing body size estimates than to varying estimates of absolute brain size.

The problem has been dealt with in a variety of ways. Attempts to avoid estimating body size, by relating absolute brain size to foramen magnum area instead (Radinsky, 1967, 1970), have been questioned because the high correlation between foramen magnum area and body weight drops considerably when the effects of brain size (also correlated with foramen magnum area) are mathematically removed (Radinsky, 1977; Martin, this volume).

Comparisons of skeletal material of modern and fossil primates (only the skull for *Tetonius*, *Necrolemur*, and *Rooneyia*; more complete skeletons for the adapids) have resulted in the use (for a fossil form) of the body weight of a modern prosimian with matching skeletal dimensions, or an interpolation based on several modern forms that appear to bracket the size of the fossil form. Radinsky (1977) has used this method.

The relationship of a linear dimension (such as body length or skull length) to body weight in a modern prosimian has been used as a model for the same relationship in the fossil forms, so that the expected body weight (P_f) of the fossil form can be calculated from that linear measurement of the fossil, using the equation:

$$P_f = (L_f/L_m)^3 \ (P_m)$$

where L_f and L_m are homologous linear measurements in the modern and fossil forms and P_m is the body weight of the modern form (Jerison, 1979). Because this relationship (the ratio of the cubed linear measurement to body weight) varies in modern forms, choice of a modern model with body proportions and size similar to the fossil is important.

Another approach is to use linear-skeletal-dimension-to-body-weight relationships in a group of modern forms to predict body weights for the fossil genera. A regression equation representing the average linear-dimension-to-body-weight relationship in a modern group can be used to predict body weights for the fossil forms from that linear dimension. Again, choice of an appropriate modern model (group) is important. Body weight estimates used most currently by Jerison (1979) in calculations of relative brain size for early tarsiiform prosimians *Tetonius*, *Necrolemur* and *Rooneyia* were derived from a regression equation representing the average skull length - body weight relationship in modern ceboid monkeys, based on the assumption that omomyids (including these fossil forms) are more closely related to other haplorhines (including ceboids) than to other prosimians. (Use of this ceboid model produced lower body size estimates and thus higher relative brain size values than would use of a prosimian model). This assumption certainly has yet to be demonstrated by the fossil record, and as Radinsky (this volume) points out, similarity in size and proportion is more important than close phylogenetic relationship in choosing an appropriate modern model group. A modern prosimian model (including species of the modern genus *Tarsius*) would seem a better choice on the basis of either phylogenetic proximity or similarity in general size and proportion to known skeletal material (skulls and few questionable postcranials) of the fossil tarsiiforms, but any clear choice of an appropriate modern model must await the discovery of more complete fossil remains.

Body size estimates obtained by different authors vary sometimes widely (Table 2), depending on choice of method and modern models. My body size estimates, which are simple interpolations between body weights published for modern prosimians (Bauchot and Stephan, 1966; Napier and Napier, 1967; Walker, 1968), which bracket the size of the fossil specimens in skull or skeletal dimensions, agree closely with Radinsky's (1977) estimates using similar methods (except for my higher estimate for *Rooneyia*; Radinsky's estimate was given in response to a previous (Jerison, 1973) very low estimate, and he cautioned that it was probably still too low). In addition, to bracket "actual" body size values, I made improbably high and low body size estimates based on comparisons of the fossil prosimians with modern forms that were larger and smaller in available skeletal dimensions. These values provide a range of estimates for each of the fossil forms, and include in every case the estimates made by Radinsky (1977). Two estimates recently used by Jerison (1979), however, still fall outside this range (*Tetonius and Rooneyia*).

RELATIVE BRAIN SIZE

Two indices of relative brain size are currently in common use. The encephalization quotient (EQ) (Jerison, 1973) expresses brain weight (E_i) for any species i as a proportion of that expected in an average living mammal (determined by regression analysis and by fitting of a minimum convex polygon principal axis) of equal body weight (P_i), using the equation:

$$EQ_i = \frac{E_i}{0.12\ P_i^{\ 2/3}}$$

For fossil primates, measurements of single specimens are used in place of species means. This practice is treated below in a discussion of individual variation.

Similarly, the index of progression (IP) expresses brain weight for any species i as a proportion of that expected in a typical "basal" insectivore (a group of living insectivores considered most primitive by Stephan (1972) and colleagues) of equal body weight.

Both indices make use of average brain-body relations in their respective groups of living mammals as represented by lines fitted to

TABLE 2. Body Size Estimates Used in the Calculation of Relative Brain Size in Early Prosimians

Early Prosimian	Source of Estimate[a]	Estimated Body Size in Grams	Minimum Body Size Estimate	Maximum Body Size Estimate
Adapis parisiensis	(JG)	2500	1400	3100
	(LR)	2540		
	(HJ)	1700		
Smilodectes gracilis				
YPM 12152[b]	(JG)	2500	1400	3100
USNM 23276, USNM 17997	(JG)	2400	1400	3100
	(LR)	2540		
	(HJ)	1700		
Notharctus tenebrosus	(JG)	2500	1400	3100
	(LR)	2540		
Tetonius homunculus	(JG)	165	85	245
	(LR)	160		
	(HJ)	74		
Necrolemur antiquus	(JG)	265	185	345
	(LR)	300		
	(HJ)	228		
Rooneyia viejaensis	(JG)	700	500	900
	(LR)	500+		
	(HJ)	326		

[a]Abbreviations: JG, estimates of J. Gurche (this paper); LR, estimates of L. Radinsky (1977); HJ, estimates of H. Jerison, (1979).
[b]Derived from YPM 12152, and similarly sized composite USNM specimens.

graphed brain and body weight points on a double logarithmic scale. Because the slopes used with each index are similar (2/3 for EQ; 0.64 for IP), the use of either index results in identical comparisons of relative brain size for data discussed here.

Table 3 shows EQs calculated from my estimates of brain volume (translated to brain weights by multiplication by 1.036 following methods of Stephan et al., 1970) and body size and those of Radinsky (1977) and Jerison (1979). Included are the minimum and maximum values calculable from my ranges of estimates. My results are presented graphically in Figure 1.

Although differences in estimation of brain size and especially body size have resulted in differing EQ values, studies by Jerison (1979), Radinsky (1977), and me agree in suggesting that the Eocene lemuriforms (*Smilodectes*, *Adapis* and *Notharctus*) were relatively smaller brained than any modern prosimian. (My upper limit EQ values for *Notharctus* and one specimen of *Smilodectes* are high because of the incompleteness of both specimens, but their probable EQ values are well below the range for modern prosimians.) All three studies show an increase in encephalization through time for the tarsiiforms, with *Tetonius* (Early Eocene) being relatively smaller brained than any other known fossil or living tarsiiform (on the lower edge or below the range for modern prosimians). *Necrolemur* (Middle to Late Eocene) is intermediate, and *Rooneyia* and *Tarsius* are the largest brained. No evolutionary progression is implied here; brain size may simply have been increasing in several different lineages. All three studies place *Rooneyia* somewhere within the range of modern prosimians.

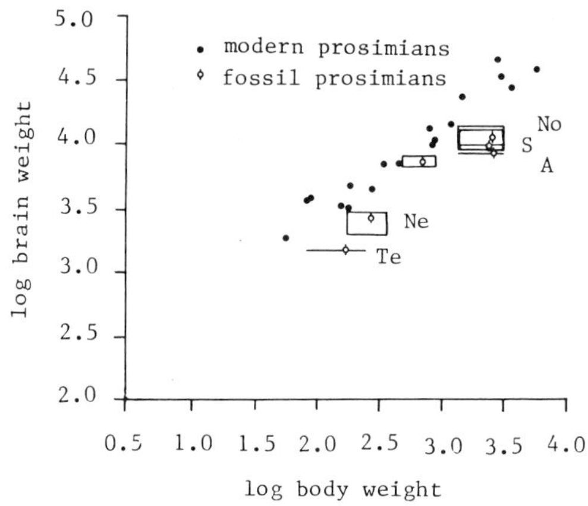

Fig. 1. Brain weight in relation to body weight in fossil and living prosimians on a double logarithmic scale. Rectangles indicate ranges of body and brain weight estimates for the fossil forms. Points within rectangles indicate more probable values. Data on modern prosimians are from Stephan et al. (1970). Abbreviations: S, mean value for two specimens of *Smilodectes*; No, *Notharctus*; A, *Adapis*; Te, *Tetonius*; Ne, *Necrolemur*; R, *Rooneyia*.

TABLE 3. Encephalization Quotients (EQ) for Early and Modern Prosimians

Early Prosimian	Source of Estimate[a]	EQ	Minimum EQ	Maximum EQ
Adapis parisiensis	(JG)	0.39	0.34	0.57
	(LR)	0.39		
	(HJ)	0.53		
Smilodectes gracilis				
YPM 12152[b]	(JG)	0.47	0.35	0.86
USNM 23276	(JG)	0.44	0.36	0.64
	(LR)	0.41		
	(HJ)	0.53		
Notharctus tenebrosus	(JG)	0.49	0.36	0.92
Tetonius homunculus	(JG)[c]	0.43	0.33	0.67
	(LR)	0.42		
	(HJ)	0.71		
Necrolemur antiquus	(JG)	0.56	0.35	0.76
	(LR)	0.79		
	(HJ)	0.94		
Rooneyia viejaensis	(JG)	0.81	0.60	1.07
	(LR)	0.97		
	(HJ)	1.23		
Modern Prosimian[d]				
Tarsius spectrum		1.14		
Tarsius syrichta		1.35		

[a]Abbreviations: JG, values calculated from volumetric measurements and body size estimates by J. Gurche (this paper); LR, values calculated by Radinsky (1977); HJ, values calculated by Jerison (1979).

[b]Endocast identification.

[c]My value for *Tetonius* is based on a brain volume of 1.5, used by both Radinsky (1977) and Jerison (1979) in their calculations.

[d]Range for modern prosimians: 0.67-1.89; mean: 1.09. Values for modern prosimians were calculated from data presented by Stephan et al. (1970). The mean EQ for modern prosimians was based on mean log body weight and mean log brain weight. Values for modern tarsiiforms are given for comparisons discussed in text.

Rooneyia has sometimes been aligned with anthropoids, and Jerison's results (1973; 1979) using graphic double integration estimates and low body-size estimates suggest that by the Early Oligocene, *Rooneyia* already had a relative brain size within the upper range for modern prosimians. Both Radinsky's values and mine indicate a relative brain size for *Rooneyia* within the lower range for modern prosimians.

No Paleocene primates are represented in my tables or figures. Absolute and relative brain size estimates have been published for *Plesiadapis* (Radinsky, 1977; Gingerich, 1976; Jerison, 1979), but these estimates are based on the external appearance of crushed skulls, and with the exception of fragments revealing olfactory bulb anatomy, no endocasts of Paleocene primates are known.

OLFACTORY BULB SIZE

Volumetric estimates for components of the brain have not been made previously, but are possible in some cases. Olfactory bulb representations are often delineated from the rest of an endocast by a sharp constriction, making it possible to estimate the posterior extent of the olfactory bulbs (using modern prosimian brains and endocasts as guides). Olfactory bulb representations were cast and restored where necessary (assuming they were symmetrical) for four of the fossil forms. In addition, to bracket actual volumes, improbably large and small restorations were made by varying the posterior extent of the olfactory bulbs and by assuming an asymmetrical shape.

These restored casts were recast in a material of known density, enabling calculations of their volumes from their weights. An object of known volume (again, a ball bearing) was measured in the same way as a check, with a maximum error of 1.3% of its actual volume.

The resulting values (Table 4) are graphed in Figure 2, along with data points for modern prosimians (data from Stephan et al., 1970). Index of progression values for olfactory bulbs were calculated similarly to those for relative brain size, expressing olfactory bulb weight (again obtained by multiplication of volumes by 1.036) for a fossil form as a proportion of that expected in an average basal insectivore (determined by Model II regression) of identical body weight. The resulting indices of relative olfactory bulb volume appear in Table 5, with the entire range for modern prosimians and an "average" value for modern prosimians, calculated from the means of logarithmically transformed body and brain weights.

Much discussion of apparent olfactory bulb size has appeared in the literature. Eocene prosimians have been widely reported to have large or unreduced olfactory bulbs compared with those of modern prosimians (indicating post-Eocene olfactory reduction). This has specifically been suggested for *Adapis*, *Tetonius*, and *Necrolemur* (Le Gros Clark, 1959;

TABLE 4. Measurements[a] of the Volumes of Olfactory Bulbs From Early Prosimian Endocasts[b]

Early Prosimian	Endocast Identification	Volumes of Restoration		
		Probable	Minimum	Maximum
Adapis parisiensis	BM. M. 20192	266.6	264.5	334.4
Notharctus tenebrosus	Endocast A	236.6	178.2	375.2
Tetonius homunculus	AMNH 4194	32.8	30.5	42.2
Necrolemur antiquus	Montauban 9	51.3	35.1	70.2

[a]In cubic centimeters.
[b]Olfactory bulb casts of minimum, maximum, and probable volume were made for each specimen using a material of known density. These were weighed and volumes were calculated.

TABLE 5. IP_o (Model II) Values[a] for Early and Modern Prosimians

Early Prosimian	Endocast Identification	IP_o		
		Probable	Minimum	Maximum
Adapis parisiensis	BM.M 20192	0.34	0.29	0.64
Notharctus tenebrosus	Endocast A	0.30	0.19	0.72
Tetonius homunculus	AMNH 4194	0.29	0.20	0.60
Necrolemur antiquus	Montauban 9	0.33	0.18	0.58
Modern Prosimian				
Range: 0.09-1.25; Mean: 0.43				

[a]These values are calculated from olfactory bulb volumes in Table 4 and body size estimates in Table 2 using the equation

$$I\,P_o\ (\text{Model II}) = \frac{0_i}{2.8948\ P_i^{0.71657}}$$

where O_i is olfactory bulb volume for species i in cubic millimeters and P_i is body weight in grams. Values for modern prosimians were calculated from data presented by Stephan et al. (1970). The mean IP_o for modern prosimians was based on mean log body weight and mean log olfactory bulb volume.

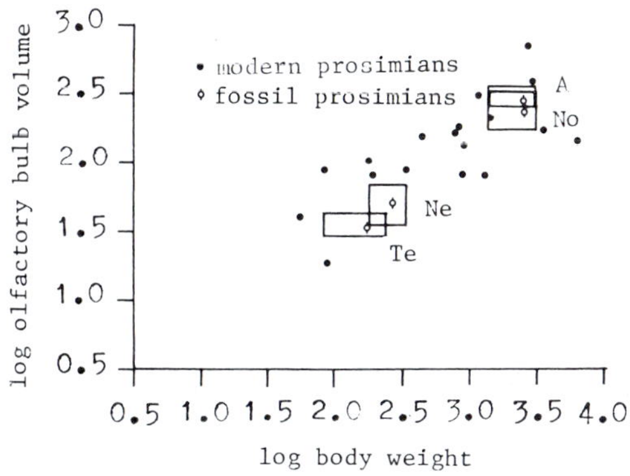

Fig. 2. Olfactory bulb volume in relation to body weight on a double logarithmic scale for fossil and living prosimians. Rectangles indicate ranges of body weight and olfactory bulb estimates for the fossil forms. Points within rectangles indicate more probable values. Abbreviations are as in Figure 1. Data on modern prosimians are from Stephan et al. (1970).

Radinsky, 1967; 1970; Simons, 1972). My measurements indicate that the olfactory bulbs of those Eocene prosimians that can be measured, including these three genera, all had relative sizes well within the range for modern prosimians. Even the improbably extreme body size and olfactory bulb size estimates for the fossil forms yield IP_0 values that fit well within the range for modern prosimians. Although the olfactory bulbs may appear to be large in relation to the rest of the endocast, the proper comparison is with body size. Eocene tarsiiforms *Tetonius* and *Necrolemur* may have had slightly larger olfactory bulbs relative to body size than their modern relative *Tarsius* (data for *Tarsius syrichta* from Stephan et al. (1970) yield an IP_0 value of 0.25), but even here the apparent differences are slight.

Martin (1973) suggested that the olfactory bulbs of prosimians may have undergone no reduction, but instead may simply have arrested as other portions of the brain enlarged. The results presented here indicate that there is still no good evidence of olfactory bulb reduction in prosimians. Olfactory reduction may have occurred before the Eocene, but this possibility has yet to be confirmed by the fossil record.

SIZE OF THE TELENCEPHALON AND CEREBELLUM

Endocasts appear to offer little in terms of other brain portions that can be directly measured volumetrically. Telencephalic and cerebellar volumes, however, can be predicted from brain size in primates and insectivores. Combined logarithmically transformed brain weight and

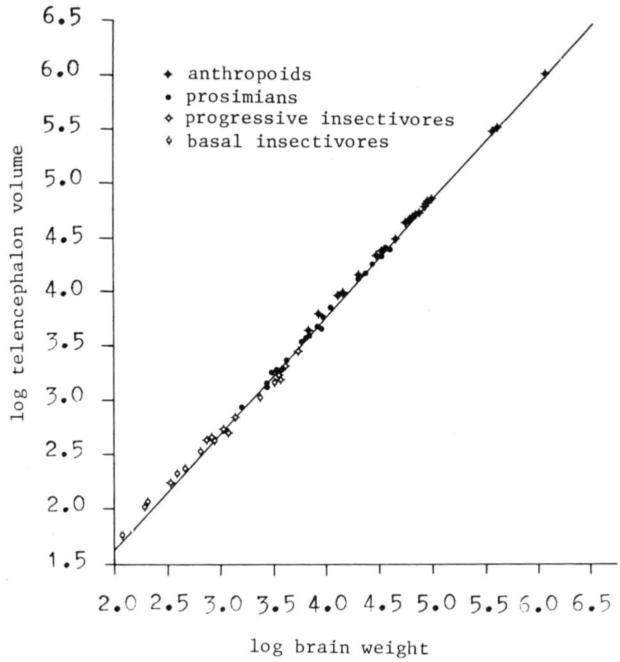

Fig. 3. Telencephalic volume in relation to brain weight on a double logarithmic scale for 22 insectivores and 41 prosimians. A Model II regression line is fitted to the data. Data are from Stephan et al. (1970).

telencephalic volumetric data for 22 species of living insectivores, 20 prosimians, and 21 anthropoids yield a product-moment correlation coefficient of 0.999. The equation for the regression line (Fig. 3) fitting these data

$$\hat{T}_i = 0.3830\,(E_i^{1.055988})$$

can be used to predict telencephalic volume from endocast volume. (E_i is the brain weight of species i and $\hat{T}_i$ is the predicted telencephalic volume. The equation has been transformed into a non-logarithmic form.) This, of course, assumes that the same relationship held for prehistoric forms, but this seems likely as it appears to hold uniformly in widely separated lineages in both orders.

Logarithmically transformed brain weights and cerebellar volumes for these same species yield a correlation coefficient of 0.996 and cerebellar volume can similarly be predicted using the equation for the regression line in Figure 4. Transformed into non-logarithmic form this equation is:

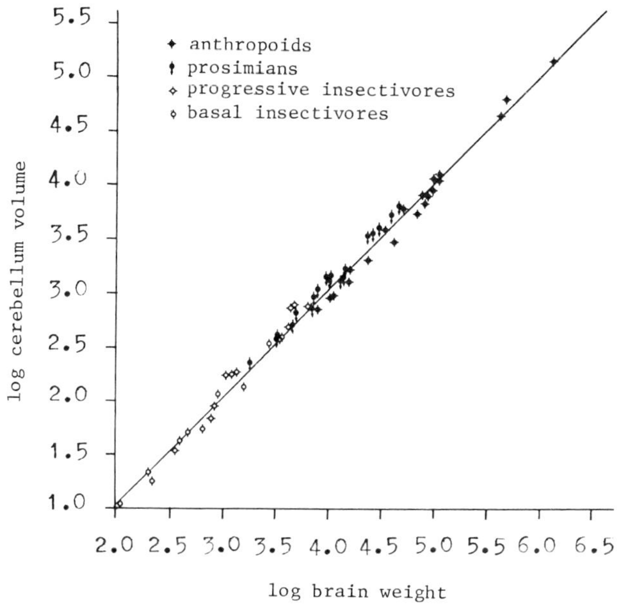

Fig. 4. Cerebellar volume in relation to brain weight on a double logarithmic scale for 22 insectivores and 41 primates. A model II regression line is fitted to the data. Data is from Stephan et al. (1970).

$$\hat{C}_i = 0.1333\ (E_i^{0.986717})$$

where $\hat{C}_i$ is the predicted cerebellar volume. The information obtained from these predictions is somewhat redundant, as comparisons will mimic those of brain size, but it may be useful for studies that concentrate on these particular parts of the brain and for attempting to understand how selection has acted to change the size of the brain and its components.

Jerison (1977, 1980) discusses other brain statistics that are highly correlated with and might be predicted by brain size, such as cortical volume, cortical surface area and neuron/glia ratio (for further discussion, see Holloway, 1980; Armstrong, this volume; Jerison, this volume).

COMPARATIVE SIZES OF NEOCORTICAL AREAS

No volumetric estimates are possible for neocortical areas, but comparisons of fossil endocasts with endocasts of modern prosimians of similar body size sometimes suggest marked differences. These comparisons should be used cautiously; they are qualitative and cannot be subjected to allometric scaling.

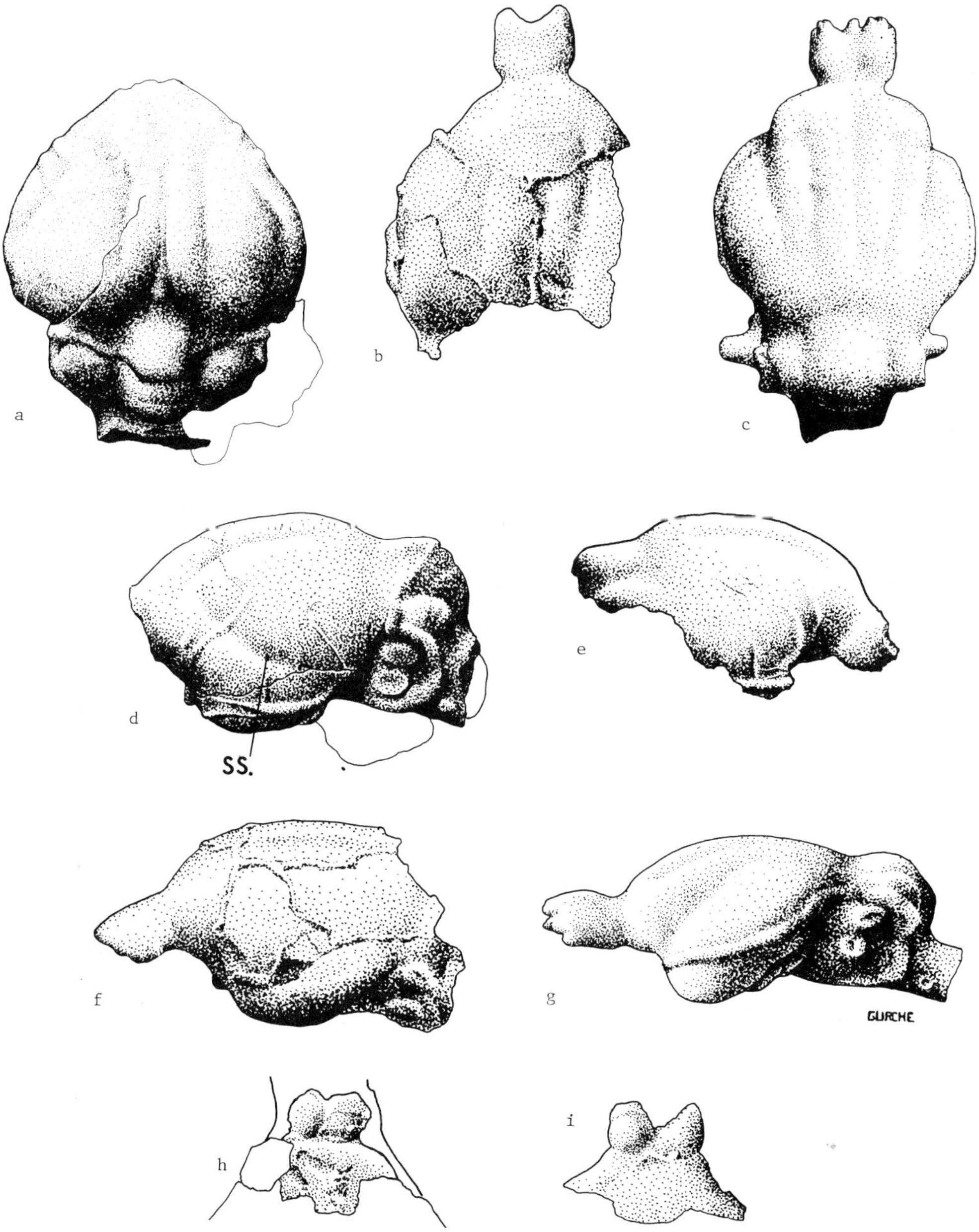

Fig. 5. Endocasts of early lemuriform prosimians: Dorsal and left lateral views of (a and d) *Smilodectes gracilis* (USNM 23276), (b and f) *Notharctus tenebrosus* (Endocast A), and (c and g) *Adapis parisiensis* (BM.M. 20192); e, right lateral view of *Smilodectes gracilis* (YPM 12152); dorsal views of anterior endocasts of (h) *Notharctus tenebrosus* (USNM 244365) and (i) Notharctus sp. (Endocast B); ss., suprasylvian sulcus? Not to scale. Nonshaded areas represent portions of the skull.

The temporal lobes are broad in endocasts of all six fossil forms for which this region is known (Figs. 5 and 6), and this area appears to have expanded early in primate evolution in comparison with other neocortical lobes (see Allman, this volume). The position of the rhinal fissure, as indicated by a vascular sinus that follows it in modern prosimians, indicates that the temporal neocortex probably expanded to its modern ventral extent very early (Early Eocene or before) in both lemuriform and tarsiiform lineages.

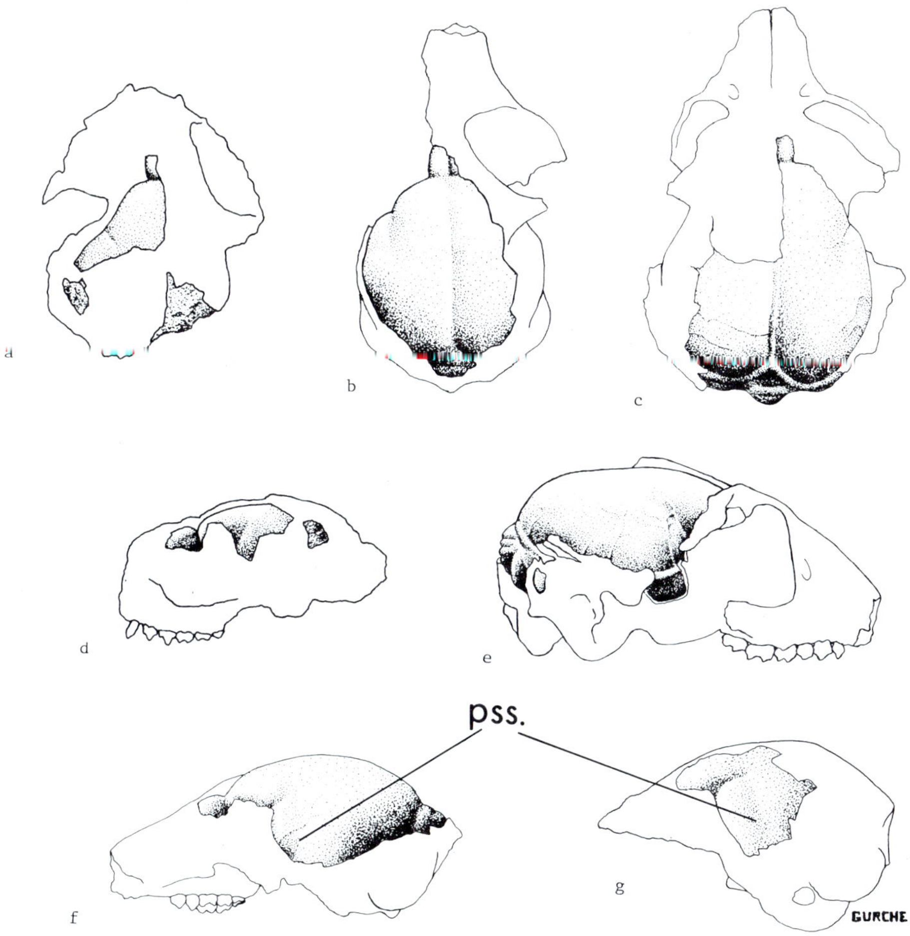

Fig. 6. Endocasts of early tarsiiform prosimians: a and d, dorsal and left lateral view of *Tetonius homunculus* (AMNH 4194); b and f, dorsal and left lateral view of *Necrolemur antiquus* (Montauban 9); c and e, dorsal and right lateral view of *Rooneyia viejaensis* (UT 40688-7); g, left lateral view of *Necrolemur antiquus* (BM.M 4490); pss., postsylvian sulcus? Not to scale. Nonshaded areas represent portions of the skull.

The frontal lobes appear very small in Middle Eocene lemuriforms *Smilodectes* and *Notharctus* (Fig. 5), somewhat larger in the Late Eocene *Adapis*, and larger still in modern lemuriforms of similar body size.

Endocasts of fossil and living tarsiiforms are characterized by small frontal lobes and moderate to advanced overlap of the cerebellum by the occipital lobes (Fig. 6).

SULCAL PATTERNS

The meningeal membranes are thin enough in most prosimians to enable good endocast representation of the sulcal patterns. All of the more complete fossil endocasts, except for those of *Smilodectes*, clearly show the Sylvian sulcus, a characteristic primate feature. It has been suggested (Radinsky, 1970) that *Smilodectes* may have lacked a Sylvian sulcus. If so, it is the only known primate, living or fossil, without it. In the two *Smilodectes* endocasts exhibiting this region, there are faint dorsoventral ridges where the Sylvian sulcus is expected (Fig. 5). These appear to be representations of subdural blood vessels. Obscuring of fissures by blood vessels is common in endocasts and it is quite possible that the Sylvian sulcus was present in the brain of *Smilodectes* but is obscured in the endocast.

All three of the Eocene lemuriforms exhibit a coronolateral sulcus (Fig. 5) in a similar position to those of modern prosimians. In modern prosimians, however, this sulcus does not run parallel to the midline throughout its entire length (as in the fossil forms), but swings ventrolaterally, possibly allowing for larger forelimb and hindlimb regions in the primary somatosensory and motor cortices than found in the fossil forms.

A shallow sulcus of unknown homology runs along the dorsolateral surface of one endocast of *Smilodectes* (USNM 23276). It is variable in its expression on the two sides of this specimen (Fig. 5), and is absent on the only other specimen (YPM 12152) exhibiting this region. This has been tentatively suggested to be a homologue of the suprasylvian sulcus (Gazin, 1965), and Radinsky (1970) suggested that it might be undeflected in the absence of a Sylvian sulcus.

No other sulci are visible on early prosimian endocasts, except for a faint indentation running roughly parallel to the anteroventral border of the temporal lobe in two specimens of *Necrolemur* that exhibit this region (Fig. 6). It is similar in position to the postsylvian sulcus in *Tarsius*, giving the impression of a slight lip on the anterior face of the temporal lobe. If this indentation is homologous with the postsylvian sulcus, its position may indicate the posterior extent of the auditory cortex in *Necrolemur*.

INDIVIDUAL VARIATION

Knowledge of individual variation within species is useful in properly evaluating the significance of differences in specimens belonging to different species. At present, sample sizes for each of the fossil species are extremely small and do not permit estimates of ranges of individual variation. Studies have resorted to treating single specimens as representative of their species, a fact that emphasizes the tentative nature of our current knowledge of early prosimian brains.

The fossil record allows only hints of the extent of variability in some features. Three endocasts of *Adapis parisiensis* differ widely in overall shape and especially in the shape of the frontal lobes. Three specimens of *Notharctus* show variation in olfactory bulb form (Fig. 5), as do x-rays of several *Smilodectes* specimens. Two specimens of *Smilodectes* are similar volumetrically, but vary in the expression of sulci.

EARLY PRIMATE BRAIN EVOLUTION: A MODEL

Our incomplete knowledge of primate phylogeny further blocks any clear picture of early primate brain evolution, but the following hypothetical model is advanced from completed work, to be tested as further evidence comes in.

Sometime before the Early Eocene the following events occurred in lineages ancestral to tarsiifrom and lemuriform primates. The occipital lobes expanded over the midbrain, the temporal lobes became comparatively broad, and the temporal neocortex expanded ventrally, reaching levels in both groups similar to those in their modern relatives. There may or may not have been olfactory bulb reduction before the Eocene. Whether these events occurred before or after the tarsiiform-lemuriform split is not known.

Subsequent tarsiiform evolution was characterized in at least two lineages by increases in relative brain size, and probably in the size of the temporal and occipital lobes. There may have been slight (if any) reduction in the olfactory bulbs.

There is little evidence of any change in the brains of lemuriforms during the Eocene, except for an apparent increase in the size of the frontal lobes in the line leading to *Adapis* (over those of earlier adapids). Post-Eocene lemuriform brain evolution is marked by increases in brain size and an apparent enlargement of frontal and occipital lobes. The hindlimb and forelimb areas of the primary somatosensory and primary motor cortices appear to have expanded, and a variety of new sulci appear to have developed. There was little change in olfactory bulb size.

In the introduction to this paper I set a goal for this type of study: eventually we might hope to identify and quantify evolutionary changes in the primate brain and understand the selection pressures involved. The present evidence allows only a tentative picture of some of the changes that may have occurred. Complete fossil specimens are needed for more species, particularly more postcranial remains for body size estimates. More specimens per species are needed to better indicate ranges of variation. A better understanding of primate phylogeny would be helpful. Any real understanding of selection pressures must await a clearer picture of what changes actually occurred, an increased knowledge of the ecology of early prosimians, and a better understanding of the functional significance of added amounts of brain tissue.

Fossil endocasts from early primates do provide some information on sulcal patterns, on brain size, and on the size of the olfactory bulbs, telencephalon, cerebellum, and various neocortical areas. New results reported in this study include new direct volumetric measurements of endocasts, volumetric measurements of olfactory bulbs indicating that there is currently no evidence of olfactory bulb reduction in prosimians since the Early Eocene, and evidence of possible expansion of the forelimb and hindlimb areas of the primary somatosensory and primary motor cortex in post-Eocene lemuriform evolution. In addition, a hypothetical model of early primate brain evolution is presented for testing as new fossil evidence comes in.

Acknowledgments: For comments on methods and available specimens I thank Dr. Leonard B. Radinsky, Department of Anatomy, University of Chicago; Dr. Ralph L. Holloway, Department of Anthropology, Columbia University; Dr. David W. Frayer, Department of Anthropology University of Kansas; Dr. James H. Mielke, Department of Anthropology, University of Kansas; Dr. Larry Martin, Department of Systematics and Ecology, University of Kansas; Dr. H. O. Hofer, Institut fur Zoologie und Vergleichende Anatomie, Universitat Kassel; and Dr. Norman A. Slade, Department of Systematics and Ecology, University of Kansas. I am grateful to Ms. Carolyn Gamble for editorial comments and help in preparing the manuscript.

REFERENCES

Bauchot, R., and Stephan, H., 1966. Données nouvelles sur l'encéphalisation des insectivores et des prosimiens. Mammalia, 30:160-196.

Bauchot, R., and Stephan, H., 1967. Encéphales et moulages endocraniens de quelques insectivores et primates actuels. Problèmes actuels de paléontologie. Colloq. Inter. Centre. Nat. Rech. Sci., 163:575-586.

Gazin, C.L., 1965. An endocranial cast of the Bridger Middle Eocene primate *Smilodectes gracilis*. Smithsonian Misc. Coll., 149:1-14.

Gingerich, P.D., 1976. Cranial anatomy and evolution of Early Teritary Plesiadapidae. University of Michigan Museum of Paleontology, Papers on Paleontology, No. 15.
Holloway, R.L., 1968. The evolution of the primate brain: some aspects of quantitative relations. Brain Res., 7:121-172.
Holloway, R.L., 1969. Some questions on parameters of neural evolution in primates. Annals N.Y. Acad. Sci., 167:332-340.
Holloway, R.L., 1976. Paleoneurological evidence for language origins. Annals N.Y. Acad. Sci., 280:330-348.
Holloway, R.L., 1980. Brain size, allometry, and reorganization: toward a synthesis. In, Development and Evolution of Brain Size: Behavioral Implications, M.E. Hahn, C. Jensen, B.C. Dudek, eds., Academic Press, New York, pp. 59-88.
Jerison, H.J., 1973. Evolution of the Brain and Intelligence, Academic Press, New York.
Jerison, H.J., 1975. Fossil evidence of the evolution of the human brain. Am. Rev. Anthrop., 4:27-58.
Jerison, H.J., 1976. Discussion paper: The paleoneurology of language. Annals N.Y. Acad. Sci., 280:370-382.
Jerison, H.J., 1977. The theory of encephalization. Annals N.Y. Acad. Sci. 299:146-160.
Jerison, H.J., 1979. Brain, body and encephalization in early primates. J. Human Evol., 8:615-635.
Jerison, H.J., 1980. The evolution of diversity in brain size. In, Development and Evolution of Brain Size: Behavioral Implications, M.E. Hahn, C. Jensen, B.C. Dudek, eds., New York, Academic Press, pp. 29-57.
Le Gros Clark, W.E., 1959. The Antecedents of Man, Edinburgh University Press, Edinburgh.
Martin, R.D., 1973. Comparative anatomy and primate systematics, Symp. Zool. Soc. Lond., 33:301-337.
Napier, J.R. and Napier, P.H., 1967. A Handbook of Living Primates; Morphology, Ecology, and Behavior of Nonhuman Primates, Academic Press, New York.
Radinsky, L.B., 1967a. The oldest primate endocast. Am. J. Phys. Anthrop., 27:385-388.
Radinsky, L.B., 1967b. Relative brain size: a new measure. Science, 155:836-838.
Radinsky, L.B., 1970. The fossil evidence of prosimian brain evolution. In, The Primate Brain: Advances in Primatology, Vol. 1, C.R. Noback and W. Montagna, eds., Appleton-Century-Crofts, New York, pp. 209-224.
Radinsky, L.B., 1977. Early primate brains: facts and fiction. J. Human Evol., 6:79-86.
Simons, E.L., 1972. Primate Evolution, Macmillan, New York.
Stephan, H., 1960. Methodische Studien über den quantitativen Vergleich architektonischer Struktureinheiten des Gehirns. Z. wiss. Zool., 164:143-172.
Stephan, H., 1972. Evolution of primate brains: a comparative anatomical investigation. In, The Functional and Evolutionary Biology of Primates, R. Tuttle, ed., Aldine, Chicago, pp. 157-174.

Stephan, H., Bauchot, R. and Andy, O.J., 1970. Data on the size of the brain and of various parts in insectivores and primates. In, The Primate Brain, C.R. Noback and W. Montagna, eds., Appleton-Century-Crofts, New York, pp. 289-297.

Walker, E.P., 1968. Mammals of the World, Vol. 1, John Hopkins Press, Baltimore.

A STUDY OF CEREBRAL VASCULAR EVOLUTION IN PRIMATES

Glenn C. Conroy*

The intelligent use of nonhuman primates in evolutionary studies requires an appreciation for both primate anatomy and systematics. The choice of an experimental animal often depends not only on an evaluation of the comparative anatomical systems involved, but also on the phenetic relationship of the animal to man. An understanding of the anatomical variations found within the order Primates make it possible to design experimental research programs on anatomical systems of widely varying complexities within a group of animals closely related to our own species.

Few studies of comparative anatomy have assumed more importance than those dealing with the vascularization of the brain. Such studies are basic to most researchers in the neurosciences who use primates as part of their experimental work. For primatologists, analyses of the carotid circulatory patterns have assumed particular importance in both phyletic and phenetic studies (Gregory, 1920; Le Gros Clark, 1959; Szalay and Katz, 1973; Saban, 1963; Cartmill, 1975; Conroy and Wible, 1978; Conroy and Packer, 1981). Such studies have played a prominent role in the systematics of other mammalian groups as well (Bugge, 1974, 1978; Guthrie, 1963; Patterson, 1965).

Indeed, the relationship of the carotid circulatory patterns to intracranial structures in living and fossil primates is pertinent to any discussion concerning the evolution of the primate brain. Of particular importance in this regard are the relative contributions of the promontory and stapedial branches of the internal carotid artery in supplying intracranial and facial structures. Studies on the phylogenetic history of arterial blood supply *to* the brain in several nonhuman primates have been undertaken, but as yet no one has presented a similar analysis of cerebral venous blood return *from* the brain.

*From the Division of Biology & Medicine, Box G, Brown University, Providence, Rhode Island 02912

Not surprisingly, studies of the arterial system have taken precedence over those of the venous system in primatology and mammology. Arteries are usually easier to visualize through various injection and corrosion techniques, due in part to their more muscular walls. Also, dural venous sinuses are often ripped on removal of the skull cap since the outer layer of the dura is the endosteal layer of the cranial bones. Veins are sometimes overlooked in medical school dissecting rooms where they are often excised to expose deeper arterial channels. Such cavalier attitudes toward the venous sytem have led to anatomical inaccuracies that are still propagated in some anatomy texts. For example, the deep, radiating grooves found on the endocranial surface of the parietal bone owe their presence more to the meningeal veins than to the meningeal arteries. This fact was known to anatomists 300 years ago, but it is sometimes overlooked in modern texts (see Jones, 1912). Veins are also notorious for their variability, and yet certain veins (e.g., the long saphenous vein at the ankle; the major intracranial venous sinuses) have a remarkably constant position in the human body.

There are definite advantages in studying the intracranial venous drainage patterns in primates. First, the venous sinuses often leave distinct, measurable impressions on the endocranial surface of the cranial bones and in this regard are often much easier to assess than the tiny grooves for the carotid branches within the middle ear. Second, and more important, there is a distinctly different morphology between the intracranial venous outflow patterns in strepsirhines (and many mammals) and most haplorhines.

The pattern of venous drainage from the brain of man and his anthropoid relatives seems unique among primates (Conroy, 1980a). This report will review this unique intracranial morphology within an evolutionary and functional framework.

The cranial dural sinuses are subject to variations and anomalies like any other anatomical system. Some of the more extreme anomalies, such as complete absence of the sigmoid sinus or displacement of the jugular bulb, result in cerebral venous drainage patterns that are reminiscent of the embryonic condition. These anomalies can produce clinical symptoms such as hearing loss, tinnitus, vertigo, and facial nerve palsy (Knott, 1882; Laff, 1930; Waltner, 1944; Kraus et al., 1965; Hadziselimovic et al., 1970; Ziv et al., 1973). Accordingly, the role of anomalous venous sinuses in these disorders cannot be overlooked in clinical evaluations. Unfortunately, the presence of such anomalous structures is sometimes only noted due to the "unpleasant disclosure in the course of routine ear surgery" (Kraus et al., 1965).

The anatomical arrangement of the arterial and venous network around the otic region is an important consideration in the pathogenesis of the disorders mentioned above. Interference with venous outflow in this region can affect not only structures around the internal acoustic meatus (e.g., cranial nerves VII and VIII), but also intracranial pressure levels.

To properly understand the normal, and abnormal anatomy of the dural venous sinuses in the otic region, a comparative embryological and anatomical approach is essential. This review particularly emphasizes the evolutionary perspective as well. Such a perspective is important for elucidating factors that affected the morphological differences one finds in cerebral venous drainage among primates. The results will contribute to a more comprehensive understanding of the evolution of the primate brain.

Various lines of evidence can be used to trace the evolutionary history of cerebral venous drainage patterns in primates. Data from comparative anatomy, embryology, paleontology, and venography can be integrated to explain this evolutionary process. The main themes from each of these lines of evidence are summarized below.

VERTEBRAL AND CAROTID ARTERIAL PATTERNS IN PRIMATES

The carotid circulatory system has been studied in several nonhuman primates (both living and fossil). In strepsirhine primates the promontory branch of the internal carotid is normally minute. Thus, in lemurs (e.g., *Lemur catta*) the brain is supplied almost entirely by the vertebral artery whereas in lorises and cheirogaleids it is supplied mainly by the vertebral and ascending pharyngeal arteries (Bugge, 1974; Cartmill, 1975). The internal carotid artery is not normally a significant supplier of arterial blood to the brain in any strepsirhine form.

The predominance of the vertebral artery over the internal carotid in strepsirhines is not surprising, given that the vertebral artery even in man supplies numerous intracranial structures including the cervical spinal cord (via anterior and posterior spinal arteries), medulla, pons, midbrain, cerebellum, posterior parts of the diencephalon, and parts of the occipital and temporal lobes of the cerebrum. A labyrinthine branch of the basilar artery also supplies the cochlea and vestibular apparatus (Truex and Carpenter, 1969). The internal carotid artery, via its cerebral branches, is mainly responsible for supplying the enlarged telencephalon of man and his anthropoid relatives. In many mammals lacking an expansive telencephalon, the internal carotid artery is normally small and the vertebral artery is generally the main source of blood supply to the brain (Struthers, 1930; Bugge, 1974). There are some exceptions to this general rule; certain myomorphs, for example, have an internal carotid supplying the brain whereas other rodents, e.g., certain sciuromorphs, do not.

The relative size of the stapedial artery also assumes importance in primatological studies. Developmentally, this is the artery of the second branchial arch; however, because the artery of the first branchial arch regresses so early in embryogenesis, its area of distribution is rapidly taken over by the primitive stapedial artery (LaTorre and Netsky, 1960). Thus, the stapedial artery in the early stages of development (and in some adult mammals) has branches that correspond to the divisions of the trigeminal nerve (nerve of the first branchial arch). In such cases the stapedial artery would normally have supraorbital, infraorbital, and mandibular branches (Padget, 1957). No primate still retains this primitive distribution of the artery, the branches having regressed to varying degrees in the different primate taxa. Even in strepsirhines the distribution of the stapedial artery supply is greatly diminished from the "primitive" condition and is largely supplanted by the external carotid system (Bugge, 1974; Hill, 1969).

It is reasonable to hypothesize that in primate evolution a critical size must have been reached whereby the developing masticatory apparatus could not have been adequately supplied by an artery (stapedial) whose diameter was limited by the dimensions of the intracrural foramen of the stapes. The tremendous effect of arterial size in determining rate of blood flow through a vessel is expressed in Poiseuille's Law whereby rate of flow is directly proportional to the fourth power of the radius of the tube (vessel in this case). Interestingly, in those rodents with functionally similar carotid patterns to *Lemur* (i.e., well developed stapedial artery and small medial entocarotid artery), few seem to average over 7 kg in weight and most weigh under 500 grams (e.g., *Tamia*, *Eutamias*, *Tamiasciurus*, *Callosciurus*, *Glaucomys*). By the way of contrast, some of the larger rodents, with a body weight between 9-32 kg (*Castor*, *Pedetes*), have carotid patterns more similar to the haplorhine condition (large internal cartoid artery and small stapedial artery), with the external carotid artery supplying the masticatory apparatus (Bugge, 1974; Walker, 1968). Thus the size of the body and brain may influence the arterial pattern.

The relative importance of the promontory and stapedial branches of the internal carotid artery in extant and fossil primates is often deduced mainly from the size of the grooves or bony tubes that are thought to house these vessels (Gingerich, 1976; McKenna, 1963). The "primitive" primate condition is said to be characterized by a larger stapedial canal than promontory canal. Such a morphology, however, is not found in the Paleocene skull of *Plesiadapis* (Gingerich, 1976; McKenna, 1963). In lorisids and cheirogaleids neither promontory nor stapedial arteries are significantly developed (Cartmill, 1975). In haplorhines the promontory artery is larger than the stapedial.

It is presently unknown when the changes in the carotid circulatory patterns that now generally distinguish strepsirhines from haplorhines evolved, i.e., when (and why) did the internal carotid artery assume dominance over the vertebral artery in intracranial blood supply?

Unfortunately, determining the relative size of the promontory and stapedial arteries in fossil primates is fraught with difficulties for several reasons: a) the auditory bulla and middle ear structures are not often preserved in fossil skulls and, if they are, postmortem crushing often makes assessment of these tiny grooves impractical; b) it is often difficult to establish whether or not grooves near the promontorium are for vessels or nerves (Conroy and Wible, 1978); and c) the grooves are often so minute and inaccessible that accurate measurements are difficult to record.

Ideally, then, what is needed is a different approach or technique that would also provide evidence for a major shift in intracranial circulatory patterns between strepsirhines and haplorhines and that would complement interpretations based on carotid patterns in fossil primate specimens. This technique should also provide evidence for deducing intracranial circulatory patterns in fossils even in the absence of any data concerning carotid circulation. Such a technique would be a powerful interpretative tool for the paleoprimatologist. I have suggested that a study of the dural venous sinus patterns is such an approach (Conroy, 1980a).

CEREBRAL VENOUS DRAINAGE PATTERNS IN PRIMATES

A marked alternation in cerebral venous drainage patterns in primates is generally found between strepsirhine and haplorhine primates (Conroy 1981a for a discussion of platyrrhines). The differentiation is as pronounced as that which took place in the carotid arterial patterns during the course of primate evolution. Thus, in haplorhines a new arterial channel assumes dominance in supplying blood to an expanded telencephalon and a new venous channel (internal jugular vein) assumes dominance in draining blood from the same structure. In strepsirhines and most other mammals, the main venous outflow from the brain is via the vertebral veins and the *external* jugular system (Padget, 1957; Dom et al., 1970; Hegedus and Shackelford, 1965). The major conduit for intracranial venous drainage in most haplorhines (including man) is via the vertebral plexus and the *internal* jugular system. The external jugular system in haplorhines is mainly involved with extracranial venous drainage (except for small amounts of intracranial blood that reach the external jugular veins via the pterygoid plexus). Exceptions to this general rule occur in some platyrrhine genera, for example, *Cebus* (Conroy, 1980b, 1981).

Because the external jugular system in strepsirhines is directly continuous with the cranial venous sinuses via the postglenoid foramen, the relative size of the foramen should be an important anatomical landmark for interpreting intracranial venous drainage in fossil and extant primate specimens (see below).

EMBRYOLOGICAL DATA ON CEREBRAL VENOUS DRAINAGE IN MAN AND OTHER MAMMALS

A brief discussion of the embryogenesis of certain dural venous sinuses will illustrate this continuity between the intracranial venous outflow and the external jugular system via the postglenoid foramen.

The primary head vein in the 28-48 day old human embryo consists of the vena capitis medialis and lateralis, which are united just posterior to the trigeminal ganglion by a post-trigeminal vein. The vena capitis medialis is formed by the union of the anterior cerebral, maxillary, and pituitary veins. The vena capitis lateralis, in turn, is continuous with the anterior cardinal vein just caudal to the vagal ganglion. The junction of the latter two veins is approximated by the presence of a large posterior rhombencephalic vein. Between the trigeminal ganglion and the otic capsule the vena capitis lateralis receives on its dorsal aspect several prootic veins and from its ventral surface several prootic emissary veins. These small prootic emissary veins connect the primary head vein (vena captis lateralis) to the subectodermal venous plexus, which is the forerunner of the external jugular venous system in the otic region. The junction of the vena capitis lateralis and the anterior cardinal vein forms an angle of about 90 degrees and receives the large posterior rhombencephalic veins; this juncture corresponds to the origin of the internal jugular vein (Padget, 1957; Butler, 1957; Streeter, 1918).

At about the 16-18mm crown-rump length stage (still in the embryonic period), a large anastomotic channel that connects the posterior rhombencephalic veins to the prootic veins is formed dorsal to the otic capsule. The vena capitis lateralis (which passes ventral to the otic capsule) assumes less importance after this new anastomotic channel is formed.

By the 21-mm crown-rump length embryonic stage, a new or "secondary" head vein is completed by a new anastomotic channel connecting the anterior cerebral vein to the prootic veins. Thus, this new "lateral sinus" is composed of the transverse sinus (formed from the anterior cerebral vein and the new anastomotic channel between the anterior cerebral vein and the prootic vein) and the sigmoid sinus (formed by the supraotic anastomosis) and the rhombencephalic vein.

With caudal expansion of the cerebral hemispheres, the transverse sinus reaches a relatively vertical course (in relation to the cranial base) by the 23-mm stage. The transverse sinus drains ventrally toward the otic capsule and divides into two equal-sized channels, the petrosquamous sinus (to the external jugular vein) and the sigmoid sinus (to the internal jugular vein). By the fetal period (9-25 weeks), the petrosquamous sinus is as large as the sigmoid sinus. As the cerebral cortex expands further during the fetal period the tentorium cerebelli and its contained transverse sinus reach nearly the adult horizontal position by the 175-mm stage. By this time in human fetal development, the petrosquamous sinus is regressing and the sigmoid sinus and internal jugular vein attain their importance and dominance as channels of intracranial venous outflow (Padget, 1957; Butler, 1957).

The development of the dural sinuses in other mammals generally follows the same pattern as that outlined above for the human fetus. There are several important distinctions, however. In the human fetus, when the transverse sinus has a more or less vertical course (due to the still incomplete telencephalic development caudally over the cerebellum), the sigmoid sinus and petrosquamous sinus were noted to be about equal in size. Naturally the orientation of the transverse sinus depends to a large degree on the growth of the cerebral hemispheres, and in catarrhines it normally lies more horizontally after complete cerebral growth. In most mammals without a caudally expanded cerebral cortex, the transverse sinus retains the more vertical position relative to the skull base and consequently the petrosquamous sinus remains as a large channel that empties via the postglenoid foramen into the external jugular system (Butler, 1967). By the 18- to 26-mm embryonic stage, an important crossroads is reached in mammalian development whereby the relative caudal expansion of the cerebral cortex will determine whether the internal or external jugular system will be the dominant channel of intracranial venous drainage.

Another distinction between haplorhines and most other mammals concerns the development of the vertebral veins. The general mammalian condition is exemplified in the 15-mm crown-rump stage of the rat embryo. By this stage the basilar plexus drains by large vertebral veins, which in turn become connected to the terminal part of the sigmoid sinus. From this stage onward in the rat (and most mammals) the vertebral veins will assume the major role in draining blood from the sigmoid sinus instead of the internal jugular vein. This basic mammalian pattern has been demonstrated for rats, hedgehogs, dogs, and bats. Monotremes have a somewhat different pattern from eutherian mammals in that the petrosquamous sinus runs down the anterior face of the temporal bone and then courses through the facial canal to exit at the stylomastoid foramen (with the facial nerve) to form an internal jugular vein. There is apparently no sigmoid sinus in these forms. Birds and reptiles also show differences from the eutherian pattern by lacking a definitive external jugular vein (Padget, 1957; Butler, 1957; Klintworth, 1968; Bull, 1969).

Rathke (1838) was the first to emphasize the importance of the external jugular vein as a main channel of venous drainage in the embryo. He correctly pointed out the pathway of this drainage as passing through a foramen located between the tympanic ring and the temporomandibular joint. Luschka (1862) named this the "foramen jugular spurium," and it can only be this foramen that is homologous to the "postglenoid" foramen in man and other primates. The petrosquamous squamous sinus enters the external jugular system in mammals through this foramen. Incidentally, the dominance of the internal jugular vein in man is a retention of the early embryonic condition because the development of the definitive external jugular vein in other mammals occurs after the formation of the internal jugular vein in the fetal period (Padget, 1957).

INTRACRANIAL VENOUS DRAINAGE IN FOSSIL AND EXTANT PRIMATE SKULLS

To interpret intracranial venous drainage patterns in paleoprimatological studies, one must have some way of determining from fossil material the extent to which the external jugular venous system participated in this drainage. The significance of the postglenoid foramen in this regard has already been noted. Thus, the presence or absence (and relative size) of this foramen in fossil basicrania should be an important morphological landmark. Also, there should be some measure of confidence that the presence or absence of this foramen can separate extant prosimian and catarrhine crania, thus reflecting the difference between external and internal jugular dominance.

Evidence to support this generalization comes from several sources. Boyd's (1930, 1934) studies on the emissary foramina of primate crania are the most exhaustive for this kind of data. In *Gorilla* (n=159 skulls), *Pan* (n=137 skulls), *Hylobates*, and *Symphalangus* (n=113 skulls), the postglenoid foramen was never observed. In 98 skulls of *Pongo* the foramen was present in only one specimen. The foramen is present as an anomaly in about 1% of human crania. In contrast, of the 104 cases of what Boyd described only as "true lemurs," the foramen was found in 99% of the cases and in all seven *Tarsius* skulls examined. Unfortunately, his data for monkeys cannot be properly assessed because they were not identified as platyrrhines or catarrhines. There is little consistent evidence of a postglenoid foramen in catarrhine primates, but certain platyrrhines such as *Cebus* still retain a large foramen (in some of the larger cebids such as *Alouatta* and *Ateles* the foramen is absent) (Madeira and Watanabe, 1973; Conroy, 1980b, 1981). The postglenoid foramen is normally present and well developed in most lemurids but tends to be relatively smaller in lorisids. In a few instances, e.g., *Arctocebus*, it is virtually absent (this is another similarity in the cranial region of this small lorisid with anthropoids--the other is the similar sulcal pattern). I suspect that in those prosimians with a small postglenoid foramen, most of the intracranial venous outflow is via the vertebral veins rather than the internal jugular system, since the latter still appears to be a rather small vessel in these forms (McClure and Silvester, 1909). In any event, preliminary analyses indicate that *most* catarrhine and hominoid primates (including man) are clearly distinct from most prosimian primates (*Tarsius* as well) in the dominance of the internal jugular over the external jugular system.

Other more direct evidence for determining venous channel patterns is provided by fossil endocasts. Many of the impressions for the dural venous sinuses are well preserved either in the natural endocasts or on artifically prepared endocranial casts. Endocasts of the following fossil primates show the presence of a petrosquamous sinus leading to the external jugular system via the postglenoid foramen: *Smilodectes gracilis*, *Adapis parisiensis*, *Rooneyia viejaensis*, *Archaeolemur majori*, *Mesopropithecus pithecoides*, *Megaladapis edwardsi* (Gazin, 1965; Hofer and Wilson, 1967; Tattersall, 1971, 1973, 1974; Radinsky, 1968, 1970, 1974a, 1974b, 1975). Indirect fossil evidence for the presence of the petrosquamous sinus in primates is of course related to the presence of the postglenoid foramen. As far as I am aware, all early Tertiary primates up to and including the Oligocene forms *Aegyptopithecus*, *Apidium* and *Rooneyia* retain the postglenoid foramen (e.g., *Phenacolemur, Necrolemur, Notharctus, Smilodectes*) (Gazin, 1958; Gingerich, 1973; Szalay, 1972; Simpson, 1955; Simons and Russell, 1960; Simons, 1961; Szalay and Wilson, 1976; Wilson, 1966). This fossil evidence spans a time range of about 50 million years and a geographical area that includes North America, Europe, North Africa, and Madagascar.

VENOGRAPHIC STUDIES

With venography, cerebral venous outflow from the brain to the heart can be visualized along its entire course and the relative amounts of venous blood returning by the different available vascular channels (e.g., vertebral veins, internal jugular, external jugular, emissary veins) can be determined. The effects that certain variables such as body posture and head position have on cerebral venous drainage patterns are significant. Work done on the rhesus monkey clearly shows the importance of these variables in that experimental animal (Epstein et al., 1970; DiLenge and Perey, 1973).

Until recently the internal jugular system was assumed to be the only major cerebral venous outflow tract in man. Indeed, most descriptive and clinical anatomy texts ignored the role of the vertebral venous plexus as a major cerebral venous pathway. Although the role of the vertebral venous plexus in metastatic disease processes was described by Batson (1957), little attention was given to its fundamental role in cerebral venous circulation until the 1960's when Eckenoff and his colleagues demonstrated its importance (Eckenoff, 1971; Epstein et al., 1970). Their experimental work was motivated by the observation that in patients undergoing radical neck surgery, the internal jugular veins were often collapsed when the head was tilted up at about 30 degrees. Venous blood was obviously returning from the brain by a route other than the internal jugular system.

In the supine monkey, breathing at ambient pressure, blood from the brain returned to the heart via both the internal jugular veins and the vertebral venous plexus. In the upright animal, however, virtually all blood left the brain via the vertebral plexus.

The vertebral venous plexus is now clearly established as a major cerebral venous tract in the upright human (DiLenge and Perey, 1973). These findings are certainly not intuitively obvious from gross anatomical studies. The paradox is apparently resolved by noting that in the standing position the cerebrospinal fluid pressure decreases in the cervical region whereas it increases in the lumbar region. This lowered pressure of the cervical cerebrospinal fluid is transmitted to the adjoining vertebral veins. Any study of the evolution of the cerebral venous system must take into account the affects of posture on venous drainage.

ASYMMETRY OF CEREBRAL VENOUS FLOW

It has been known for centuries that the human right jugular foramen is usually larger than the left (DiChero et al., 1964; Shapiro, 1972). This cerebral venous asymmetry is known from both anatomical and radiological studies (Frenckner, 1940; Solter and Paljan, 1973; Hochberg and LeMay, 1975). There is still no consensus as to why this right-sided jugular dominance has evolved in the human brain, nor how widespread it is among primates. I have documented the phenomenon in several platyrrhine genera (Conroy, 1981). The first functional explanation for this asymmetry was put forward by Von Meyer (1877). He surmised that in the human fetus the right side became the shortest and most direct route to the heart after the development of the left-right shunt between the anterior cardinal veins draining the head. The left side of the fetal head was drained by the long, obliquely placed brachiocephalic vein.

In his comparative studies of human and ape brains, Elliot-Smith (1907) noted that the left occipital pole (area striata) projected further posteriorly than the corresponding right pole in humans. He suggested that this left-sided "bulge" pushed the sagittal sinus more towards the right side, thus explaining the venous asymmetry. He also considered this venous asymmetry as an "advanced" feature of the human brain (or at least of certain human races) and considered the more asymmetrical venous flow found in apes to be a "primitive" feature.

Bluntschli (1908) also considered the symmetrical venous flow of "lower apes" to be primitive and the right-sided jugular dominance of man to be an advanced feature. Like his predecessor Von Meyer, Bluntschli also turned to embryology to explain the cerebral venous asymmetry. He was the first to consider how the developing occipital lobes would affect blood flow through the transverse sinus. I discussed how this fetal growth phenomenon might affect patterns of cerebral blood flow, but it is difficult to see how it could account for differential flow predominantly to the right side unless one could show differential growth rates of the telencephalon between left and right cerebral hemispheres (Galaburda et al., 1978).

Zeiger (1922-23) considered the relationship of "handedness" to cerebral venous asymmetry. His case was significantly weakened by his belief that the asymmetry in cerebral flow did not occur until after birth, concomitant with asymmetrical growth patterns in the upper extremity. Unfortunately for Zeiger, this cerebral asymmetry is already apparent in the 20-mm fetus (Streeter, 1915). The relationship of "handedness" and cerebral asymmetry has been thoroughly investigated by Hochberg and LeMay (1975) and LeMay (1976, this volume). These authors have found no significant relationships between transverse sinus flow and handedness, sex, or Sylvian point angle differences.

Many earlier authors (e.g., Elliot-Smith, 1907; Bluntschli, 1908) assumed that "lower apes" (monkeys) had a more symmetrical cerebral drainage than did "higher apes." There are very little quantitative data bearing on this point, however.

Using a modification of Poiseuille's Law, I calculated that eight of ten New World primate genera studied display statistically significant right-sided jugular dominance (Conroy, 1980b, 1981). Only six of the ten genera, however, show significant venous asymmetry in total cerebral venous drainage, indicating that two genera (*Cebus* and *Saimiri*) have "compensated" for right jugular dominance by preserving other well developed emissary foramina on the contralateral side. I speculated that some new world monkeys (e.g., *Cebus*) display a type of cerebral venous morphology which is "transitional" between the "primitive-symmetrical" and "advanced-asymmetrical" condition (Conroy, 1980b).

SUMMARY

Primate evolution has witnessed two major cerebral vascular alterations--the ascendency of the internal carotid artery and the internal jugular vein. Whether or not these two events are causally related is still unclear. The paleontological record has not been examined with this question in mind. I suspect that they are causally related and that both systems were fully "in place" by the Oligocene.

I have reviewed the pertinent anatomical, embryological, and paleontological data bearing on this question in the sections above. From the data presented, three primary factors emerge that may provide the key to understanding the evolution of the internal jugular vein and internal cartoid artery: 1) the caudal expansion of the telencephalon over the cerebellum and its affect on the hemodynamics of blood flow through the dural sinuses, 2) the relative size increase of the masticatory apparatus and other facial structures, and 3) alterations in body posture and associated head positions. None of these factors alone could explain the changes in the cerebral vascular system that evolved through primate history; clearly some combinations of these factors are involved. Relevant data on primates is meager at best; however, the identification of these factors should provide a "jumping off" point for future experimental work on this issue.

The correct "solution" to the puzzle of human brain evolution requires an input from many specialists and at least some "generalists" who dare a synthesis of seemingly unrelated data. This contribution represents the spirit of the latter!

Acknowledgments: This review is based on research supported by the National Science Foundation and Brown University.

REFERENCES

Batson, O., 1957. The vertebral vein system. Am. J. Roentgenol., 78:195-212.

Bluntschli, H., 1908. Ueber die Asymmetrie der Sinus transversi Durae matris bei Mensch und Affe. Verhandl. d. Geselsch. deutsch Naturf. u. Aerzte., 2:361.

Boyd, G., 1930. The emissary foramina of the cranium in man and the anthropoids. J. Anat., 65:108-121.

Boyd, G., 1934. The emissary foramina of the cranium in primates. J. Anat., 69:113-117.

Bugge, J., 1974. The cephalic arterial system in insectivores, primates, rodents, and lagomorphs, with special reference to the systematic classification. Acta. Anat., 87 (Suppl. 62):1-160.

Bugge, J., 1978. The cephalic arterial system in carnivores, with special reference to the systematic classification. Acta. Anat., 101:45-61.

Bull, J., 1969. Tentorium cerebelli. Proc. Roy. Soc. Med., 62:1301.

Butler, H., 1957. The development of certain human dural venous sinuses. J. Anat., 91:510-526.

Butler, H., 1967. The development of mammalian dural venous sinuses with especial reference to the postglenoid vein. J. Anat., 102:33-56.

Cartmill, M., 1975. Strepsirhine basicranial structures and affinities of Cheirogaleidai. In, Phylogeny of the Primates, W. Luckett and F. Szalay, eds., Plenum, N.Y., pp. 313-354.

Conroy, G., 1980a. Evolutionary significance of cerebral venous patterns in paleoprimatology. Z. Morph. Anthrop., 71:125-134.

Conroy, G., 1980b. Cerebral venous hemodynamics and the basicranium of *Cebus*. Am. J. Phys. Anthrop., 53:37-42.

Conroy, G., 1981. Cranial asymmetry in Ceboid primates: The emissary foramena. Am. J. Phys. Anthrop., 55:187-194.

Conroy, G., and Packer, D., 1981. The anatomy and phylogenetic significance of the cartoid artery and nerves in Strepsirhine primates, Folia Primatol., 35:237-247.

Conroy, G., and Wible, J., 1978. Middle ear morphology of *Lemur variegatus*: some implications for primate paleontology. Folia Primatol., 29:81-85.

DiChero, G., Fisher, R., and Nelson, K., 1964. The jugular foramen. J. Neurosurg., 21:447-460.

DiLenge G., and Perey, B., 1973. An angiographic study of the meningorachildian venous system. Radiology, 108:333-337.

Dom, R., Fisher, B., and Martin, G., 1970. The venous system of the head and neck of the opposum. J. Morphol., 132:487-496.

Eckenhoff, J., 1971. The vertebral venous plexus. Canad. Anaesth. Soc. J., 18:487-49.

Elliot-Smith, G., 1907. On the asymmetry of the caudal poles of the cerebral hemispheres and its influence on the occipital bone. Anat. Anz., 30:574-578.

Epstein, H., Linde, H., Crampton, Ciric, I., and Eckenhoff, J., 1970. The vertebral venous plexus as a major cerebral venous outflow tract. Anesthesiology, 32:332-337.

Frenckner, P., 1940. The value of roentoenography in estimating the degree to which the lateral sinus and jugular vein allow emptying of the venous blood from the skull. Acta. Otol., 28:107-135.

Galaburda, A.M., LeMay, M., Kemper, T.L., and Gerschwind, N., 1978. Right-left asymmetries in the brain. Science, 199:852-856.

Gazin, C., 1958. A review of the middle and upper Eocene primates of North America, Smithsonian Misc. Coll., 126:1-112.

Gazin, C., 1965. An endocranial cast of the Bridger Middle Eocene primate *Smilodectes gracilis*. Smithsonian Misc. Coll., 149:1-15.

Gingerich, P., 1973. Anatomy of the temporal bone in the Oligocene anthropoid *Apidium* and the origin of Anthropoidea. Folia Primatol., 19:329-337.

Gingerich, P., 1976. Cranial anatomy and evolution of eary Tertiary Plesiadapidae. Univ. Michigan Papers on Paleontology, 15:1-141.

Gregory, W., 1920. On the structure and relationships of *Notharctus*, an American Eocene primate. Mem. Am. Mus. Natl. Hist., 3:49-243.

Guthrie, D., 1963. The carotid circulation in the Rodentia. Bull. Mus. Comp. Zool., 128:455-481.

Hadziselimovic, H., Cus, M., and Tomic, V., 1970. Appearance of the sigmoid groove and jugular foramen in relation to the configuration of the human skull. Acta Anat., 77:501-507.

Hegedus, S., and Shackelford, R., 1965. A comparative-anatomical study of the craniocervical venous system in mammals, with special reference to the dog. Am. J. Anat., 116:375-386.

Hill, W., 1969. The vascular supply of the face in long-snouted primates (*Papio* and *Mandrillus*). Rec. Adv. Primatol., 2:155-159.

Hochberg, F., and LeMay, M., 1975. Arteriographic correlates of handedness. Neurology, 25:218-225.

Hofer, H., and Wilson, J., 1967. An endocranial cast of an early Oligocene primate. Folia Primatol., 5:148-152.

Jones, F., 1912. On the grooves upon the ossa parietalia commonly said to be caused by the arteria meningea media. J. Anat. Physiol., 46:327-352.

Klintworth, G., 1968. The comparative anatomy and phylogeny of the tentorium cerebelli. Anat. Rec., 160:635-642.

Knott, J., 1882. On the cerebral sinuses and their variation. J. Anat., Physiol., 16:27-42.

Kraus, P., Leventon, G., and Man, A., 1965. Symptom producing anomalies of the sigmoid sinus. Acta Otol., 60:315-321.

Laff, H., 1930. Unilateral absence of sigmoid sinus. Arch. Otol., 11:151-157.

LaTorre, E., and Netsky, M., 1960. Study of persistent primitive maxillary artery in human fetus: some homologies of cranial arteries in man and dog. Am. J. Anat., 160:185-195.

Le Gros Clark, W.E., 1959. Antecedents of Man, Edinburgh Univ. Press, Edinburgh.

LeMay, M., 1976. Morphological cerebral asymmetry of modern man, fossil man, and nonhuman primates. Ann. NY Acad. Sci., 280:349-366.

Luschka, H., 1862. Die Venen des menschlichen Halses. Denskschr. Akad. Wiss. Wien, 20:5 (cited in Butler, 1967).

Madeira, M., and Watanabe, I., 1973. Superficial venous arrangement on the face and neck of the tufted capuchin, *Cebus apella*. J. Morph., 140:397-404.

McClure, C., and Silvester, C., 1909. A comparative study of the lymphaticovenous communications in adult mammals. Anat. Rec., 3:534-551.

McKenna, M., 1963. Early tertiary primates and their ancestors. Proc. XVI Int. Cong. Zool., Washington, D.C., pp. 69-74.

Padget, D., 1957. The development of the cranial venous system in man from the viewpoint of comparative anatomy. Contr. Embryol. 36:81-139.

Patterson, B., 1965. The auditory region of the borhyaenid marsupial Cladosictis. Breviora., 217:1-9.

Radinsky, L., 1968. A new approach to mammalian cranial analysis, illustrated by examples of prosimian primates. Am. J. Phys. Anthrop., 124:167-180.

Radinsky, L., 1970. The fossil evidence of prosimian brain evolution. In, The Primate Brain, C.R. Noback and W. Montagna, eds., Appleton-Century-Crofts, N.Y., pp. 209-224.

Radinsky, L., 1974a. The fossil evidence of anthropoid brain evolution. Am J. Phys. Anthrop., 41:15-28.

Radinsky, L., 1974b. Prosimian brain morphology: functional and phylogenetic implications. In, Prosimian Biology, R. Martin, G. Doyle, and A. Walker, eds., Duckworth, London, pp. 781-798.

Radinsky, L., 1975. Primate brain evolution. Am. Sci., 63:656-663.

Rathke, M., 1838. Über den Bau und Entwicklung des Venensytem der Wirbelthiere (cited in Butler, 1967).

Saban, R., 1963. Contribution a l'étude de l'os temporal des Primates. Mem. Mus. Natl. Hist. Natl. (Paris), Ser. A. 29:1-378.

Shapiro, R., 1972. Compartmentation of the jugular foramen. J. Neurosurg., 36:340-343.

Simons, E., 1961. Notes on Eocene tarsioids and a revision of some Necrolemurinae. Bull. Brit. Mus. Natl. Hist. (Geol.), 5:45-75.

Simons, E., and Russell, D., 1960. Notes on the cranial anatomy of *Necrolemur*. Breviora, 127:1-14.

Simpson, G., 1955. The Phenacolemuridae, new family of early primates. Bull. Am. Mus. Natl. Hist., 105:415-441.

Solter, M., and Paljan, D., 1973. Variations in shape and dimensions of sigmoid groove, venous portion of jugular foramen, jugular fossa, condylar and mastoid foramina classified by age, sex, and body size. Z. Anat. u. Entwickl. Gesch., 140:319-335.

Streeter, G., 1915. The development of the venous sinuses of the dura mater in the human embryo. Am. J. Anat., 18:145-178.

Streeter, G., 1918. The developmental alterations in the vascular system of the brain of the human embryo. Contr. Embryol., 8:7-38.

Struthers, P., 1930. The aortic arches and their derivatives in the embryo porcupine. J. Morph. Physiol., 50:361-392.

Szalay, F., 1972. Cranial morphology of the early Tertiary *Phonacolemur* and its bearing on primate phylogeny. Am. J. Phys. Anthrop., 36:59-76.

Szalay, F., and Katz, C., 1973. Phylogeny of lemurs, galagos, and lorises. Folia Primatol., 19:88-103.

Szalay, F., and Wilson, J., 1976. Basicranial morphology of the early tertiary tarsiform. Rooneyia from Texas. Folia Primatol., 25:288-293.

Tattersall, I., 1971, Revision of the subfossil Indriinae. Folia Primatol., 16:257-269.

Tattersall, I., 1973. Cranial anatomy of the Archaeolemurinae. Anthrop. Pap. Am. Mus. Natl. Hist., 52:1-110.

Tattersall, I., 1974. Facial structure and mandibular mechanics in *Archaeolemur*. In, Prosimian Biology, R. Martin, G. Doyle, and A. Walker, eds., Duckworth, London, pp. 563-577.

Truex, R., and Carpenter, M., 1969. Human Neuroanatomy, Williams & Wilkins, Baltimore.

Von Meyer, H., 1877. Die Sinus transversus des Hinterhauptbeins. Arch f. Anat. u. Physiol., 1:271-272.

Walker, E., 1968. Mammals of the World, Johns Hopkins Press, Baltimore.

Waltner, J., 1944. Anatomic variations of the lateral and sigmoid sinuses. Arch. Otol., 39:307-312.

Wilson, J., 1966. A new primate from the earliest Oligocene, West Texas, preliminary report. Folia Primatol., 4:227-248.

Zeiger, K., 1922-23. Ueber die Ursachen der Asymmetrie der Sinus transversi und sigmoidei beim Menschen. Beit. a. Anat., Physiol., Path., u. Therap. d. Ohres, 19:184.

Ziv, M., Leventon, G., and Nathan, H., 1973. Ossification in the jugular foramen in an ancient skull. Arch. Otol., 97:406-409.

ASYMMETRIES OF THE BRAINS AND SKULLS OF NONHUMAN PRIMATES

Marjorie LeMay*
Michael S. Billig
Norman Geschwind

Morphological asymmetries in the bodies and facies of modern man have long been recognized, but only during the past 10 years has attention been given to the numerous anatomical asymmetries that are present in the human brain. In the past few years both functional and specific morphological cerebral asymmetries have also been noted in nonhuman primates and other vertebrates (Dewson, 1977; Glick et al., 1977; Nottebohm, 1970; Webster, 1972; Webster and Webster, 1974). The present study reviews the cerebral and cranial asymmetries that have been reported in the past and also presents some new findings made on endocranial casts of a number of nonhuman primate skulls.

BRAIN ASYMMETRIES

Temporoparietal Region

Length of the Sylvian fissure. In human brains the length of the Sylvian fissure (SF) has been found to be commonly greater on the left than the right. Cunningham (1892), studying a few nonhuman primate brains, found the left fissure longer in the chimpanzee and macaque. Yeni-Komshian and Benson (1976) and Fisher (cited by Yeni-Komshian and Benson) found the left fissure longer than the right in a group of chimpanzee brains.

*From the Department of Radiology (ML), Massachusetts General Hospital and Harvard Medical School, Boston, MA 02114; the Department of Anthropology (MSB), Harvard University, Cambridge, MA 02138; and the Department of Neurology (NG), Beth Israel Hospital and Harvard Medical School, Boston, MA 02215.

Falk (1978), in a study of casts of Old World monkeys, found the right SF to be more often longer than the left.

Height of the end of the Sylvian fissure. In *Homo sapiens* the end of the SF is commonly higher on the right than the left. This is especially true in right-handers. The posterior end is more often nearly equal in height, and higher on the left than on the right, in left-handers than in right handers (Hochberg and LeMay, 1974; LeMay and Culebras, 1972).

Cunningham (1892), measuring the angle formed by the SF and a line through the anterior end of the SF perpendicular to the greatest anteroposterior (AP) diameter of the hemisphere, found the posterior end of the SF to be higher on the left in a macaque, but higher on the right in some brains of baboon, mangaby, orang, and chimpanzee.

In a study of photographs of brains of the great apes, we found asymmetries in the position of the posterior end of the SF similar to those found in humans (LeMay and Geschwind, 1975). On the brains of 28 higher apes, the posterior end of the right SF was higher in 16 whereas the left was higher in only 1.

Superior temporal sulcus (STS). In lower primates the posterior ends of the superior temporal sulcus and Sylvian fissure are joined and with increasing size of the hemispheres they become separated. Ingalls (1914) found them to be separated more often on the left than the right. Studying these in the brains of 30 New and Old World monkeys we found the STS and SF to be either separated bilaterally or joined bilaterally in 25 (LeMay and Geschwind, 1979). In 5 they were separated on the left but joined on the right. Although the number of brains is small, our findings, in conjunction with those of Ingalls, suggest greater development in the superior temporal gyrus on the left than on the right.

Frontal Lobe Asymmetries

Size of the frontal lobe. Significant differences in the surface areas of the frontal lobes have not been reported in human brains. The upper end of the central sulcus reportedly shows more variation in position than does the lower end. Studying photographs of the surfaces of a series of nonhuman primate brains, we found the lower end of the central sulcus to lie more posteriorly on the left than on the right in 9 of 12 *Pongo* brains and in 5 of 7 *Papio* brains, suggesting a larger frontal lobe on the left (LeMay and Geschwind, 1979).

Falk (1978) made a large number of endocranial casts of Old World monkey skulls, and in her measurements of the casts she reported that the anterior margins of the frontal lobes, overlying the orbits, were significantly longer on the right than on the left in *Presbytis* and *Cercopithecus*. She reported no species showing longer left-frontal margins.

Sulci. In studying the sulcal patterns in 144 endocranial casts of Old World monkeys, Falk found the rectus sulcus touched the arcuate in 21, and in 20 casts this was on the left side. Also, in 12 *Presbytis* and 8 *Papio* casts the space between the rectus and arcuate was shorter on the left than on the right, and she suggested the longer rectus sulcus on the left may be a result of greater growth of the left prefrontal cortex.

Asymmetries in Brain Shape

Although human brains vary considerably in size and shape, the posterior portion of the left hemisphere and, less striking, the anterior portion of the right hemisphere have been found to be wider and to protrude farther more often than their contralateral counterparts (Galaburda et al., 1978; LeMay, 1976).

Nonhuman primates: Endocranial casts. We studied endocranial casts in a series of ape and New and Old World monkey skulls (24 apes and 5 monkeys from the collection at the Museum of Comparative Zoology, Harvard University; most of the endocasts of the New and Old World

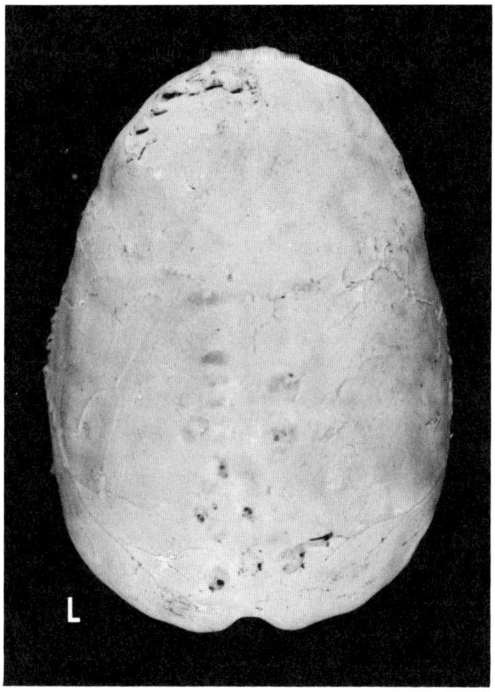

Fig. 1. Telephotograph of superior surface of a *Gorilla gorilla* endocast. The cast is oriented to give the greatest AP diameter. The anterior end of the right hemisphere and the posterior end of the left hemisphere project slightly beyond their counterparts.

monkeys were from the collection of D. Falk) for asymmetries and compared them with those found in human brains. Measuring specific variations was difficult on some of the casts, partly because of their small size (particularly for some of the New World monkey casts) and also because occasionally the true median borders of the hemispheres could not easily be determined. Telephotographs were made of superior and base views, with the casts oriented so as to give the greatest AP diameter (Fig. 1). The widths of the forward ends of the hemispheres were measured at the junction of the frontal lobe with the anterior end of the temporal lobe. Width measurements of the hemispheres' posterior ends were compared at about 1/10 the length of the hemisphere from the occipital poles. By means of program SAGIT (Neurosciences Computer Facility-Children's Hospital, Boston), the perimeters of the photographed casts were traced, and the outer distances from the midlines computed (Fig. 2). Because in some cases

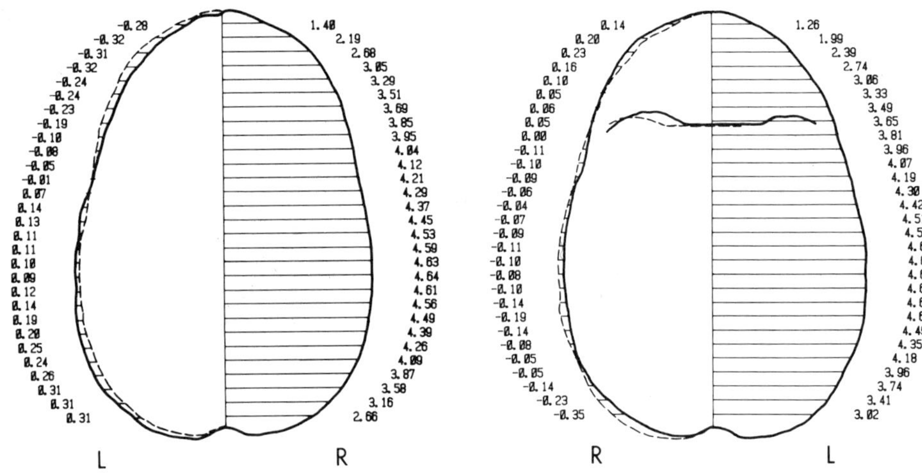

A B

Fig. 2. Computerized printout of tracings from the telephotographs of the specimen shown in Figure 1. The outline of the photograph is printed in black and the figures on the right give the measurements from the midline to the outer margin of the cast at 5 mm intervals. The interrupted line on the left is the outline of the right portion of the cast projected onto the left side computer printout. The figures on the left give a difference in measurements between the two halves of the hemispheres. In the original computer printouts, the interrupted lines are continuous red lines. They were changed to interrupted lines for clearness in the present illustration. (A) Superior surface. The anterior end of the right hemisphere and posterior end of the left hemisphere are wider. (B) Base view of the same cast. The broken lines mark the margins of the left hemisphere that is projected onto the outline of the right hemisphere. The curved lines over the central portion of the printout represent the anterior ends of the temporal lobes.

establishing the correct angulation of the casts for photographing was difficult, the computer measurements were not always similar on the top and base views. In such cases, width measurements were determined by using a stainless steel contour gauge and/or calipers.

For determining the frontal and occipital petalia, a polar cordinate, clear plastic graph was centered over the midpoint of each cast on each photograph and the forward and posterior extension of the hemispheres compared.

X-rays of the skulls of apes were taken with the greatest AP diameter of the inner table of the vault parallel to the film. The petalia of the inner tables of the vault was determined with the polar coordinate graph and compared with that found on the photographs of the endocranial casts.

Table 1 shows the findings on the casts as judged by the measurements described above. Asymmetries appear to be present in the ape endocasts, which tend to resemble those usually seen in human skulls. Less striking asymmetries, but in a similar direction, are also seen in Old World monkeys. Significant asymmetries were not noted on the endocast measurements on the New World monkeys, with the exception that left occipital petalia was more common than right occipital petalia.

TABLE 1. Asymmetries of the Endocranial Casts

	Width of Hemispheres						Petalia					
	Frontal			Occipital			Frontal			Occipital		
	L>	=	R>	L>	=	R>	L↑	=	R↑	L↓	=	R↓
APES												
Gorilla		2	4	4	2			2	4	4	2	
Pan	1	4	1	3	3		1	2	3	3	3	
Pongo		3	3	2	4			1	5	4	2	
Symphalangus			1		1				1		1	
Hylobates	1	3	1	4	1			1	4	2	3	
	2	12	10	13	11	0	1	6	17	13	11	0
OLD WORLD MONKEYS												
Simias (4), *Presbytis* (6), *Macaca* (5) *Papio* (3), *Mandrillus* (1), *Cercocebus* (2)							5	10	6	7	11	1
NEW WORLD MONKEYS												
Lagothrix (4), *Ateles* (3), *Alouatta* (4), *Cebus* (5), *Callicebus* (3), *Aotus* (6), *Cacajao* (2), *Chiropotes* (3), *Pithecia* (5), *Callimico* (1), *Leontopithecus* (2), *Sanguinas* (2), *Cebuella* (1), *Callithrix* (2)							11	22	10	15	22	6

Venous Channels

In the human brain, the main flow of the superior venous channels has been seen to drain into the right transverse sinus as early as in the 20-mm fetus (Streeter, 1915). In a series of 100 routine carotid arteriograms performed on adult humans during life (personal data, ML), the right transverse sinus was *larger* than the left in 57%. In 27% of the arteriograms, the main flow of the sagittal sinus appeared to be into the left transverse sinus. In the same study, the left transverse sinus was *lower* than the right in 52% of the arteriograms. In 32%, the right transverse sinus was lower.

The size and position of the transverse sinuses were studied on the series of endocranial casts of nonhuman primates mentioned above. The drainage was mainly to the right in the endocasts of 61% of the apes, 40% of the Old World monkeys, and 51% of the New World monkeys (Table 2). A main flow to the left was seen on the endocasts in 17% of the ape skulls, 16% of those of the Old World monkeys, and 23% of the New World monkey skulls. In 70% of the endocasts of the greater and lesser apes, the left transverse sinus was lower than the right, and lowering of the left transverse sinus was found in 68% of the endocasts of the Old World monkey skulls and 35% of those of the New World monkey casts. In only 4 of 92 endocasts was the right transverse sinus lower than the left (see also Conroy, this volume).

SKULL ASYMMETRIES

Interest in skull asymmetries in nonhuman primates was renewed when Groves and Humphrey (1973) noted asymmetries in the facial area of some mountain gorilla skulls. They described measuring from the anterior point of the temporal fossa on either side to the gonion and commonly found the left side to be longer than the right. They questioned whether the asymmetries were more likely to be caused by a difference in size of the cerebral hemispheres or because of a tendency for the gorilla to chew more on the left side than the right. Other scientists earlier in this century also described skull and facial asymmetries in gorilla skulls. Gyldenstolpe (1938) reported that in the skulls of some mountain gorillas the right facial portion was more prominent than the left and the teeth were worn more on the right side. These were skulls obtained in a hunting expedition by Prince William of Sweden. Henschen (1926) noted that on 6 of these skulls the left posterior portion was larger than the right. That investigator, along with G. Elliot-Smith, also noted that the left hemisphere in a gorilla skull at the University Museum in London was wider, and they concluded that the gorilla was probably right-handed.

X-rays of the skulls of the great apes used in the present study were taken with the greatest AP diameter of the inner tables parallel to the film. The forward and posterior extension of the inner table of one hemisphere

TABLE 2. Position and Size of Transverse Sinus

	Major Transverse Sinus Flow			Height of Transverse Sinus		
APES	to Lt	Rt = Lt	to Rt	Lt ↓	=	Rt↓
Gorilla	1	2	3	4	2	
Pan	1		5	4	1	1
Pongo		2	3	2	3	
Symphalangus	1			1		
Hylobates	1	1	3	5		
TOTAL	4	5	14	16	6	1
OLD WORLD MONKEYS						
Simias		2	2	2	2	
Nasalis			2	2		
Presbytis	2	1	3	4	1	1
Macaca	1	3	3	4	3	1
Papio	1	2		2	1	
Mandrillus		1		1		
Cercocebus		2		2		
TOTAL	4	11	10	17	7	1
NEW WORLD MONKEYS						
Lagothrix		1	3	3	1	
Ateles		1	2	2	1	
Alouatta	1	1	2	1	2	1
Cebus	2		3	3	1	1
Saimiri	1	1	1		3	
Callicebus	1	1	1		3	
Aotus			5	2	2	a
Cacajao			2	2		
Chiropotes	2	1		1	2	
Pithecia	1	2	2	2	3	
Leontopithecus	1		1		2	
Sanguinas		1			1	
Cebuella		1			1	
Callithrix	1		2		2	b
TOTAL	10	10	24	16	24	2

[a]On one brain the flow was entirely to the left.
[b]Because of an artifact, it was not possible to compare the heights of the sinuses on one cast.

beyond the other was determined in the skulls by using a polar coordinate graph. The graph was centered over the foramen magnum of the skull x-rays and the anterior end of the vertical line centered on the midline of the inner table of the frontal bone. The asymmetries of the inner tables of the frontal and occipital regions were compared along the angle lines. Measurement of the forward extension of one brow ridge beyond the other was also determined.

Variations in measurements of skulls of nonhuman primates are more difficult to determine than in those of humans because the skulls are smaller and also because the large bony excrescences on the outer tables (particularly the adult male skulls) make it difficult to project the skull so that the greatest AP diameter of the inner table of the skull is parallel to the x-ray film. In human skulls, the right frontal and left occipital bones have been found to extend beyond their counterparts more often than vice versa (LeMay, 1977). This finding was also true for the skulls of the apes studied at this time. In the past we found right frontal petalia more often in the skulls of chimpanzees than in orangs, but in the present group of skulls studied, right frontal petalia was also obvious in the orang skulls studied. In the past, left frontal petalia was rarely found in the ape skulls and this was also true in the present study. The brow ridge was also found to be more often forward, associated with the more forwardly located inner table of the frontal bone on the same side (Table 3, Fig. 3).

DISCUSSION

Variations in length of limb bones and asymmetries of the skulls and brains of nonhuman primates, and some other vertebrates, have been described in the past (Mollison, 1896; Schultz, 1926), but only in recent years has there been a more active interest in morphological asymmetries of the brain in the hope of defining structural changes relating to some of the functional and biochemical asymmetries now being found in nonhuman vertebrates (Dewson, 1977; Glick et al., 1977; Webster and Webster, 1974). The functional significance of the numerous asymmetries noted in the human brain are not known, and it may be of value to attempt to correlate cerebral asymmetries in nonhuman primates with functional asymmetries found in the animals.

The most constant asymmetries seen in the present study are the main flow of the superior sagittal sinus to the right and the lower position of the left transverse sinus. The flow pattern of the sagittal sinus is established in early life (Streeter, 1915), but the lowering of the transverse sinus must be influenced to some degree, particularly in individuals with large brains, by the size of the posterior portions of the hemispheres overlying them (see Conroy, this volume).

Human brains and endocranial casts often show a prominence along the orbital margins of the frontal lobe just anterior to its junction with the

TABLE 3. Relationship of Browridge to the Forward Extension of the Hemisphere

		Frontal Petalia			Browridge		
		Lt↑	=	Rt↑	Lt↑	=	Rt↑
GORILLA							
gorilla	750			x			x
"	162		x			x	
"	850			x		x	
"	048		x			x	
"	261			x			x
"	325			x			x
PAN							
paniscus	018			x			x
troglodytes	163			x			x
"	166	x				x	
"	847		x			x	
"	849		x			x	
"	416			x			x
PONGO							
pygmaeus	364		x			x	
"	365			x		x	
"	516			x		x	
"	518			x			x
"	959			x			x
"	960			x			x
SYMPHALANGUS							
syndactylus				x			x
HYLOBATES							
lar	383			x			x
concolor	116		x		x		
lar	531			x			x
"	538			x		x	

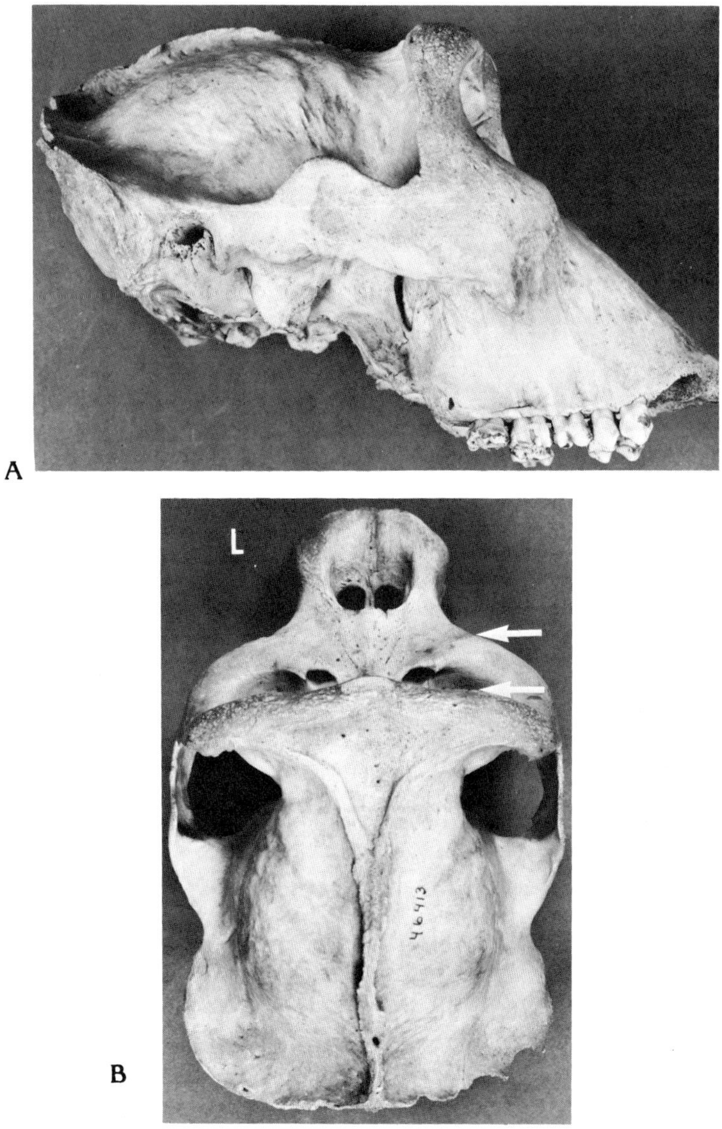

Fig. 3. Skull of a male *Gorilla gorilla.* (A) Lateral view. The skull is oriented to give the greatest AP diameter of the inner table of the vault. (B) Superior view of the same skull oriented as shown in 3A. The brow ridge and maxilla are forward on the right. The large occipital crest extends further posteriorly on the left.

temporal lobe that Kochetkova (1978) referred to as a "tuber." These prominences are often more apparent on the right and are also commonly present on the casts of nonhuman primate skulls (Fig. 4). In the literature, prominence in the frontal lobe region associated with a protuberance along the posterior orbital margin of endocranial casts is sometimes referred to as "Broca's area" (Kochetkova, 1978; Heim, 1976). Historically, Broca's area refers to the area in the brain in which a lesion produces a persistent Broca's aphasia; it is not known at present whether this region is defined by anatomical landmarks or by architectonic fields. The lesion is most often in the posterior half of the inferior frontal gyrus and commonly involves the opercular portion of the frontal lobe, the posterior portion of the pars triangularis, and the lower end of the motor complex (i.e., the hand and face area) (Dejerine, 1908; Lassen et al., 1978). The orbital portion of the frontal lobe, where the "tuber" is most prominent, is not commonly involved in aphasia-producing lesions and it would seem misleading to refer to this "tuber" as Broca's area. The amount of neocortex increases greatly from lower primates to humans. The principal sulci are present in lower primates, but the position of sulci change with the development of the convolutions. The "tuber" seen on the human casts, therefore, probably does not represent the same neurological structure as that on the nonhuman primate casts. Pandya and Galaburda (1980) (see Galaburda and Pandya, this volume) have shown clearly by cytoarchitectonic studies that one can be misled in attempting to relate too closely the morphological changes seen in nonhuman primate brains and endocasts to those seen in humans.

The development of the orbital ridges, as particularly shown in the brow ridge of adult males, seems from the present study to be influenced by the asymmetry of the anterior margins of the cerebral hemispheres. Clearly, in humans the shape of the skull usually reflects the shape of the brain. Even the basicranium is influenced by the mass of the brain (Moss, 1975), and in humans the main influence on the development of forehead shape has been shown to be the brain mass (Moss, 1964). The finding in the present study that the brow ridge is more often prominent on the right, as the anterior margin of the right hemisphere is more often forward, would suggest that the brain size and shape--and not external factors--are the main influence on the variations in shape of the brow ridge. In humans, the anterior cranial fossa has nearly completed its growth by the third year of life, but the facial growth continues throughout adolescent life. The basic shape of the face, however--including the angle of the mandible--does not change but remains constant during its late prenatal and postnatal growth (Salentijn and Moss, 1971). In anthropoids, the brain is comparatively larger at birth than it is in man, and rapidly gains mature size (see also Leutenegger, this volume). One could logically assume that the basic shape of the skull, including the brow ridge is also mainly programmed in fetal life in nonhuman primates as well as in humans.

The question must arise as to the significance of the asymmetries in nonhuman primates. We know that in humans both the gross and cytoarchitectonic asymmetries of the upper surface of the temporal lobe (Galaburda

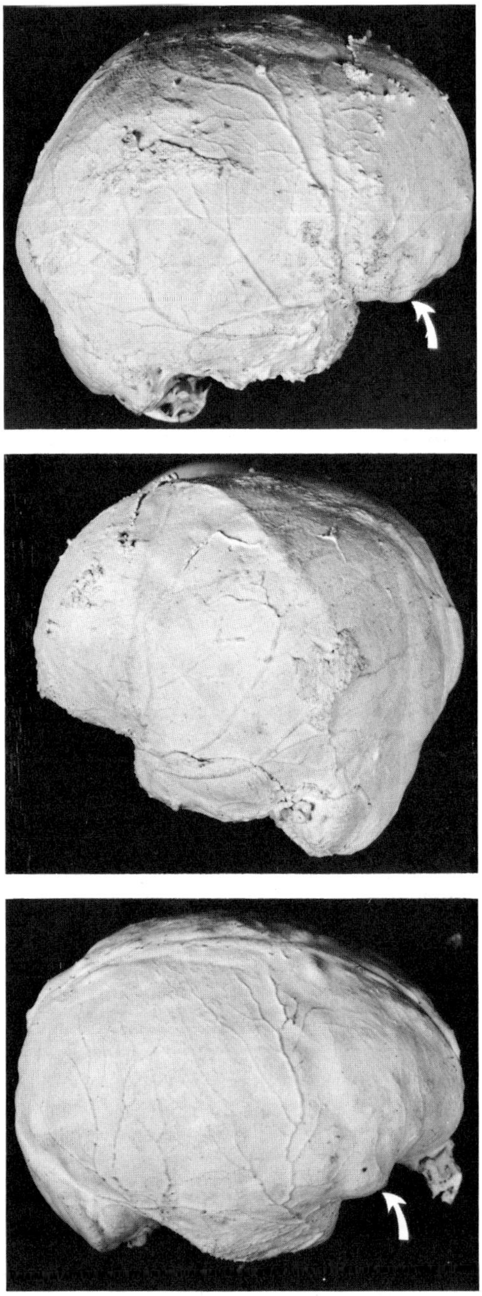

A

B

C

Fig. 4. Endocranial casts showing projections from the orbital ridges of the frontal lobes. (A) Right hemisphere of North American Indian. Arrow points to prominent tuber. (B) Left hemisphere of same skull. No obvious "tuber." (C) Right hemisphere of a cast of *Pan troglodytes*. Arrow points to prominent "tuber."

et al., 1978a,b) are in regions that, on the basis of studies of lesions, constitute part of Wernicke's area, one of the major regions devoted to language. We also know that there are statistically different distributions of several of the asymmetries, discussed earlier in this paper, on the right and left sides of the brain. On the bases of their locations it seems likely that at least some of the human asymmetries are related to dominance for language although there is as yet no direct proof of this. On the other hand, despite the different distributions of some of the asymmetries (e.g., height of the posterior ends of Sylvian fissures) in right and left handers, there may be serious question as to whether the relationship is a direct one. The relationship between language dominance and handedness is very complex; less than 3% of aphasias in right handers result from right hemisphere lesions whereas in left handers the lesions are distributed much more symmetrically (about 60% left, 40% right). Furthermore, there is much evidence that many, if not most, left handers develop aphasia with lesions on either side of the brain whereas aphasia associated with right-hemisphere lesions is rare in right handers.

The meaning of the asymmetries in the nonhuman primates is even more obscure at present. They are almost certainly not related to handedness since although the present studies show that the distribution of the asymmetry of the Sylvian fissures in many apes resembles that in humans there is no evidence of a predominant type of handedness in any population of nonhuman primates.

If we were to speculate, which is all that one can do at present, we would suggest that the primate anatomical asymmetries are indeed related to asymmetrical functions. It seems reasonable to assume that at least some of these asymmetries are related to right-hemisphere dominance for limbically related and spatial functions since Denenberg and his co-workers (1978) have, in the last few years, advanced considerable evidence that there is right-hemisphere dominance for these functions in rats. Furthermore it is conceivable that some of the asymmetries seen in the brains of nonhuman primates may be located in areas that subserve functions that are forerunners of language. Dewson (1977) has brought evidence to support the view that the left temporal lobe of monkeys is specialized for certain auditory tasks. Future studies can be expected to add greatly to our knowledge of the relationship between cerebral asymmetries and function.

Acknowledgments: We thank M. Edith Rutzmoser, of the Museum of Comparative Zoology, Harvard University, for her help with the selection of the skull specimens for the study and Gene Belch, Computer Analyst at the Neuroscience Department at Children's Hospital in Boston, for designing the computer program to measure the asymmetries of the endocasts. This study was supported by a grant from NINCIDS #5R01-NS14018 and the Milton Fund, Harvard University.

REFERENCES

Cunningham, D.J., 1892. Contribution to the Surface Anatomy of the Cerebral Hemispheres, Royal Irish Academy, Dublin.

Dejerine, M.L., 1908. Discussion sur l'aphasie, Société de neurologie de Paris. Rev. Neurol., 16:974-1024.

Denenberg, V.H., Garbanati, J., Sherman, G., Yutzey, D.A., and Kaplan, R., 1978. Infantile stimulation induces brain lateralization in rats. Science, 201:1150-1152.

Dewson, J.H., 1977. Some behavior effects of removal of superior temporal cortex in the monkey. In, Proc. 6th Congr. Intl. Primatol. Soc., Academic Press, New York.

Falk, D., 1978. External neuroanatomy of old world monkeys (cercopithecoidea). In, Contributions to Primatology, Vol. 15, F.S. Szaley, ed., S. Karger, Basel-New York.

Fisher, E., 1921. Reported by Yeni-Komishian G.H., and Benson, D.A., 1976.

Galaburda, A.M., LeMay, M., Kemper, T.L., and Geschwind, N., 1978a. Right-left asymmetries in the brain. Science, 199:852-856.

Galaburda, A.M., Sanides, F., and Geschwind, N., 1978b. Human brain: cytoarchitectonic left-right asymmetries in the temporal speech region. Arch. Neurol., 35:815-817.

Glick, S.D., Zimmerberg, G., and Jerussi, T.P., 1977. Adaptive significance of laterality in the rodent. In, Evolution and Lateralization of the Brain, D.A. Blizard, and S.J. Dimond, eds., The New York Academy of Science, New York, pp. 180-185.

Groves, C.P., and Humphrey, N.K., 1973. Asymmetry in gorilla skulls: Evidence of lateralization brain function? Nature, 244:53-54.

Gyldenstolpe, N., 1928. Zoological results of a Swedish expedition to Central Africa. Arkiv. Zool., 20A:46.

Heim, J.L., 1976. Les hommes fossiles de la Ferrassie. Arch. de Institute de Paleontologie Humaine Mem. Vol. 35, Masson, New York.

Henschen, S.E., 1926. On the function of the right hemisphere of the brain and its relation to the left in speech, music, and calculation. Brain, 49:110-123.

Hochberg, F.H., and LeMay, M., 1974. Arteriographic correlates of handedness. Neurology, 25:218-222.

Ingalls, N.W., 1914. The parietal region in the primate brain. J. Comp. Neurol., 24:291-341.

Kochetkova, V.A., 1978. Paleoneurology, H.J. Jerison, and I. Jerison, trans., J. Wiley & Sons, New York-Toronto.

Lassen, N.A., Ingavar D.H., and Skinhoj, E., 1978. Brain function and blood flow. Sci. Am., 239:62-71.

LeMay, M., 1976. Morphological cerebral asymmetries of modern man, fossil man and nonhuman primate. Ann. NY Acad. Sci., 280:349-366.

LeMay, M., 1977. Asymmetries of the skull and handedness. Phrenology revisited. J. Neurol. Sci., 32:243-253.

LeMay, M., and Culebras, A., 1972. Human brain-morphological differences in the hemispheres demonstrable by carotid angiography. N. Engl. J. Med., 287:168-170.

LeMay, M., and Geschwind, N., 1975. Hemispheric difference in the brains of great apes. Brain Behav. Evol., 11:48-52.

LeMay, M., and Geschwind, N., Morphological asymmetries of the brain and skull in primates, presented at the International Symposium of Primatology, Bangalore, India, January 1979.

Mollison, T., 1911. Die Körperproportionen der Primaten. Morphol. Jahrb., 42:79-304.

Moss, M.L., 1964. The capsular nature. Am. J. Orthod., 56:474-490.

Moss, M.L., 1975. New studies of cranial growth. In, Morphogenesis and malformation of face and brain, D. Bersma, ed., Alan R. Liss, Inc., New York.

Nottebohm, F., 1970. Ontogeny of bird song. Science, 167:950-956.

Pandya, D.M., and Galaburda, A.M. The role or architectonics and connections in the study of primate brain evolution, presented at Am. Assoc. Phyl. Anthrop., Niagara Falls, April, 1980.

Salentijn, L., and Moss, M.L., 1971. Morphological attributes of the logarithmic growth of the human face: Gnomonic growth, Acta Anat. (Basel), 78:185-199.

Schultz, A.H., 1926. Studies on the variability of platyrrhine monkeys. J. Mamm. 7:286-305.

Streeter, G.H., 1915. The development of the venous sinuses of the dura mater in the human embryo. Am. J. Anat., 18:145-178.

Yeni-Komshian, G.H., and Benson, D.A., 1976. Anatomical study of cerebral asymmetry in the temporal lobe of humans, chimpanzees, and rhesus monkeys. Science, 192:387-389.

Webster, W.G., 1972, Functional asymmetry between the cerebral hemispheres of the cat. Neuropsychologia, 10:75-87.

Webster, W.G., and Webster, T.H., 1974. Anatomical asymmetries of the cerebral hemispheres of the cat brain. Physiol. Behav., 14:867-869.

NEUROBIOLOGICAL ASPECTS IN THE PHYLOGENETIC ACQUISITION OF SPEECH

Charles R. Noback*

The ability to articulate words and to communicate verbally is a major hallmark unique to man compared with the living primates from whose ancestors man evolved. Even this dichotomy may not be absolute. Recent evidence demonstrates that vervet monkeys in Kenya, East Africa, use semantic communication in their alarm calls (Seyfarth et al., 1980). These monkeys use three different alarm calls for three different predators and respond differently and appropriately according to whether the call (live or tape-recorded) means a leopard, a martial eagle, or a python. To a leopard alarm the monkeys run into the trees, to an eagle alarm the monkeys look up, and to a snake alarm the monkeys look down.

Three theories are current concerning the relation of vocalizations in nonhuman mammals to speech in man. The most widely accepted view is that animal vocal activities evolved into the speech of man (Hockett, 1960). Another concept theorizes that human speech evolved from gestures and body movements and not from the vocalization of mammals (Hewes, 1973). Still another concept states that human speech arose *de novo* and has little relation with mammalian vocal responses (Lancaster, 1968; Myers, 1976).

Man's acquisition of speech is the culmination of evolutionary changes in the anatomy of the entire voice organ coupled with a finely tuned integration with the nervous system. Man's voice organ was attained after a number of critical modifications occurred in the preexisting anatomy of the respiratory system and the oral, nasal, pharyngeal, and laryngeal structures associated with chewing, swallowing, breathing, and sound production. The voice organ comprises the lungs, larynx, pharynx, nose, and mouth. The lungs function as the power supply to produce an excess of air pressure

*From the Department of Anatomy, College of Physicians and Surgeons, Columbia University, New York, New York 10032

(airstream) through the activity of the abdominal and thoracic musculature. The larynx, called the voice source, contains the elastic vocal folds (cords), which act as the oscillators; the vibrating vocal folds chop up the airstream and thereby produce the basic elements for speech. The larynx, pharynx, mouth, and nose constitute the vocal tract or resonators; they form the resonant chamber. The actions of the velum (soft palate), tongue, and lips are also major contributors to the formulation of voiced sounds. The phylogenetic and ontogenetic development of the voice organ were discussed by many authors, including Lieberman (1975), DuBrul (1976), Laitman and Crelin (1976), and Wind (1976).

The exquisitely coordinated activities of the musculature associated with the production of speech are activated and regulated by central motor programs that are, in turn, modulated by other complex neural circuit systems in the higher levels of the brain. These programs and systems have their roots in man's ancestral primates, mammals, and even nonmammals where they exerted their influences on the musculature involved with respiration, swallowing, and sound production. Although precise knowledge of where in the brain the orders arise and precisely how the neural systems are linked to produce the voluntary movements of the voice organ is incomplete at the present time, information is available concerning many neural substrates relevant to the motor control of speech. In this discussion emphasis will be directed to brief accounts of some of the concepts associated with the formulation of speech.

CONCEPT OF THE NEURONS

The cardinal functional property of each neuron is the relative constancy of its output function. The inputs to a given neuron stimulate it to express itself in a programmed way and this expression is "known" to the neuron. For example, the neuron receives inputs--both of an excitatory and inhibitory nature--to alter its on-going activity. The neuron responds either by increasing its firing rate (excitation) or decreasing its firing rate (inhibition). Consider the lower motor neurons that innervate the voluntary muscles of the voice organ. A lower motor neuron is a neuron whose cell body is located in either the brain stem or spinal cord and whose axon terminates by branching peripherally and innervating from several to many muscle cells. If excited, a lower motor neuron evokes its activity by stimulating the muscle cells of its motor unit to respond by contracting (obligatory contraction). The excitatory input merely sets off the functional feature of that motor neuron. This input carries the message "to whom it may concern." If the neuron is influenced to respond, it responds in a predictable and relatively predetermined way. For speech, the assemblage of lower motor neurons (both alpha and gamma neurons) innervating the muscles activating the voice organ are given "to whom it may concern" signals that are exquisitely coordinated for the production of sounds.

CONCEPT OF NEURONAL CIRCUITS

The activities of the nervous system are expressed by organized sequences of neurons known as neuronal circuits. Their functional properties are expressed on pools of neurons and ultimately on output neurons of the central nervous system--namely, the lower motor neurons. In essence, the neuronal circuits are the products of complex neuronal interactions, with each circuit expressing the properties of its neurons and their connections. In turn, many circuits influence other circuits. The results of the processing activities within these circuits are ultimately directed (focused) to the lower motor neurons. Each circuit is, in a way, similar to a neuron or pool of neurons in that it receives the "to whom it may concern" excitatory and inhibitory message as input. If that circuit "chooses" to respond, it responds in its own predetermined way on other circuits or lower motor neurons.

Each lower motor neuron has an intrinsic activity, which can be stimulated (1) by tonic circuits, which result in giving tone to the vocal organ musculature, and (2) by phasic circuits, which produce contractions superimposed on the pre-existing tone. It is through the control of conformational changes of the musculature of the vocal organ that the different sounds of speech are produced.

CONCEPT OF CENTRAL CONTROL PROGRAMS (CENTRAL CONTROL THEORY AND PERIPHERAL CONTROL THEORY)

Historically the coordinated activity of the musculature (motor theory) was conceived as operating according to either (1) the peripheralist concept or (2) the centralist concept (DeLong, 1972; Delcomyn, 1980, and others). According to the peripheralists, motor activity is based primarily on the reflex arcs and on the idea that the rhythmic pattern of movements are an expression of the use of feedback from peripheral receptors in the structures in motion. A crucial feature of this concept is that the sensory cues of the prior phase of the motion is essential for the proper timing of the succeeding phase. According to the centralists, motor activity is said to be intrinsically capable of generating the proper timing and coordination of muscle activation without the need of sensory feedback. In this centralist concept, a central pattern generator (pacemaker neuron, pacemaker neural network, or neural oscillator) is presumed to produce the neural activity resulting in a repetitive rhythmic output to the musculature innervated (Edgerton et al., 1976). A central pattern program involving a rhythm generator is the basis of the concept of a triggered movement such as swallowing. In a coordinated movement this central pattern generator (1) stimulates the appropriate muscles (spatial aspect), (2) activates or inactivates each participating muscle in the proper temporal relation to each other (temporal aspect), and (3) excites and inhibits to the appropriate amount each muscle of the activity (quantitative aspect) (DeLong, 1972). In a highly complex action such as speech, several central pattern generators are presumed to be coordinated by higher level circuitry. In its precise formulation the central pattern generator of the central control system may not require peripheral feedback for its basic activity. Such feedback, however, can modulate the intrinsic central pattern (Delcomyn, 1980). The current concept of a central programmed behavioral action does not mean that an action is generated wholly within the central nervous system without reference to sensory feedback, but that the essential features of their activation are centrally determined (MacKay, 1980). Currently, neurobiologists favor the view that both concepts are operative in motor activity to a greater or lesser degree.

In mammals, the vital activities of swallowing and respiration lack specific spatial orientation to the environment. Hence, a reasonable postulation is that sensory feedback is minimal for carrying out their roles. In contrast, the musculature involved in movements of the head and eyes are spatially oriented. Hence peripheral input likely has a more significant role in these activities. Thus differential degrees of sensory feedback are features involved in different movements.

Spatially, the appropriate muscles in an action are similarly selected in both the peripheral and the central control theories by precisely defined neural circuits that convey neural information to the appropriate lower motor neurons. This is accomplished either by reflex circuits according to the peripheralists or by centrally located neural circuits according to the centralists. The basic difference between the peripheral and central concepts is in the way in which the temporal (timing) and quantitative aspects are fulfilled. According to the peripheralists, these two aspects are a primary function of the differential activities of the feedback loops whose connections provide the excitory and inhibitory influences in the different phases of musculature activity after appropriate temporal delays. According to the centralists, the central circuitry has within it the critical information essential to the spatial, temporal, and quantitative aspects of movement. The central control theory does allow for a role for peripheral input for modifying the details of the output and for modulating the excitability within the central circuitry.

The lower brain stem (pons and medulla) and spinal cord contain the neural circuits within which are located the central pattern generators for the central motor programs for such activities as respiration and swallowing. These basic circuits and programs are present and functioning in mammals and even in nonmammals. Thus the primary control of the coordination of the musculature regulating the activity of the voice organ is exerted by intrinsic circuits and programs embedded in the central nervous system. The central programs in the lower brain stem and spinal cord have neural substrates that are capable of stimulating synchronized motor activity even when denied the neural influences from the higher centers and circuits in the brain. Such central control is exhibited in the control and regulation of respiration. For example, the central pattern generator for respiration, which is located in the medulla, sets the basic rhythm of inspiration (Wyman, 1976). This generator, called the pneumotaxic center, also integrates input from the pulmonary stretch receptor feedback that normally terminates inspiration. Rostral to the generator are other centers responsive to several variables such as blood gases; these centers act to modulate the pneumotaxic center, but not to change its basic mode of rhythm generation. Another central center in the pons, called the apneustic center, is responsible for the rhythmic inhibition of the central pattern generator for respiration. Breathing is an example of a cyclic motor act that is governed by a central pattern generator that does not require phasic afferent feedback from the moving parts for the production of a rhythmic motor output (von Euler, 1980).

Suprasegmental Feedback Circuits Involving the Cerebrum

The central control motor programs are, in turn, modulated and biased by a number of complex circuits involving the cerebrum. Two will be discussed as they relate to motor activities. Schematically the *first* is the feedback circuitry of cerebral neocortex to cerebellum and back to motor neocortex. This process involves the core sequence commencing with (a) the neocortex of many lobes of the cerebrum (frontal, parietal, occipital and temporal lobes) to (b) the pontine nuclei to (c) the cerebellum (deep nuclei and cortex) to (d) the thalamus (ventral lateral and ventral anterior nuclei) and finally to its termination in (e) the motor neocortex (areas 4, 6, and 8 of the frontal lobe). The *second* is the feedback circuitry of cerebral neocortex to basal ganglia and back to motor neocortex. This process involves the core sequence with (a) the neocortex (frontal, parietal, occipital and temporal lobes) to (b) the basal ganglia (striatum and globus pallidus) to (c) the thalamus (ventral lateral and ventral anterior nuclei) and finally to (d) the premotor and motor cotex (frontal lobe). Further details of these circuits and other aspects of neural pathways are outlined in Noback and Demarest (1981).

Several important points concerning these suprasegmental circuits are relevent to this discussion. (1) In both circuits, the neocortex of the lobes of the cerebrum gives rise to efferent projections (corticopontine and corticostriate fibers). (2) Both circuits have projections terminating in the ventral lateral and ventral anterior thalamic nuclei. (3) The ventral lateral and ventral anterior nuclei project to a limited area of the neocortex--namely the motor and premotor cortices (areas 4, 6 and 8). Because of the nature of these projections, these nuclei and their projections are known as the afferent nuclei of the motor cortex, motor thalamus, and afferents to the motor cortex. (4) With respect to volitional movements by subjects responding to cues, neurons of the basal ganglia, cerebellum, and motor thalamus fire well in advance of the muscular action (DeLong, 1974; Thatch, 1975). Thus these circuits are critically important in exerting influences to the motor cortex before the discharge from the motor cortex activating the movement. The circuit involving the basal ganglia is apparently essential to the "earliest stages of initiating movement, the stages when, by processes not yet understood, an abstract thought is translated into a concrete motor action" (Evarts, 1979). These feedback circuits are continuously active in the subsequent phases of any movement.

Thus both the basal ganglia and the cerebellum receive input from many areas of the cerebral neocortex and both project a major portion of their output to the premotor and motor cortex (areas 4, 6 and 8) via the motor thalamus. In essence, both the basal ganglia and cerebellum are wired in parallel from many areas of the neocortex at the beginning of the circuits to the premotor and motor cortex at the end of the circuits. It has been theorized that the basal ganglia and cerebellum influence are largely responsible for the basic patterns of activity in the motor cortex and, to a large extent, in motor control and motor planning. They may have roles in virtually all the components of motor function (DeLong, 1974). In this respect, the results of cognitive functions in the neocortex may possibly exert influences through these complex feedback circuits that function in the formulation of speech patterns.

CEREBRAL NEOCORTEX

In the hierarchy of the nervous system, the neocortex is considered to be at the highest command level. Within the fabric of the neocortical neuronal elements and their circuitry are the substrates that initiate the events leading both to speech and to language. Because man is the only species capable of speech, as the term is commonly used, mammals are not suitable subjects for analyzing the command levels of speech. Thus, knowledge concerning the relation of the neocortex to speech has been determined from humans primarily in two ways: (1) from persons with localized penetrating injuries (Luria, 1970) and (2) from persons whose various cortical areas have been stimulated during surgery (Milner, 1974). From studies of these subjects, information has been obtained from which current interpretations have been formulated (Adams and Victor, 1977; Geschwind, 1970, 1979; Marin et al., 1977; Goodglass, 1980, and others). As yet, the formulation of a neorophysiology of speech and language that is based on objective evidence is not possible--that is, a theory that characterizes the manner in which the linguisitic structure is coded by the brain (Zuriff, 1980). The following abbreviated account outlines several aspects of the relation of the neocortex to speech.

The executive region of the neocortex concerned with the motor aspects of speech is thought to be Broca's area (areas 44 and 45, the anterior speech area) located in the posterior third of the inferior frontal gyrus just in front of the head and neck region of the motor cortex (precentral gyrus near the lateral fissure). This area of the motor cortex contains the motor representation for the larynx, pharynx, tongue, lips and face--structures of the voice organ. In about 95 percent of humans, speech, per se, is a function of Broca's on the left side of the brain (major hemisphere). Focal lesions of this area on the left side result in the aphasic of Broca's type, called Broca's aphasia (Adams and Victor, 1977). In the most advanced form of this syndrome, patients lose all power of speaking aloud. In milder forms, patients are able to speak to some degree but do so with poor articulation. In effect, the primary defect after a lesion in Broca's area is an alteration in speech production. Electrical stimulation of Broca's area results in the arrest of speech. It is still reasonable to assume, as Wernicke did in 1874, that Broca's area contains the rules by which language is coded into its articulating form (Geschwind, 1970).

By convention, speech is used as a measure of cerebral hemispheric dominance, with the left hemisphere being dominant for speech in most persons. Speech dominance by the left side of the cerebrum is apparently involved with some subtle, as yet to be delineated, neurophysiological difference between the cerebral hemispheres. Geschwind and Levitsky (1968) showed an anatomical asymmetry between the "temporal speech regions" of the superior surface of the two temporal lobes. This planum temporale, located just behind the transverse gyri of Heschl or primary auditory cortex, is larger on the left side in 65 percent of brains and larger on the right side in 11 percent of brains. It may represent a functional asymmetry. Functional aspects of lateralization of higher cortical functions including speech are discussed by Whitaker and Ojemann (1977) and Rasmussin and Milner (1977). Anatomical asymmetries in primate and human cerebral cortex were reviewed and analyzed by Whitaker and Selnes (1976), Rubens (1977), and LeMay et al., this volume.

Linguistic competence may be conceived as being the resultant of tho cooperative activities of several cortical areas. The neural activity within these areas are presumably relayed either directly and indirectly to and via other areas before being relayed to Broca's area where the central program (or programs) is evoked for a specific articulation (see Galaburda and Pandya, this volume). In turn, the code from this program is then projected to the motor cortex (precentral gyrus), where the appropriate neurons of descending motor pathways projecting to the brain stem and spinal cord are stimulated.

The nature of the sequential linkage in the cortex from the initiating sites to the motor cortex is not fully known. The following has been presumed to occur. When a word is seen, the visual pathways from the eyes convey coded messages to the primary visual cortex (occipital lobe area 17). Furthur neuronal processing occurs in the visual association cortex (areas 18 and 19) before being projected to the angular gyrus (area 39) where the visual form of the word is then conveyed to Wernicke's area (area 22) of the posterior temporal lobe. From Wernicke's area, coded information is relayed by association fiber (arcurate fibers) to Broca's area. Similarly, when a word is heard, the auditory pathways from the ears convey coded messages to the primary auditory cortex (area 41 of the temporal lobe) and then further processing occurs in the auditory association cortex (area 42) before projection to Wernicke's area, from which the coded information for that word is relayed to Broca's area.

Based on observations of the effects of brain injury in human subjects, multistage sequential models have been proposed (Goodglass, 1980). In a three-stage model, Goodglass proposed a linkage from (1) the locale of the origin of a concept to (2) a cortical locale of the phonological representation of its name (of the object), and finally to (3) Broca's area for the presumed conversion of the phonological representation into a chain of articulatory gestures.

Another significant contribution to the formulation of speech is, in all probability, performed by the suprasegmental feedback circuits previously outlined.

OUTPUT FROM THE MOTOR CORTEX

The output from the motor cortex is projected downstream to the brainstem and spinal cord via several pathways, including (1) the cortico-reticular, corticobulbar pathways to the brain stem reticular formation and directly to some lower motor neurons of cranial nerves and (2) the corticospinal, corticorubrospinal, and corticoreticulospinal pathways to the spinal cord. These descending pathways modulate and bias the activity of the neuronal circuits of the central control programs located in the brain stem and spinal cord, which, in turn, influence the lower motor neurons innervating the musculature of the voice organ.

The neurons of the corticospinal and other tracts project to many levels of the neuraxis, including the cerebral cortex itself (via recurrent collateral fibers), thalamus, pontine nuclei, brain stem reticular formation, posterior column nuclei, and spinal cord. The functional role of these tracts is to coordinate normal movements. The pyramidal cells comprising the corticospinal tracs are thought to be the motor command elements of the cerebral cortex.

Among the neurons of the motor cortex are large pyramidal neurons called Betz cells. The axon of these Betz cells descend in the corticospinal and corticobulbar tracts and terminate by synapsing directly with lower motor neurons (Evarts, 1979). These lower motor neurons, including those innervating respiratory muscles, have a significant functional role with the most precise of movements, such as those in speech. According to Phillips and Porter (1977), these direct projections from the motor cortex to the lower motor neurons innervating the respiratory musculature are probably related not to the act of respiration but rather to the use of the respiratory musculature in the skilled activities of speech and song.

CONCLUSION

A concept of hierarchial motor organization has been conceived as comprising three levels (Evarts, 1976): (1) the peripheral apparatus that gives rise to the actual movements and the generation of sensory feedback, (2) the pattern generator to which the command system sends signals and from which peripheral apparatus is influenced, and (3) the command system that initiates the neural events leading to the motor activity. The neural organization regulating and modulating the musculature of the voice organ expressed these three levels.

The Peripheral Apparatus comprises two elements. The lower motor neurons innervating the musculature of the vocal organ are components of the cranial and spinal nerves. The afferent neurons in the peripheral nerves transmit sensory feedback to the central nervous system.

The Pattern Generator. Several central control generators (systems) and suprasegmental feedback systems are integrated to coordinate and to modulate the activity of the lower neurons innervating the vocal organ.

The Command System. The neocortex and its circuitry that is eventually directed to influence Broca's area and the motor cortex comprise the major components of the command system. The commands are projected from the motor cortex via descending motor pathways primarily to the central control systems and, to a lesser extent, to the appropriate lower motor neurons in the brain stem and spinal cord. The suprasegmental feedback systems, because of their relations to the neocortex, may be considered to be components of the command system.

The phylogenetic acquisition of the motor aspects of speech by *Homo* is the culmination of evolutionary changes that follow the principle that evolution is basically conservative. In this respect, the human vocal organ evolved by modification of the anatomy of the preexisting respiratory and swallowing systems and the nasal and oral regions, while the neural control systems regulating the vocal organ also evolved by subtle changes and additions within the circuitry of the nervous system.

Acknowledgment: This project was supported, in part, by NIH Grant NS 12436.

REFERENCES

Adams, R.D., and Victor, M., 1977. Principles of Neurology, McGraw-Hill Book Co., New York, pp. 318-333.

Delcomyn, F., 1980. Neural basis of rhythmic behavior in animals. Science, 210:492-498.

DeLong, M.R., 1972. Central patterning of movement. Neurological Research Program Bulletin, 9:10-30.

DeLong, M.R., 1974. Motor functions of the basal ganglia: Single-unit activity during movement. In, The Neurosciences--Third Study

Program, F.O. Schmitt and F.G. Sorder, eds., MIT Press, Cambridge, Massachusetts. 28:319-325.

DuBrul, E. L., 1976. Biomechanics of speech sounds. In, Origin and Evolution of Language and Speech, S.R. Harnad, H.D. Steklis and J. Lancaster, eds., Annals of the New York Academy of Sciences, 280:631-642.

Edgerton, V.R., Grillner, S., Sjostrom, A., and Zangger, P., 1976. Central generation of locomotion in vertebrates. In, Neural Control of Locomotion, R.M. Herman, S. Grillner, P.S. Stein and D.G. Stuart, eds., Plenum Press, New York, pp. 631-632.

Euler, C. von, 1980. Central pattern generation during breathing. Trends in Neuroscience, 3:275-277.

Evarts, E., 1976. The interaction of central commands and peripheral feedback in pyramidal tract neurons of the monkey. In, Neural Control of Locomotion. R.M. Herman, S. Grillner, P.S Stein and D.G. Stuart, eds., Plenum Press, New York, pp. 808-817.

Evarts, E.V., 1979. Brain mechanisms of movement. Scientific American, 240:164-179.

Geschwind, N., 1970. The organization of language and the brain. Science, 170:940-944.

Geschwind, N., 1979. Specializations of the human brain. Scientific American, 240:180-201.

Geschwind, N., and Levitsky, W., 1968. Human brain. Left-right asymmetrics in temporal speech region. Science, 161:186-187.

Goodglass, H., 1980. Disorders of naming following brain injury. American Scientist, 68:647-655.

Hewes, G.W., 1973. Primate communication and gestural origin of language. Current Anthropology, 14:5-24.

Hockett, C.F., 1960. The origin of speech. Scientific American, 203:88-96.

Laitman, J.T., and Crelin, E.S., 1976. Postnatal development of the basicranium and vocal tract region in man. In, Development of the Basicranium, J.F. Bosma, ed., National Institute of Dental Research. DHEW Pub. No. (NIH) 76-989, Bethesda, Maryland, pp. 206-220.

Lancaster, J. B., 1968. Primate communication systems and the emergence of human language. In, Primates, J.C. Jay, ed., Holt, New York, 439-457.

Lieberman, P., 1975. On the Origins of Language. Macmillan, New York, pp. 153-155.

Luria, A.R., 1970. Traumatic Aphasia, Mouton, The Hague.

MacKay, W.A., 1980. The motor program: back to the computer. Trends in Neuroscience, 3:97-100.

Marin, O.S., Saffram, E.M., and Schwartz, M.F., 1977. Dissociations of language in aphasia: implications for normal functions. In, Origin and Evolution of Language and Speech, S. Harnad, H.D. Stekis and J. Lancaster, eds., Annals of the New York Academy of Science, 280:868-884.

Milner, B., 1974. Hemispheric specialization: Scope and limits. In, Neurosciences, F.O. Schmitt and F.G. Worden, eds., The MIT Press, Cambridge, pp. 75-89.

Myers, R.E., 1976. Comparative neurology of vocalization and speech: Proof of a dichotomy. In, Origins and Evolution of Language and Speech, S.R. Harnad, H.D. Steklis and J. Lancaster, eds., Annals of the New York Academy of Sciences, 280:745-757.

Noback, C.R., and Demarest, R.D., 1981. The Human Nervous System. Basic Principles of Neurobiology, McGraw-Hill, New York.

Phillips, C.G., and Porter, R., 1977. Corticospinal Neurons: Their Role in Movement. Academic Press, London.

Rasmussen, T., and Milner, B., 1977. The role of early left-brain injury in determining lateralization of cerebral speech. In, Evolution and Lateralization of the Brain, S.J. Diamond and D.A. Blizard, eds., Annals of the New York Academy of Sciences, 299:355-369.

Rubens, A.B., 1977. Anatomical asymmetrices of human cerebral cortex. In, Localization in the Nervous System, S. Harnad, R. Doty, J. Jaynes, L. Goldstein and G. Krauthamer, eds., Academic Press, New York, pp. 503-516.

Seyfarth, R.M., Cheney, D.L., and Marler, P., 1980. Monkey responses to three different alarm calls: Evidence of predator classification and semantic communication. Science, 210:801-803.

Thatch, W.T., Jr., 1975. Timing of activity in cerebellar dentate nucleus and cerebral motor cortex during prompt volitional movement. Brain Res., 88:233-241.

Whitaker, H., and Selnes, O., 1976. Anatomic variations in the cortex, individual differences and the problem of the localization of language functions. In, Origins and Evolution of Language and Speech. S.R. Harnad, H.D. Steklis and J. Lancaster, eds., Annals of the New York Academy of Sciences, 280:844-854.

Whitaker, H.A., and Ojemann, G.A., 1977. Lateralization of higher cortical functions: A critique. In, Evolution and Lateralization of the Brain. S.J. Dimond and D.A. Blizard, eds., Annals of the New York Academy of Sciences, 299:474-476.

Wind, J., 1976. Phylogeny of the human vocal tract. In, Origins and Evolution of Language and Speech. S.R. Harnad, H.D. Steklis and J. Lancaster, eds., Annals of the New York Academy of Sciences, 280:612-630.

Wyman, R.J., 1976. Neurophysiology of the motor output pattern generator for breathing. Fed. Proc. Fed. Am. Soc. Exp. Biol., 35:2013-2023.

Zuriff, E.B., 1980. Language mechanism: A neuropsycholinguistic perspective. American Scientist, 68:305-311.

ON THE ORIGIN AND PROGRESSIVE EVOLUTION OF THE TRIUNE BRAIN

Paul D. MacLean*

That the cerebral hemispheres are requisite for the spontaneous, directed activities of terrestrial vertebrates has been well known since the last century. As Ferrier (1876) noted, if a decerebrated animal "be left to itself, undisturbed by any form of external stimulus, it remains fixed and immovable on the same spot, and unless artificially fed, dies of starvation...." As has since been repeatedly confirmed, the neuraxis below the level of the hemispheres contains the neural apparatus required for posture and locomotion and the integrated performance of bodily actions involved in self-preservation and procreation. Since the cerebral hemispheres are essential for psychological functions, they may appropriately be referred to as the *psychencephalon.*

A comparison of the brains of extant vertebrates together with an examination of the fossil record indicates that the human psychencephalon has evolved and expanded to its great size while retaining the features of three basic evolutionary formations that reflect an ancestral relationship to reptiles, early mammals, and recent mammals (Fig. 1). Radically different in structure and chemistry, and in an evolutionary sense countless generations apart, the three formations constitute a hierachy of three brains in one--a triune brain (MacLean, 1970, 1973a). This situation suggests that psychological and behavioral functions are under the joint direction of three different mentalities. For human beings, there is the added complication that the two older formations do not have the capacity for verbal communication.

Just as the three basic evolutionary formations of the brain can be distinguished on the basis of their anatomy and chemistry, they can also be

*From the Laboratory of Brain Evolution and Behavior, National Institute of Mental Health, Poolesville, Maryland 20837

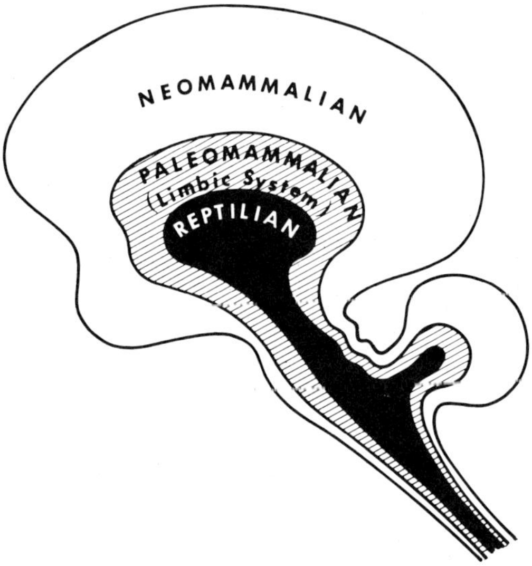

Fig. 1. The triune brain. In its evolution the primate forebrain expands along the lines of three basic formations that anatomically and biochemically reflect an ancestral relationship to reptiles, to early mammals, and to late mammals. The three formations are labeled at the level of the forebrain, which may be regarded as the *psychencephalon.* (From MacLean, P.D., 1967.)

shown to account for certain different functions. In this abbreviated account, I will deal successively with the protoreptilian, paleomammalian, and neomammalian formations.

THE PROTOREPTILIAN BRAIN

Developments in histochemistry have been of great help in identifying corresponding parts of the brain in reptiles, birds, and mammals. In regard to the protoreptilian counterpart, two illustrations will suffice. As illustrated in Figure 2, a Koelle stain for cholinesterase brings into sharp contrast the greater part of ganglionic structures that form the base of the forebrain in reptiles, birds, and mammals. With the application of the histofluorescence technique of Falck and Hillarp (Falck, 1962; Falck et al., 1962; Carlsson et al., 1962), the same structures shown in the figure glow bright green because of the presence of dopamine, a neural sap that appears to be requisite for bringing into play the total energies of the organism. These same regions are rich in serotonin (Paasonen and Vogt, 1956; Paasonen et al., 1957) and in the recently discovered opiate receptors and endorphins (Pert and Snyder, 1973). In reptiles and birds these ganglionic structures are

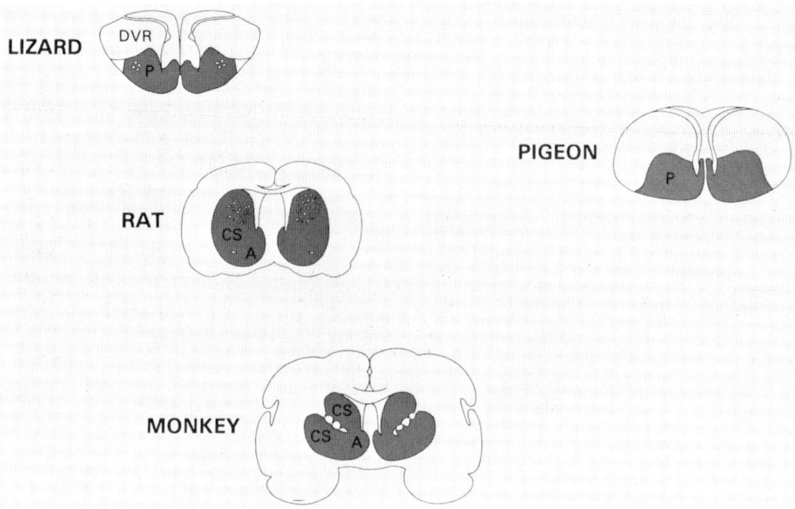

Fig. 2. Picture obtained on applying a stain for cholinesterase to brain sections of a lizard, rat, monkey, and pigeon. Combined with other evidence, the stain reveals that the striatal complex is a common denominator of the basal forebrain of terrestrial vertebrates (redrawn from Parent and Olivier, 1970, with substitution of lizard for turtle) DVR = Dorsal ventricular ridge, CS = corpus striatum, P - paleostriatum, A = accumbens septi.

identified as the olfactostriatum and paleostriatum (*paleostriatum augmentatum* and *paleostriatum primitivum*) and lie beneath the dorsal medullary lamina. In mammals the striatal (literally, striped) structures are referred to as the olfactostriatum and corpus striatum. The latter comprises the caudate nucleus and putamen, which project to the globus pallidus and satellite collections of gray matter. Because no term generally applies to the whole striatal complex, I will simply refer to it as the R-complex (reptilian complex) (MacLean, 1973b).

Someone not engaged in brain research might find it odd that after more than 150 years of investigation, the functions of the R-complex have not yet been identified (Crosby et al., 1962). In clinical neurology the structures in question are usually referred to as the basal ganglia. Because motor disorders occur after disease of related ganglia in the thalamus and midbrain, the R-complex has been inferred to be primarily implicated in motor functions. Consequently in textbooks and neurological teaching the R-complex is traditionally discussed as part of the motor system dominated by the motor cortex of the neocortex. This view has prevailed despite the recognition that large destructions of different parts of the R-complex may result in no notable motor deficit. Moreover, electrical stimulation of extensive parts of the R-complex results in no movement.

The Mammal-Like Reptiles

Unfortunately for comparative research, no existing reptiles are directly in line with mammals. Might some living form provide an appropriate experimental substitute? If the reading of the fossil record is correct, the lineage of mammals can be traced to the therapsids, the mammal-like reptiles. Long before the dinosaurs, they profusely populated the world. Two hundred fifty million years ago, when there was but one continent, Pangaea, they roamed every part of it (Fig. 3). Today their remains are found on every continent. A few years ago fossils of mammal-like reptiles similar to those in the Karroo Beds were found in Anarctica, which was once joined to Africa and formed part of the massive southern continent that Edward Suess called Gondwanaland (see Colbert, 1972). Robert Broom calculated that the Karroo Beds of South Africa hide the

Fig. 3. Pangaea. At the time of the earliest mammal-like reptiles the land masses forming most of the present continents appear to have been gathered together into one megacontinent known as Pangaea. The remains of mammal-like reptiles are found on every continent (see text). Note where India and Madagascar (M) were located. India has since moved through the Sea of Tethys and come to rest against the Himalayan Mountains. (Drawing based on a reconstruction by Irving, 1977; from MacLean, in press.)

remains of more than 800 *billion* mammal-like reptiles (Broom, 1932, p. 308). There were two main varieties--carnivores and herbivores (Fig. 4). Present-day mammals derive from the carnivores. Some forms resembled dogs and wolves (Fig. 5). In body carriage and structure of the jaws and teeth, the advanced forms closely approached the condition of mammals. Late in Triassic times the mammal-like reptiles mysteriously became extinct. As much as paleontologists would welcome information about brain development in these animals, there are no good endocranial casts of the most advanced mammal-like reptiles. Morphological considerations based on casts of less advanced mammal-like reptiles, along with other evidence, indicate that the brains were not too unlike present-day lizards.

In the evolution from reptiles to mammals, three key changes were the development of (1) a warm-blooded condition, (2) nursing, and (3) parental care. Did the advanced mammal-like reptiles lay eggs; did they care for their young? Or, like the offspring of today's giant Komodo dragons, did the young have to escape to the trees to avoid being cannibalized (Auffenberg, 1972)?

In the evolution of mammals, vocalization and hearing became of utmost importance for maintaining parent-offspring relationships; vocal communication would have helped to insure contact among the diminutive early mammals living in the dark floor of the forest. The so-called

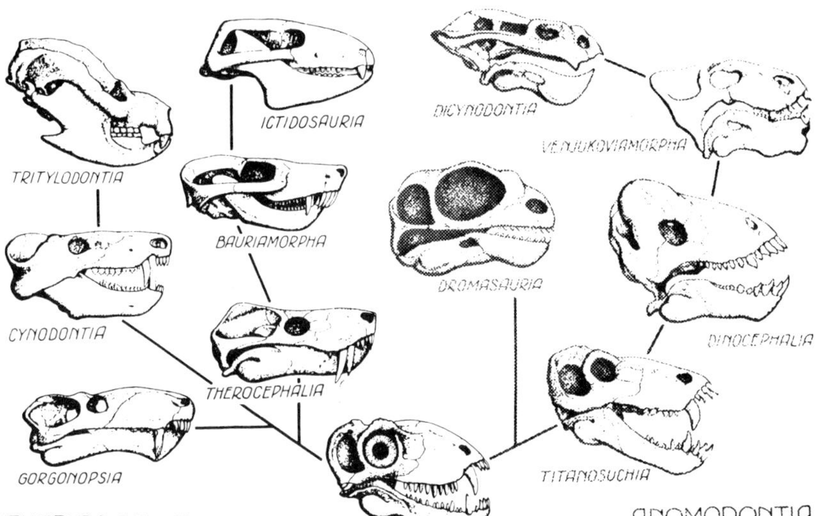

Fig. 4. The "family tree" of therapsids. Romer considers the Phthinosuchia as the parent stock leading to two great groups of mammal-like reptiles--one, carnivorous and the other, herbivorous. Here the herbivorous line has been partly shaded to focus attention on the two main lines of carnivores, either of which might have led to the mammals (from Romer, A.S., 1966. Courtesy of the University of Chicago Press).

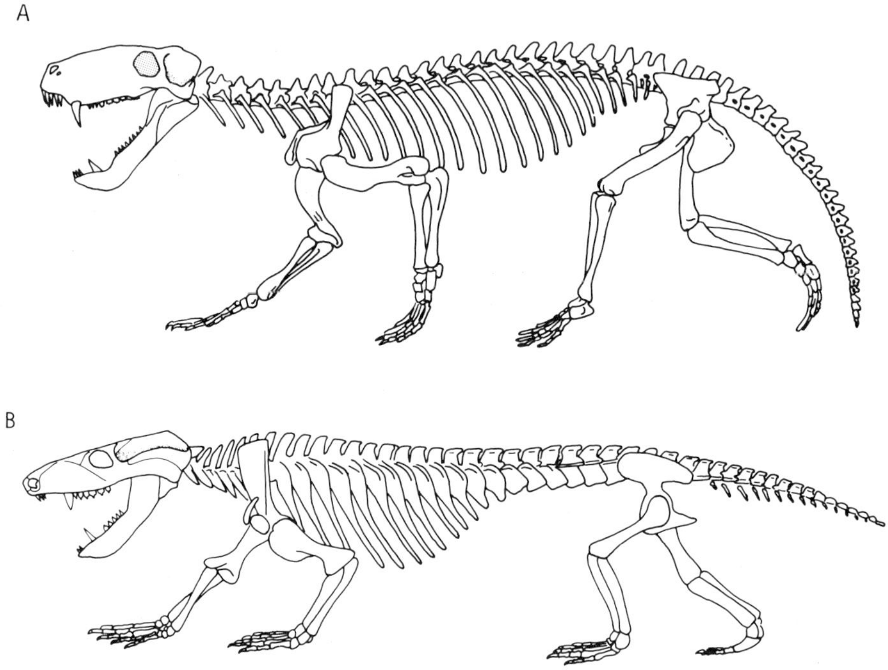

Fig. 5. Examples of two carnivore types of mammal-like reptiles. Note that the teeth and carriage of the body resemble the mammalian condition. **A.** A gorgonopsian called *Lycaenops* because of its wolf-like appearance. **B.** A cynodont named *Thrinaxodon.* The name Cynodontia for the mammal-like reptile of this type refers to its dog-like teeth. (Upper and lower figures, respectively, redrawn after Romer, 1966, and Brink, 1958, with the difference that the jaws are shown agape.)

"isolation call" characteristic of mammals is probably the most basic of mammalian vocalizations, serving to maintain maternal-offspring contact and familial affiliation (Newman and MacLean, 1981). Could the mammal-like reptiles hear and vocalize, or were they hard of hearing and mute like most existing lizards? In the advanced forms, two small bones of the jaw joint (the quadrate and articular) had become smaller, but were far from being transformed into the hammer and anvil of the mammalian internal ear (e.g., Romer, 1966, p. 85).

Reptilian Behavior

The so-called stem reptiles are considered to be the ancestors of the mammal-like reptiles. The bony structure of one of the stem reptiles appeared sufficiently lizard-like for that one order to be called *Varanosaurus* (Colbert, 1969), the same name given to today's monitor lizards, of which the Komodo dragon is one variety. Curiously, from an evolutionary standpoint, ethologists have paid little attention to reptiles, concentrating instead on fishes and birds. Of existing reptiles, lizards would probably bear the closest resemblance to the mammal-like reptiles, with the giant Komodo dragon perhaps being the best prototype.

In neurobehavioral work, as in other scientific pursuits, advances in knowledge depend partly on the recognition of similarities and differences. In analyzing the behavior of lizards one can identify more than 20 forms of behavior that are also characteristic of mammals (Table 1). All such behavior can be further reduced to individual components that form distinctive constructs and sequences of constructs (MacLean, 1975a, 1977). The sum total of all the building blocks of behavior and their various assemblages constitute the so-called behavioral profile, or ethogram, of an animal. Exclusive of verbal behavior, one recognizes in human beings and other terrestrial vertebrates two main aspects of the behavioral profile. For descriptive purposes, they can be compared to the profiles of two main mountain ranges. In one range are the distinctive peaks representing the chain of activities in an animal's daily master routine and subroutines. In the other range are four main peaks and subpeaks corresponding to four main types of behavioral patterns (displays) used in prosematic communication. Prosematic, meaning rudimentary signaling, applies to any nonverbal signal--vocal, bodily, or chemical--used in communication (MacLean, 1975a).

In lizards, the four main types of communicative signaling are referred to as (1) signature; (2) challenge (territorial); (3) courtship; and (4) submissive (assentive) displays.

The displays that variously incorporate static and dynamic modifiers may be illustrated by those of the common green anolis lizards (*Anolis carolinesis*), commonly referred to as the American "chameleon", which we have used extensively in our neurobehavioral studies. Anolian displays were described in detail by Greenberg and Noble (1944). As diagrammed in Figure 6, the signature display consists simply of three to five combined head nods and pushups together with a brief extension of a crimson throat fan occurring after the second head nod. Signature displays seem to reflect a kind of self assertion that comes into play in both nonsocial and social contexts. Among members of a group, the signature display appears to serve as a form of recognition. Challenge displays are of two types (distant and near) and are used chiefly by territorial males in establishing territory, maintaining dominance within a social group, and fending off invaders. The challenge display includes the dynamic components of the signature display combined with a number of static modifiers that, like many athletic uniforms, make the subject appear larger. The courtship display begins with the signature display followed by a number of head nods and an approach with a prancing strut toward the female. The anolian submissive display is

TABLE 1. Special Forms of Basic Behavior

1	SELECTION AND PREPARATION OF HOMESITE	DOMAIN
2	ESTABLISHMENT OF TERRITORY	
3	USE OF HOME RANGE	
4	SHOWING PLACE-PREFERENCES	
5	TRAIL MAKING	
6	MARKING TERRITORY	
7	PATROLLING TERRITORY	
8	RITUALISTIC DISPLAY IN DEFENSE OF TERRITORY, COMMONLY INVOLVING THE USE OF COLORATION AND ADORNMENT	
9	FORMALIZED INTRASPECIFIC FIGHTING IN DEFENSE OF TERRITORY	
10	TRIUMPHAL DISPLAY IN SUCCESSFUL DEFENSE	
11	ASSUMPTION OF DISTINCTIVE POSTURES AND COLORATION IN SIGNALING SURRENDER	
12	USE OF DEFECATION POSTS	
13	FORAGING	
14	HUNTING	
15	HOMING	
16	HOARDING	
17	FORMATION OF SOCIAL GROUPS	
18	ESTABLISHMENT OF SOCIAL HIERARCHY BY RITUALISTIC DISPLAY AND OTHERS MEANS	
19	GREETING	
20	GROOMING	
21	COURTSHIP, WITH DISPLAYS USING COLORATION AND ADORNMENTS	
22	MATING	
23	BREEDING AND, IN ISOLATED INSTANCES, ATTENDING OFFSPRING	
24	FLOCKING	
25	MIGRATION	

characterized by four slight up-and-down motions of the head. It is performed by members of either sex and of all ages.

Neurobehavioral Studies

In our initial behavioral studies on lizards we examined the challenge displays because they can be reliably induced in two adult territorial males.

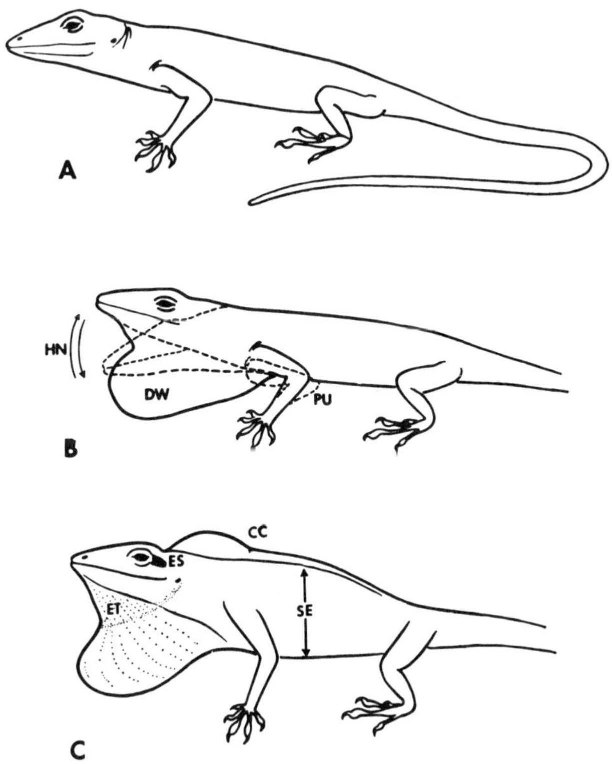

Fig. 6. Features of the signature and challenge displays of the common green anolis lizard (*Anolis carolinensis*). **A.** The usual attentive posture. **B.** Diagrammatics of the signature (assertion) display. The signature display consists of three to five head nods (HN) and pushups (PU) along with an extension of the dewlap (DW). The broken lines indicate the excursion of the head and flexion of forelimbs during pushups. Note absence of static modifiers. **C.** Diagnostic features of the challenge display of adult male lizards. In addition to the dynamic components of the signature display, the challenge display has several static modifiers. The first to appear are the extended throat (ET) and sagittal expansion (SE), followed by an elevation of the nuchal and dorsal crests (CC). A darkly pigmented eyespot (ES) may appear after two to three minutes. See text for further details. (From Greenberg et al., 1979. Brain Res., vol. 172. Courtesy of Elsevier.)

In the attempt to identify brain mechanisms underlying this display, we used the green anolis lizard. Because the optic nerves of this species are almost entirely crossed (Fig. 7), we can place a lesion in only one hemisphere of the brain and then test the animal's behavior with either of its eyes covered. In this way we avoid the possible complication of interfering with its ability to maintain an adequate bodily temperature for being active. As schematically indicated in Figure 7, we found that only hemispheric lesions largely confined to the R-complex (marked with an "X") interfered with the expression of the challenge display. Although capable of climbing, jumping, and catching crickets, the experimental animal showed no interest in the rival lizard when looking monocularly at it with the eye projecting to the injured hemisphere. Seeing the rival with the eye leading to the undamaged hemisphere, however, evoked the full challenge display (Greenberg et al., 1979).

Findings on monkeys. In a study continuing since 1961, I observed the effects of brain lesions on the display behavior of squirrel monkeys. Like lizards, squirrel monkeys have the four main forms of displays mentioned above. In both the aggressive (challenge) and courtship displays, the

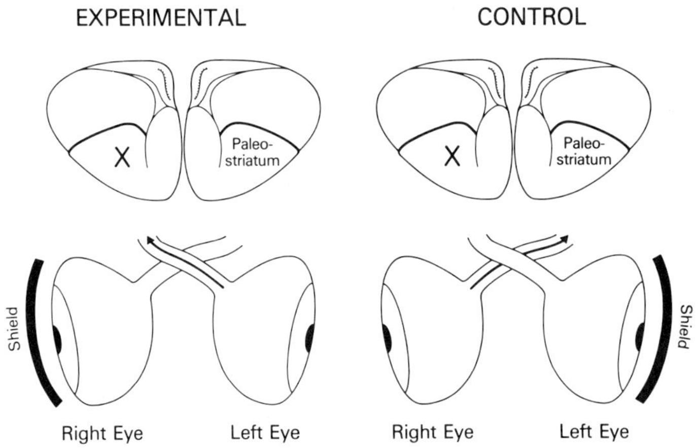

Fig. 7. Schematic of procedure for obtaining control and experimental observations in the same brain-damaged lizard. In neurobehavioral research on lizards, the possibility of interfering with the regulation of the body temperature must be minimized. Because the optic nerves are almost completely crossed in the anolis lizard, the conditions are provided for destroying nerve tissue in one hemisphere (e.g., at site "X") and then testing the animal with either eye covered by a shield. Under control conditions (right side of figure), the lizard gives its usual distant challenge display on seeing a rival territorial lizard with the eye projecting to the normal hemisphere. When the rival lizard is seen with only the eye projecting to the hemisphere in which part of the paleostriatum is destroyed (left side of figure), the test animal does not respond. Lesions elsewhere in the hemisphere are without effect (from MacLean, in press).

dominant male approaches the other animal, vocalizes, and directs the erect phallus toward the other animal. The monkeys of one subspecies regularly perform a somewhat similar greeting display to their reflection in a mirror (MacLean, 1964). As diagrammed in Figure 8, we refer to the mirror-displaying animal as the gothic type because the natural markings include an ocular patch (top left of figure) that forms a peak over the eye like a gothic arch. The so-called roman-type monkeys with the round arch (top right) are not interested in mirrors, although they perform the other kinds of displays, including aggressive, courtship, and submissive displays.

Because it allows a control of several variables, I used the mirror display test as a model for detecting what parts of the brain are implicated

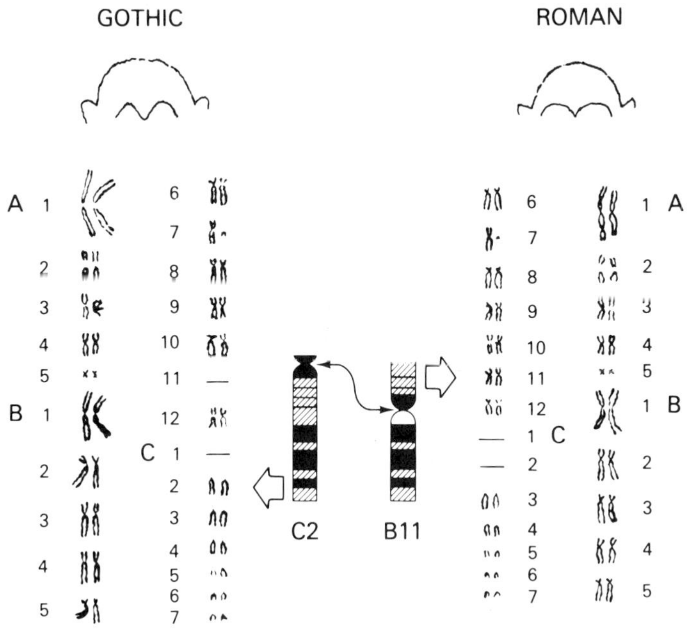

Fig. 8. Contrasting ocular patches of so-called gothic- and roman-type monkeys (shown diagrammatically in left and right upper parts of figure). Only the gothic-type monkeys will display consistently to their reflections in a mirror. In addition to differences in appearance and behavior, these two varieties of squirrel monkeys have minor karyotypic differences. Their 44 chromosomes can be classified into three groups (A, B, and C) on the basis of the length and position of the centromeres. As illustrated by the enlargement at the center of the Figure, Ma and co-workers (1974) postulate that the difference between the gothic-type monkey from Columbia and the roman-type from Peru is owing to a pericentric inversion of the B11-C2 chromosomes. Regarding geographic factors, the authors speculate that "differences in the number of acrocentric chromosomes in *Saimiri* may be due to individual migration followed by recombination and selection." (entirely redrawn after Ma et al., 1974; from MacLean, in press).

in prosematic communication. In observations on more than 120 monkeys, I found that the medial pallidal segment of the R-complex represents a region of convergence of neural mechanisms essential for the expression of the mirror display. Electrocoagulation of this region or of its projecting pathways may either eliminate or fragment the display (MacLean, 1975b, 1978a, 1981).

Comment

The results of these experiments indicate that in animals as diverse as lizards and monkeys, the R-complex plays a basic role in displays used in social communication. By inference, they also indicate that the R-complex is essential for conspecific recognition through the performance of like kinds of behavior. The performance of like kinds of behavior (for which I use the term "isopraxis," MacLean, 1975a) is what typifies a species.

Isopraxis is one of six important interoperative behaviors seen in reptiles and higher forms (Table 2). Without defining them, I shall simply

TABLE 2. General ("Interoperative") Forms of Basic Behavior

1 ROUTINIZING
2 ISOPRAXIC
3 TROPISTIC
4 REPETITIOUS
5 REENACTMENT
6 DECEPTIVE

point out that the listed behaviors find expression in such human activities as the performance of daily routines and subroutines; adherence to fashion (both social and scientific); responses to partial representations whether alive or inanimate; repetitious, obsessive-compulsive acts; slavish conformance to old ways of doing things; obeisance to precedent as in legal and other matters; ceremonial reenactments; and all manner of deception.

I mention these proclivities because of the frequent statement that all human behavior is learned (e.g., Miller and Dollard, 1941; Montagu, 1956). If that is so, one might ask why, in spite of the high degree of human intelligence and culturally learned behavior, do human beings continue to do all the ordinary things that animals do? Here we need consider but one curious example. There appear to be some carry-overs from animal to human displays that are so subtle that they have escaped even the notice of expert ethologists, making it seem all the more remarkable that, as is

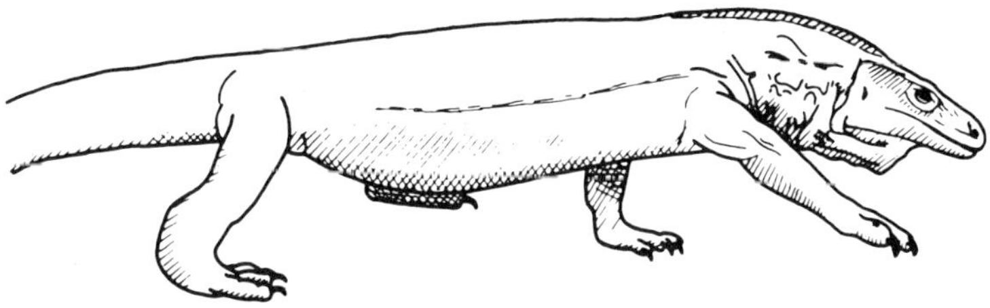

Fig. 9. Display of a Komodo dragon. The close-in agonistic display of an adult Komodo dragon is similar to that of an appeasement display of a juvenile shown here. The animal walks slowly in a stiff-legged, stilted manner. The angle of the right forelimb in this picture is reminiscent of the goose step. Note three static modifiers seen in other lizards--namely, elevated roach (nuchal and dorsal crests), extension of gular fold, and sagittal expansion. (From Auffenberg, W., 1978. DHEW Publication. Courtesy of U.S. Government Printing Office.)

claimed, everything that human beings do as human beings they have had to learn from other human beings (Montagu, 1956). The subtlety pertains to the close-in challenge display of territorial lizards. In the close-in display lacertilians rise up on all fours and present themselves sideways while stepping in a stilted, staccato manner that makes them appear off balance. Some rodents perform a similar broadside display, but it happens so rapidly that observers may fail to notice it. As Barnett has observed, two rats in a confrontation rise up with all four limbs extended, the back arched, and the flank turned toward the opponent. While in this posture each moves around the other "with short, mincing steps, still presenting his flank" (Barnett, 1963, p. 87). I had been unaware that the "challenge" display of two adult, rival gorillas incorporated lacertilian features until I saw Dian Fossey at one of our laboratory seminars perform what she refers to as the "parallel display" of two silverbacks (Fossey, 1976). When she mimicked their sideways presentation and their walking with stilted, awkward steps, one was immediately reminded of the close-in display of certain lizards. In the case of chimpanzees, Goodall (Lawick-Goodall, 1971) described a bipedal swagger that appears to correspond to the strutting display of the gorilla. Her description calls to mind the posture and movements of a Japanese wrestler.

As in the case of lizards, the stilted, staccato steps seen in the displays of the great apes seem to carry the message of a series of exclamation marks, calling to mind the goose-step of a military parade and its similarity in profile to the "Schrägstellung" gait of a Komodo dragon (Fig. 9). Among different species, the sideways presentation and stilted, staccato steps have such an uncanny resemblance that it would almost seem that the challenge display had been genetically packaged and handed up the phylogenetic tree of mammals.

THE PALEOMAMMALIAN BRAIN (LIMBIC SYSTEM)

Earlier, in considering lizards, I mentioned the cannibalism of the young. The evolution of mammals seemed to bring with it a primal commandment against cannibalism. Also of momentous significance in mammalian evolution is the development of hearing, vocalization, and protracted parental care. Indeed, one might say that the history of the evolution of mammals is to a large extent the history of the evolution of the family.

If judged by existing reptiles, the extinct forms leading up to mammals would have had only a rudimentary cortex. In the lost transitional forms between mammal-like reptiles and mammals, the primitive cortex is believed to ballooned have out and become further differentiated. The cortex presumably provides the organism with a greater capacity to learn from current experience. This function in turn depends on the ability to retain the memory of a present experience and to compare it with past experiences.

As illustrated in Figure 10, most of the phylogenetically old cortex is found in a large convolution that Broca (1878) called the great limbic lobe because it surrounds the brain stem. In 1952, I suggested the term limbic system as a designation for the limbic cortex and structures of the brain stem with which it has primary connections (MacLean, 1952). In its totality, the limbic system represents an inheritance from early mammals--hence the expression paleomammalian brain.

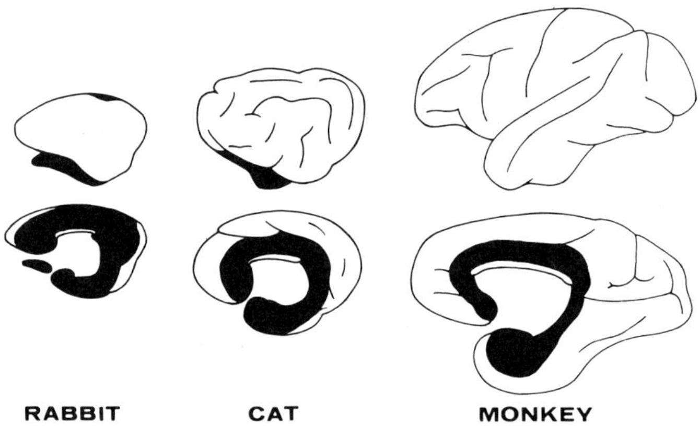

Fig. 10. The cortex of the paleomammalian brain (limbic system). Largely contained in the limbic lobe of Broca (shaded) the evolutionary old cortex appears as a common denominator in the brains of all mammals. The cortex of the neomammalian brain (shown in white) mushrooms late in evolution (after MacLean, 1954).

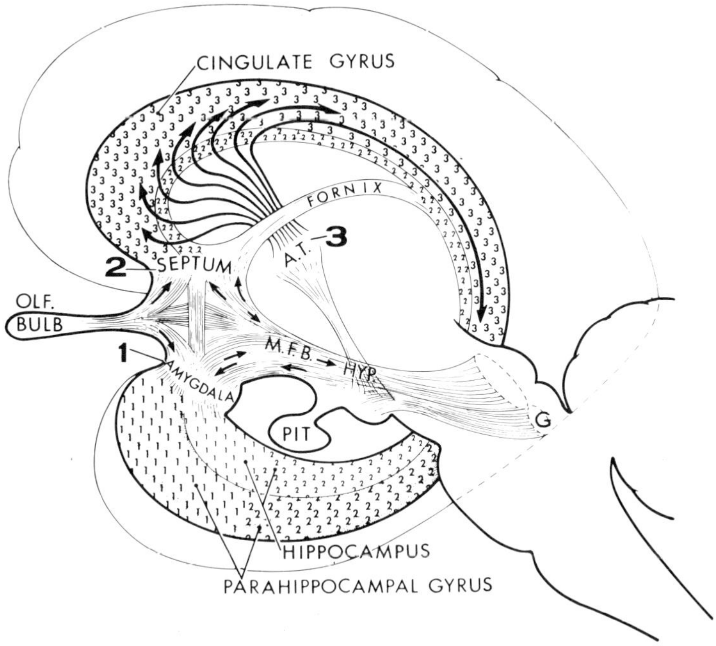

Fig. 11. A brain map of three main subdivisions of the limbic system and their major pathways identified by the overlying numbers. See text for summary of their respective functions. Abbreviations: AT, anterior thalamic nuclei; G, dorsal and ventral tegmental nuclei of Gudden; HYP, hypothalamus; MFB, medial forebrain bundle; PIT, pituitary; OLF, olfactory (after MacLean, 1958, 1973a).

Anatomical Connections

In the past decade the introduction of the autoradiographic and horseradish peroxidase methods for identifying the cellular origin and destination of nerve fibers has resulted in a spiralling increase in hodological knowledge, as well as a modification of certain classical views. These and other new techiniques have been applied with particular force in studies on the limbic system. Nevertheless, the brain diagram shown in Figure 11, and adapted from ones published in 1958, still provides a useful framework for presenting the connections of the limbic system. For purposes of simplification the afferent and efferent pathways can be indicated on a single diagram because, in general, they follow a parallel course. The medial forebrain bundle (MFB) and its continuation as the cingulum are major lines of communication.

Afferent connections. The telencephalic limbic structures receive connections from both intero- and exteroceptive systems. Comparative neuroanatomists have long recognized that the isthmus region located at the junction of the midbrain and pons is an important location for the integration of somatovisceral systems (Herrick, 1948). Herrick, for example, identified a "visceral" and "gustatory" nucleus in the isthmus region of the salamander. In mammals there are cells in this same region that form a cuff around the superior cerebellar peduncle. The use of the newer techniques has shown that cells in this area receive fibers from the (vagal) solitary nucleus (Ricardo and Koh, 1978) and the rostral extension of this nucleus innervated by gustatory fibers from the 7th and 9th cranial nerves (Norgren, 1976). The parabrachial nucleus projects to medial thalamic and amygdala nuclei (Mehler, 1980; Nomura et al., 1979; Ricardo and Koh, 1978), which in turn project to the limbic cortex in the rostral part of the hippocampal gyrus and hippocampus. Radioautographic methods show in the squirrel monkey that the posterior part of the solitary nucleus also projects to the same lateral brachial area that contains labelled cells after horseradish peroxidase injections in the region of the amygdala (unpublished observations).

The isthmic dorsal and ventral tegmental nuclei of Gudden (G in Fig. 11) are another source of visceral afferents (see Bleier, 1969, for review). The dorsal nucleus receives connections from the solitary nucleus (Morest, 1967). Some ascending fibers from the Gudden complex join the medial forebrain bundle and terminate in the septum, whereas others project to the mamillary bodies and thereby join the mamillary-anterior thalamic-cingulate circuit. Other afferents from the dorsal nucleus of Gudden ascend via the dorsal longitudinal fasciculus of Schütz to intralaminar nuclei (Morest, 1961) that have been convincingly shown by recent techniques to project to the amygdala (Mehler, 1980) and hippocampal formation (Herkenham, 1978; Amaral and Cowan, 1980).

Among recent provocative findings are those showing that aminergic afferents from cell bodies located in the isthmus region and from the midline and ventral parts of the midbrain ascend to limbic structures. Fibers from an isthmus nest of cells in locus coeruleus ascend not only to the amygdala, septum, and anterior thalamic nuclei of the limbic system, but also project directly to the hippocampal formation and cingulate gyrus (Pickel et al., 1974; Ungerstedt, 1971). Serotonin-containing fibers from cells in the dorsal raphe nucleus and from the superior central nucleus of Bechterev innervate the hippocampal formation (Moore and Halaris, 1975). Finally, dopamine-containing fibers arising from cells in the ventral tegmentum of the midbrain innervate the central nucleus of the amygdala and the caudal part of the bed nucleus of the stria terminalis (Mehler, 1980), as well as the olfactostriatum (olfactory tubercle and nucleus accumbens) (Nauta and Domesick, 1978) that receive direct connections from the hippocampal formation (Poletti et al., 1973).

In regard to afferents from exteroceptive systems, the reader is referred to an extensive review of olfactory connections by Heimer (1978) and to articles by the present author regarding the developing knowledge of auditory, somatic, and visual inputs (MacLean, 1970; 1973a; 1975a).

For the most part, the efferent projections from limbic structures run parallel to the ascending connections. One recent finding at variance with the classical picture of limbic efferents is that the subicular areas of the hippocampal formation, not areas CA4 and CA3, project to the mamillary bodies (Swanson and Cowan, 1975). In addition, significant differences in the connections of the proximo-septal and proximo-amygdala parts of the hippocampus have been observed (e.g., Meibach and Siegel, 1977; Pasquier and Reinoso-Suarez, 1977; Segal and Landis, 1974). The two parts in question are indicated in Figure 11 by the small numerals #1 and #2 overlying the hippocampus.

Functions of Limbic Subdivisions

Somewhat like a great metropolis, the limbic populations of nerve cells fall into a number of districts. In the brain map of Figure 11, the three main cortical subdivisions of the limbic system are identified by overlying small numerals 1, 2, and 3. As diagrammed, the first two subdivisions are closely associated with the olfactory apparatus. The nerve cells in the first subdivision have been shown to be concerned with activities insuring self-preservation--namely, feeding, fighting, and self-protection. The second subdivision is involved in primal sexual functions and sociosexual expression subserving procreation (see MacLean, 1973a). The intimate interrelationship between oral and sexual functions in this part of the brain can be attributed to the strong mutual interconnections of the two subdivisions with the olfactory apparatus, which spearheaded the evolution of the forebrain and which plays a primary role in both feeding and mating, as well as the fighting that may precede.

As indicated in Figure 11, the main pathway of the third subdivision bypasses the olfactory apparatus. According to some authorities (Le Gros Clark and Meyer, 1950), there is no rudimentary counterpart of the third subdivision in the brains of reptiles. Although well-developed in all mammals, this subdivision shows progressive expansion in higher primates, reaching its greatest development in the human brain (see Armstrong, this volume). Our studies of 20 years ago revealed that parts of the third division are involved in primal sexual functions (MacLean and Ploog, 1962). As will be further explained below, there is also evidence that this division is involved in maternal behavior (Stamm, 1955; Slotnick, 1967; Murphy et al., 1978) and in play.

Global Functions of the Limbic System

The clinical study of psychomotor epilepsy provides the best evidence that the limbic system is involved in emotional experience and expression. Scarring of the temporal limbic and neighboring cortex creates unpredictable bioelectrical storms that have a tendency to spread within parts or all of the limbic system. During the aura, which represents the beginning of a storm, the patient's mind lights up with vivid emotional feelings that range all the way from intense fear to ecstasy. Significantly, a storm may also spark eureka-type feelings like those associated with discovery, or free-floating feelings of belief or conviction regarding what is real, true, and important (MacLean, 1970). The last mentioned manifestations have profound epistemological implications (MacLean, 1975a). A "primitive" mind providing feelings of what is real, true, and important may be adequate for judging the authenticity of food or mate, but where do we stand if we must depend on that same mind for belief in our ideas, concepts, and theories?

There is also clinical evidence that the limbic system plays an important role in the memory of current happenings. It is one of the wonders of the brain that limbic storms tend to spread in and be confined to the limbic system. During such storms, a person may carry out highly complicated behavior and have no memory of it afterward. One might say that victims of limbic epilepsy behave like disembodied spirits.

Chronic localized discharges in the limbic system may result in symptoms resembling those of schizophrenia and other psychoses. The beneficial effects of some psychotropic drugs are believed to be due to a rather specific action on the R-complex and limbic system.

THE BASIC ANIMALITY

Before a brief consideration of the "neomammalian brain," it is pertinent to ask what an animal would be like with only the R-complex and limbic system. In animals such as rats and hamsters, experimental intervention at the time of birth can prevent the development of the neocortex and its connections. We found that hamsters growing up without the neocortex showed every form of hamster-typical behavior (Murphy et al., 1981). They appeared to grow and develop normally. They went through their daily routines like their control litter mates. They developed play behavior at the appropriate time. They mated and bred and reared their young.

Most significantly, when such animals were also deprived of tissue including the cortex of the third subdivision of the limbic system, they did not play, and in addition females showed deficits in maternal behavior (Murphy et al., 1981; MacLean 1978b). It was as though these animals had regressed toward a reptilian condition.

In summary, the results show that the two older evolutionary formations of the brain are capable, along with the rest of the neuraxis, of giving expression to most forms of species-typical behavior, whereas the evolutionary newer parts of the limbic system appear necessary for the full expression of maternal behavior and the capacity for play. Meriting reemphasis are the three cardinal behavioral advances in the evolution from reptiles to mammals: (1) nursing, (2) parental care, and (3) play.

THE NEOMAMMALIAN BRAIN

To credit the two older evolutionary formations of the psychencephalon with providing the underpinnings of basic behavior is not to downplay the importance of the neocortex. As for human beings, nothing is more neurologically certain than that the neocortex provides the neural substrate for language and speech and that we owe to it the infinite variety of ways in which we can express ourselves. The neocortex mushrooms progressively in higher mammals (cf. Fig. 10) and reaches its greatest development in human beings.

A remarkable feature of the neocortex is that it evolves primarily in relation to systems receiving and processing information from the external world--namely, the exteroceptive visual, auditory, and somatic systems. It was as though the neocortex was designed to serve as a more "objective" intelligence in coping with the external environment.

Handedness and Speech

Lack of available space prevents discussion of the protean functions identified with neocortical systems. From the standpoint of primate evolution, it would be particularly pertinent to consider the factors that have accounted for the tie-in of vocalization with handedness and speech. This subject, however, invariably ends up on the shoals of speculation. Take, for example, the question as to why most human beings are right-handed--a condition that, according to artifacts of prehistory, has existed for thousands of years (Coren and Porac, 1977). As a caveat in regard to the complexity of the problem, a predisposition to "rightness" in some form exists among some snails, flatfish, and other animals. Elsewhere I have speculated about the evolution of handedness (MacLean, 1978b), starting with Thomas Carlyle's "primitive warfare theory" and giving it a new twist in connection with Raymond Dart's osteodontokeratic culture (Dart, 1959). Given the inherited predisposition to right-handedness, one can develop an explanation of the cerebral dominance of speech, first noting how it would be neurologically advantageous for a midline organ of speech such as the tongue with its bilateral innervation to receive its commands from a single hemisphere (MacLean, 1978b). The representation of speech in the left hemisphere would provide the quickest and most effective means of coordinating speech and right-handed action. It is evident how under conditions of group hunting or of internecine strife, split-second timing in enunciating a directional signal might make the difference between life and death.

The same neurological economy with respect to dominance would apply to a written language, because whenever the idea first struck to jot things down, the right hand was ready and waiting!

That the nondominant hemisphere sits idly by without ever mastering a language has been considered a great functional waste. In the light of computer technology, I suggested that this seeming deficiency may have compensations. With computers, an insufficient "memory" presents a continuing hindrance to achieving solutions of complicated problems. One might propose that nature, in placing linguistic functions in one hemisphere, killed two birds with one stone--putting the midline organ, the tongue, under a single command and freeing the nondominant hemisphere to be used for a greatly expanded memory (MacLean, 1978b). There is evidence that the nondominant hemisphere plays a role in the registration of memories (Penfield and Milner, 1958; Milner et al., 1968).

Transcendent Speech

In an evolutionary sense, the "isolation call" is probably the oldest and most basic of mammalian vocalizations, serving to maintain maternal-offspring contact, as well as contact with other members of the same species. In squirrel monkeys we found that gray matter at the core of the forebrain (posterior periventricular gray) appears to be essential for the production of these calls (Newman and MacLean, 1981). This gray matter receives part of its connections from the limbic cortex. Interestingly, in view of the questioned ability of vocalization in reptiles ancestral to mammals, when the subhuman mammalian telencephalon is electrically stimulated, vocalization can be elicited only from limbic structures (Jurgens and Ploog, 1970). The vocalization required for speech appears to have necessitated a quantum jump to the neocortex. Electrical stimulation of the so-called four speech areas interferes with speech, but does not produce speech (Penfield and Jasper, 1954).

Curiously, the human brain attained its large size thousands of years before there was a language of words. And just as curiously, it has been only two thousand years since human beings first saw the "sunya" (the empty space, the "zero," existing between the fingers) that has since afforded a workable language of numbers (Hogben, 1937). With the soaring developments in modern communication, who is to say what other languages, what metalanguages, what transcendent speech, may still be in the making?

That one form of transcendent speech is in the making, we may be quite sure. Mention was made earlier that initially with the evolution of the neocortex there was a predominant representation of the exteroceptive systems. As though an essential ingredient had been left out, a new sector of cortex appears in the neomammalian brain that ties in with interoceptive systems. This relatively recent neocortical development turns our attention once more to mammalian evolution and the evolution of the human family. The location of the cortical development in question becomes apparent by contrasting the low brow of a Neanderthal skull and the recently evolved high brow of a Cro-Magnon skull (Fig. 12). Underneath the high brow is a greatly expanded sector of cortex that appears to be the only neocortex that establishes a strong communicative link with the internal world. This once speculative statement can now be said with some conviction, because recently we have shown in monkeys that more than 25% of the cells of the

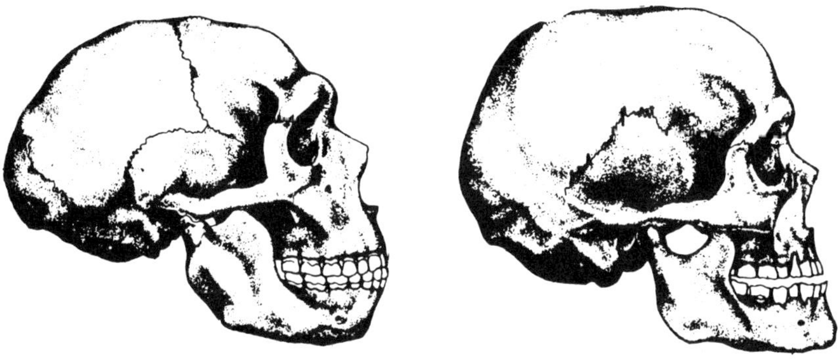

Fig. 12. Contrasting profiles of Neanderthal and Cro-Magnon skulls. See text regarding suggested implications. (From Romer, A.S., 1966. Courtesy of the University of Chicago Press.)

medial dorsal nucleus are activated by the vagus nerve--the great visceral nerve (Hallowitz and MacLean, 1977). The medial dorsal nucleus is the main nucleus projecting to the prefrontal cortex.

Clinically, there are indications that the expansion of the prefrontal cortex affords an increased capacity to relate internal and external experience and thus to identify one's inner feelings with those of other beings. The prefrontal cortex is also recognized clinically to play a fundamental role in relating past, present, and future in regard to "looking ahead" and making possible both anticipation and planning (Freeman and Watts, 1950). Presumably, through its connections with the medial dorsal nucleus (Fig. 13), the prefrontal cortex obtains the "insight" required for the foresight to plan for the needs of others as well as the self. In the flowering of our medical heritage we have lived to see the evolution of such historically unique institutions as the World Health Organization and the National Institutes of Health dedicated to the alleviation of suffering, not only of human beings everywhere, but, hopefully, of all living things. Such developments must be regarded as a 180° turnabout from what has been a reptile-eat-reptile and dog-eat-dog world.

Significantly, the prefrontal cortex is intimately geared in with the third great subdivision of the limbic system that has been found to be implicated in parental behavior and in play. As should be noted, play not only facilitates tolerance in the close quarters of the nest, but also promotes adult affiliations.

Mentioned earlier was the importance of the evolution of vocalization and hearing for maintaining parent-offspring relationships. Separation of offspring from the mother is calamitous. We can discern in this situation the evolutionary roots of unity of the family, unity of the clan, unity of

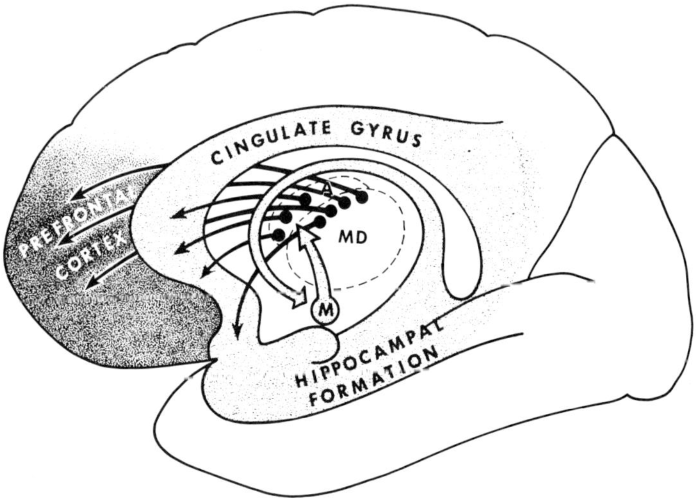

Fig. 13. A limbic-prefrontal connection. The diagram indicates how the limbic system (light stipple) and its third subdivision are anatomically related to the prefrontal cortex through the third pathway shown in Fig. 11. Abbreviations: F, fornix; M, mammillary bodies of hypothalamus; MD, medial dorsal nucleus; A, anterior thalamic nuclei. (From MacLean, 1958, 1973a)

larger societies, as well as the emotional intensity of feelings attending separation and, in its most acute form, the utter isolation of death. In evolution, a parental concern for the young eventually generalizes to other members of the species, a psychological development that amounts to a progression from a sense of responsibility to what we call conscience. Given the human capacity to recognize gradations of "yes" and "no" on various issues and to decide accordingly, one need hardly look beyond the evolving family to find a reason for being, an ethic to live by.

REFERENCES

Amaral, D.G., and Cowan, W.M., 1980. Subcortical afferents to the hippocampal formation in the monkey. J. Comp. Neurol., 189:573-591.

Auffenberg, W., 1972. Komodo dragons. Natl. Hist., 81:52-59.

Auffenberg, W., 1978. Social and feeding behavior in *Varanus komodensis*. In, The Behavior and Neurology of Lizards, N. Greenberg and P.D. MacLean, eds., U.S. Government Printing Office, Washington, D.C., DHEW Publication No. (ADM) 77-491, pp. 301-331.

Barnett, S.A., 1963. A Study in Behaviour, Methuen & Co. Ltd., London.

Bleier, R., 1969. Retrograde transsynaptic cellular degeneration in mammillary and ventral tegmental nuclei following limbic decortication in rabbits of various ages. Brain Res., 15:365-393.

Brink, A.S., 1958. Note on a new skeleton of *Thrinaxodon liorhinus*. Paleontologia Africiana, 6:15-22.

Broca, P., 1878. Anatomie comparée des circonvolutions cérébrales. Le grand lobe limbique et la scissure limbique dans la serie des mammifères. Rev. Anthrop., 1:Ser. 2, 385-498.

Broom, R., 1932. The Mammal-Like Reptiles of South Africa and the Origin of Mammals, H.F. and G. Witherby, London.

Carlsson, A., Falck, B., and Hillarp, N.A., 1962. Cellular localization of brain monoamines. Acta Physiol. Scand., 56:1-28.

Colbert, E.H., 1969. Evolution of the Verterates, John Wiley & Sons, Inc., New York.

Colbert, E.H., 1972. Antarctic fossils and the reconstruction of Gondwanaland. Natl. Hist., 81:66-73.

Coren, S., and Porac, C., 1977. Fifty centuries of right-handedness: The historical record. Science, 198:631-632.

Crosby, E.C., Humphrey, T., and Lauer, E., 1962. Correlative Anatomy of the Nervous System, Macmillan, New York.

Dart, R.A., 1959. Adventures with the Missing Link, The Institutes Press, The Institutes for the Achievement of Human Potential, Philadelphia.

Falck, B., 1962. Observations on the possibilities of the cellular localization of monoamines by a fluorescence method. Acta Physiol. Scand. Suppl., 197.

Falck, B., Hillarp, N.A., Thieme, G., and Torp, A., 1962. Fluorescence of catecholamines and related compounds condensed with formaldehyde. J. Histochem. Cytochem., 10:348-354.

Ferrier, D., 1876. The Functions of the Brain, Smith, Elder, London.

Fossey, D., 1976. The behavior of the mountain gorilla. Doctoral dissertation, University of Cambridge, 460 pp.

Freeman, W., and Watts, J., 1950. Psychosurgery in the Treatment of Mental Disorders and Intractable Pain, Charles C Thomas, Springfield, Illinois.

Greenberg, B., and Noble, G.K., 1944. Social behavior of the American chameleon (*Anolis carolinesis Voigt*). Physiol. Zool., 14:392-439.

Greenberg, N., MacLean, P.D., and Ferguson, J.L., 1979. Role of the paleostriatum in species-typical display behavior of the lizard (*Anolis carolinesis*). Brain Res., 172:229-241.

Hallowitz, R.A., and MacLean, P.D., 1977. Effects of vagal volleys on units of intralaminar and juxtalaminar thalamic nuclei in monkeys. Brain Res., 130:271-286.

Heimer, L., 1978. The olfactory cortex and the ventral striatum. In, Limbic Mechanisms, K.E. Livingston and O. Hornykiewizc, eds., Plenum Press, New York, 195-197.

Herkenham, M., 1978. The connections of the nucleus reuniens thalami: Evidence for a direct thalamo-hippocampal pathway in the rat. J. Comp. Neurol., 177: 589-610.

Herrick, C.J., 1948. The Brain of the Tiger Salamander, Chicago, University of Chicago Press.

Hogben, L., 1937. Mathematics for the Million, W.W. Norton and Co., New York.

Irving, E., 1977. Drift of the major continental blocks since the Devonian. Nature, 270:304-309.

Jurgens, U., and Ploog, D., 1970. Cerebral representation of vocalization in the squirrel monkey. Exp. Brain Res., 10:532-554.

Lawick-Goodall, J.V., 1971. In the Shadow of Man, Houghton Mifflin Co., Boston.

Le Gros Clark, W.E., and Meyer, M., 1950. Anatomical relationships between the cerebral cortex and hypothalamus. Br. Med. Bull., 6:341.

Ma, N.S.F., Jones, T.C., Thorington, R.W. and Cooper, R.W., 1974. Chromosome banding patterns in squirrel monkeys (*Saimiri sciureus*). J. Med. Primatol., 3:120-137.

MacLean, P.D., 1952. Some psychiatric implications of physiological studies on frontotemporal portion of limbic system (visceral brain). Electroenceph. Clin. Neurophysiol., 4:407-418.

MacLean, P.D., 1954. Studies on limbic system ("visceral brain") and their bearing on psychosomatic problems. In, Recent Developments in Psychosomatic Medicine, R.A. and E. Wittkower, eds., Sir Isaac Pitman and Sons, London, pp. 101-125.

MacLean, P.D., 1958. Contrasting functions of limbic and neocortical systems of the brain and their relevance to psychophysiological aspects of medicine. Am. J. Med., 25:611-626.

MacLean, P.D., 1964. Mirror display in the squirrel monkey, *Saimiri sciureus*. Science, 146:950-952.

MacLean, P.D., 1967. The brain in relation to empathy and medical education. J. Nerv. Ment. Dis., 144:374-382.

MacLean, P.D., 1970. The triune brain, emotion, and scientific bias. In, The Neurosciences Second Study Program, F.O. Schmitt, ed., The Rockefeller University Press, New York, 336-349.

MacLean, P.D., 1973a. A triune concept of the brain and behavior. In, The Hincks Memorial Lectures, T. Boag and D. Campbell, eds., University of Toronto Press, Toronto, pp. 6-66.

MacLean, P.D., 1973b. The brain's generation gap: Some human implications. Zygon J. Relig. Sci., 8:113-127.

MacLean, P.D., 1975a. On the evolution of three mentalities. Man-Environment Systems, 5:213. Reprinted in Human Evolution; Biosocial Perspectives, S.L. Washburn and C.R. McCown, eds., Benjamin Cummings, Menlo Park, California, 1978, pp. 33-57.

MacLean, P.D., 1975b. Role of pallidal projections in species-typical display behavior of squirrel monkey. Trans. Am. Neurol. Assoc., 100:29-32.

MacLean, P.D., 1977. An evolutionary approach to brain research on prosematic (nonverbal) behavior. In, Reproductive Behavior and Evolution, J.S. Rosenblatt and B.R. Komisarus, eds., Plenum Press, New York, 137-164.

MacLean, P.D., 1978a. Effects of lesions of globus pallidus on species-typical display behavior of squirrel monkeys. Brain Res., 149:175-196.

MacLean, P.D., 1978b. A Mind of Three Minds: Educating the Triune Brain, Seventy-seventh Yearbook of the National Society for the Study of Education, University of Chicago Press, Chicago, Illinois, pp. 308-342.

MacLean, P.D., 1981. Role of transhypothalamic pathways in social communication. In, Handbook of the Hypothalamus, P. Morgane and J. Panksepp, eds., Vol. 3, Marcel Dekker, New York, pp. 259-287.

MacLean, P.D., and Ploog, D.W., 1962. Cerebral representation of penile erection. J. Neurophysiol., 25:29-55.

Mehler, W.R., 1980. Subcortical afferent connections of the amygdala in the monkey. J. Comp. Neurol., 190:733-762.

Meibach, R.C., and Siegel, A., 1977. Efferent connections of the hippocampal formation in the rat. Brain Res., 124:197-224.

Miller, N., and Dollard J., 1941. Social Learning and Imitation, Yale University Press, New Haven.

Milner, B., Corkin, S. and Teuber, H.-L., 1968. Further analysis of the hippocampal amnesic syndrome: 14-year follow-up study of H.M. Neuropsychologia, 6:215-234.

Montagu, A., 1956. The Biosocial Nature of Man, Grove Press, New York.

Moore, R.Y., and Halaris, A.E., 1975. Hippocampal innervation by serotonin neurons of the midbrain raphe in the rat. J. Comp. Neurol., 164:171-184.

Morest, D.K., 1961. Connexions of dorsal tegmental nucleus in rat and rabbit. J. Anat., 95:229-246.

Morest, D.K., 1967. Experimental study of the projections of the nucleus of the tractus solitarius and the area postrema in the cat. J. Comp. Neurol., 130:277-299.

Murphy, M.R., MacLean, P.D., and Hamilton, S.C., 1981. Species-typical behavior of hamsters deprived from birth of neocortex. Science, 213:459-461, 1981.

Nauta, W.J.H., and Domesick, V.B., 1978. Crossroads of limbic and striatal circuitry: Hypothalamo-nigral connections. In, Limbic Mechanisms, K.E. Livingston and O. Hornykiewicz, eds., Plenum Press, New York, pp. 75-93.

Newman, J., and MacLean, P.D., 1981. Effects of tegmental lesions on the isolation call of squirrel monkeys. Brain Res. (in press)

Nomura, S., Mizuno, N., Itoh, K., Matsuda, K., Sugimoto, T. and Nakamura, Y., 1979. Localization of parabrachial nucleus neurons projecting to the thalamus or the amygdala in the cat using horseradish peroxidase. Exp. Neurol., 64:375-385.

Norgren, R., 1976. Taste pathways to hypothalamus and amygdala. J. Comp. Neurol., 166:17-30.

Paasonen, M.K., MacLean, P.D. and Giarman, N.J., 1957. 5-Hydroxytryptamine (serotonin, enteramine) content of structures of the limbic system. J. Neurochem., 1:326-333.

Paasonen, M.K., and Vogt, M., 1956. The effect of drugs on the amounts of substance P and 5-hydroxytryptamine in mammalian brain. J. Physiol., 131:617-626.

Parent, A., and Olivier, A., 1970. Comparative histochemical study of the corpus striatum. J. Hirnforsch., 12:75-81.

Pasquier, D.A. and Reinoso-Suarez, F., 1977. Differential efferent connections of the brain stem to the hippocampus in the cat. Brain Res., 120:540-548.

Penfield, W., and Jasper, H., 1954. Epilepsy and the Functional Anatomy of the Human Brain, Little, Brown, and Co., Boston.
Penfield, W., and Milner, B., 1958. Memory deficit produced by bilateral lesions in the hippocampal zone. AMA Arch. Neurol. Psychiat., 79:475-497.
Pert, C.B., and Snyder, S.H., 1973. Opiate receptor: Demonstration in nervous tissue. Science, 179:1011-1014.
Pickel, V.M., Segal, M., and Bloom, F.E., 1974. A radioautographic study of the efferent pathways of the nucleus locus coeruleus. J. Comp. Neurol., 155:15-42.
Poletti, C.E., Kinnard, M.A., and MacLean, P.D., 1973. Hippocampal influence on unit activity of hypothalamus, preoptic region, and basal forebrain in awake, sitting squirrel monkeys. J. Neurophysiol., 36:308-324.
Ricardo, J.A., and Koh, E.T., 1978. Anatomical evidence of direct projections from the nucleus of the solitary tract to the hypothalamus, amygdala, and other forebrain structures in the rat. Brain Res., 153:1-26.
Romer, A.S., 1966. Vertebrate Paleontology, The University of Chicago Press, Chicago.
Segal, M., and Landis, S., 1974. Afferents to the hippocampus of the rat studied with the method of retrograde transport of horseradish peroxidase. Brain Res., 78:1-15.
Slotnick, B.M., 1967. Disturbances of maternal behavior in the rat following lesions of the cingulate cortex. Behaviour, 24:204-236.
Stamm, J.S., 1955. The function of the median cerebral cortex in maternal behavior of rats. J. Comp. Physiol. Psychol., 48:347-356.
Swanson, L.W., and Cowan, W.M. 1975. Hippocampo-hypothalamic connections: Origin in subicular cortex, not Ammon's horn. Science, 189:303-304.
Ungerstedt, U., 1971. Stereotaxic mapping fo the monoamine pathways in the rat brain. Acta Physiol. Scand., 367:1-48.

LIST OF CONTRIBUTORS

John Allman
Division of Biology
California Institute of Technology
Pasadena, CA 91125

Este Armstrong
Department of Anatomy
L.S.U. Medical Center
1901 Perdido Street
New Orleans, LA 70112

Roland Bauchot
Laboratoire d'Anatomie Comparée
Université Paris 7
75221 Paris, FRANCE

Michael S. Billig
Department of Anthropology
Harvard University
Cambridge, MA 02138

C. B. G. Campbell
Department of Medical Neurosciences
Division of Neuropsychiatry
Walter Reed Army Institute of Research
Washington, D.C. 20012

Glenn C. Conroy
Division of Biology and Medicine
Brown University
Providence, RI 02912

Dean Falk
Department of Anatomy and
Caribbean Primate Research Center
University of Puerto Rico
School of Medicine
San Juan, Puerto Rico 00936

Albert M. Galaburda
The Neurological Units of Beth Israel
and Boston City Hospitals
Department of Neurology of
Harvard Medical School
Boston, MA and
The Charles A. Dana Research Institute

Norman Geschwind
Department of Neurology
Beth Israel Hospital
Harvard Medical School
Boston, MA 02215

John A. Gurche
2002 N. Kenmore
Arlington, VA 22207

Ralph L. Holloway
Department of Anthropology
Columbia University
New York, NY 10027

Harry J. Jerison
Department of Psychiatry
and Behavioral Science
UCLA Medical School
Los Angeles, CA 90024

Marjorie S. LeMay
Department of Radiology
Massachusetts General Hospital
and Harvard Medical School
Boston, MA 02114

Walter Leutenegger
Department of Anthropology
University of Wisconsin
Madison, WI 53706

Paul D. MacLean
Laboratory of Brain Evolution
and Behavior
National Institute of Mental Health
Bethesda, MD 20205

R. D. Martin
Department of Anthropology
University College
Gower Street
London WC1
ENGLAND

Charles R. Noback
Department of Anatomy
College of Physicians and Surgeons
Columbia University
New York, NY 10032

Marilee Ogren
Section of Neuroanatomy
Yale University School of Medicine
New Haven, CT 06510

Deepak N. Pandya
Edith Nourse Rogers Memorial
Veterans Hospital
Bedford, MA and
Boston University School of Medicine

David G. Post
Department of Anthropology
Columbia University
New York, NY 10027

Leonard Radinsky
Department of Anatomy
University of Chicago
Chicago, IL 60637

George A. Sacher
Division of Biological and
Medical Research
U.S. Department of Energy
Argonne National Laboratory
9700 South Cass Avenue
Argonne, IL 60439

Axel Schleicher
Anatomical Institute
University Kiel
Kiel, FRG

Heinz Stephan
Max-Planck-Institute
for Brain Research
Frankfort/M.
FRG

Karl Zilles
Anatomical Institute
University Kiel
Kiel, FRG

AUTHOR INDEX

SUBJECT INDEX